Biomass Utilization: Conversion Strategies

Nhuan Phu Nghiem · Tae Hyun Kim ·
Chang Geun Yoo
Editors

Biomass Utilization: Conversion Strategies

Springer

Editors
Nhuan Phu Nghiem
Department of Environmental Engineering
and Earth Sciences
Clemson University
Clemson, SC, USA

Tae Hyun Kim
Department of Materials Science
and Chemical Engineering
Hanyang University ERICA
Ansan-si, Gyeonggi-do
Korea (Republic of)

Chang Geun Yoo
Department of Chemical Engineering
SUNY College of Environmental Science
and Forestry
Syracuse, NY, USA

ISBN 978-3-031-05837-0 ISBN 978-3-031-05835-6 (eBook)
https://doi.org/10.1007/978-3-031-05835-6

This Springer imprint is published by the registered company Springer Nature Switzerland AG
The registered company address is: Gewerbestrasse 11, 6330 Cham, Switzerland

Preface

In many history books, 1973 was recorded as the year that the Yom Kippur War broke out, which triggered the stoppage of oil export to the United States and several other Western countries by members of the Organization of Arab Petroleum Exporting Countries (OAPEC). For those of us who have engaged in energy research, that year marked the opening of a new chapter in bioenergy development. The oil embargo sparked a sudden interest in ethanol as an alternative transportation liquid fuel to petroleum. Initially, corn and sugarcane juice were used as the main feedstocks in the ethanol fermentation industry. The focus then was shifted to lignocellulosic biomass as an alternative feedstock for ethanol production. Through the long course of biomass ethanol process technology research, many valuable lessons have been learned. Some of the most important ones include: (1) It is not economically favorable to produce ethanol as the only product; (2) In addition to ethanol, higher value-added co-products can be produced and their production will improve the economic feasibility and stability of the overall process; (3) Lignin is not simply a waste but rather should be considered as a potential feedstock for the production of fuels and value-added co-products; (4) Ethanol should not be used solely as a liquid fuel but should also be considered as a feedstock for drop-in biofuel production, which can be directly integrated into the existing petroleum refining infrastructure. These and other lessons learned gave rise to the concept of a biorefinery, which is a highly complex and integrated plant using biomass as the feedstock for the production of various fuels and a wide range of chemicals and consumer products.

Numerous research efforts have been attempted to develop the technologies and processes, which employed both biochemical and thermochemical approaches, for the conversion of all the three major components of biomass, i.e., cellulose, hemicellulose, and lignin, to fuels and chemicals. The main objective of this reference book is to review and summarize the most recent results of these research efforts in a single volume. Future research directions to improve the developed technologies and to add to the current knowledge are also suggested and discussed. The authors of the chapters have made their best efforts to present the most current developments. However, since the advances in biomass conversion research progress so fast, by the time the book is published, many new technologies and processes may have been developed. We hope, in the future, if this book is published again in its second edition, these new advances will be included. On behalf of all the chapter authors, we would like to thank the readers. We also look forward to receiving their comments, suggestions, and criticisms on the contents of the book to help us make improvements in our future projects.

Laguna Niguel, CA, USA — Nhuan Phu Nghiem, Ph.D.
Ansan, Gyeonggi-do, Korea (Republic of) — Tae Hyun Kim, Ph.D.
Syracuse, NY, USA — Chang Geun Yoo, Ph.D.

Contents

About the Editors

Dr. Nhuan Phu Nghiem received a Ph.D. in Chemical Engineering from Louisiana State University. He is a retired Lead Scientist with the Eastern Regional Research Center, USDA ARS. He has been an Adjunct Professor in the Department of Environmental Engineering and Earth Sciences at Clemson University for more than 15 years. He has 40 years of experience in industry and government on bioprocess R&D including biomass processing, biorefining and biofuels/biobased products.

Dr. Tae Hyun Kim received a Ph.D. in Chemical Engineering from Auburn University. He is currently a Professor in the Department of Materials Science and Chemical Engineering at Hanyang University ERICA. He was formerly a faculty at Iowa State University and Kongju National University. He has more than 20 years of experience in teaching and research in lignocellulosic biomass processing and conversion.

Dr. Chang Geun Yoo received a Ph.D. in Agricultural and Biosystems Engineering from Iowa State University and currently is an Assistant Professor in the Department of Chemical Engineering at State University of New York College of Environmental Science and Forestry. He has investigated biomass characterization and utilization strategies with 70+ journal articles and book chapters on biomass characterization, pretreatment, thermochemical/ biological conversion, and biorefining.

Chapter 1
Introduction

Nhuan Phu Nghiem

Abstract The inevitable depletion of fossil fuels together with the concerns for climate changes due to their production and utilization have called for a more sustainable industrial system, which will rely on renewable resources for the generation of energy and production of chemicals. This chapter introduces the concept of a biorefinery, which is similar to a petroleum refinery in many aspects, except that the feedstock is lignocellulosic biomass. The basic processing steps for the conversion of the three main components of this renewable feedstock are also introduced.

The heavy dependence of our modern society on fossil fuels, mainly oil and natural gas, is indisputable. Almost all liquid transportation fuels that are currently in use come from fossil sources. The petroleum oil refineries also generate precursors for the manufacture of plastics, which are used to make products we are using in all aspects of our lives. The inevitable depletion of fossil fuels together with the concerns for climate changes due to the production and utilization of these fuels have called for a more sustainable industrial system, which will rely on renewable resources for the generation of energy and production of chemicals. Because of many reasons, which are both political and technical in nature, the development of technology to achieve that goal progressed at a rather slow pace. The turning point probably came with the oil crisis in 1973, which led to the sudden surge in the interest in fuel ethanol produced by fermentation. The global bioethanol production increased six-fold from 18 billion liters at the turn of the century to reach 110 billion liters in 2019 (USDA 2007; US DOE 2021). Bioethanol traditionally was produced from sugar-based and starch-based feedstocks with sugarcane and corn being the two most widely used. The concern on foods vs. fuels and the limited supplies of sugarcane and corn shifted the attention to an alternative feedstock, and lignocellulosic biomass (or simply biomass) appeared as the obvious choice. Biomass feedstocks have two main advantages: (a) They can be grown in areas that are not used for food production and (b) they are

N. P. Nghiem (✉)
Department of Environmental Engineering and Earth Sciences, Clemson University, Clemson, SC, USA
e-mail: nnghiem@g.clemson.edu

N. P. Nghiem et al. (eds.), *Biomass Utilization: Conversion Strategies*,
https://doi.org/10.1007/978-3-031-05835-6_1

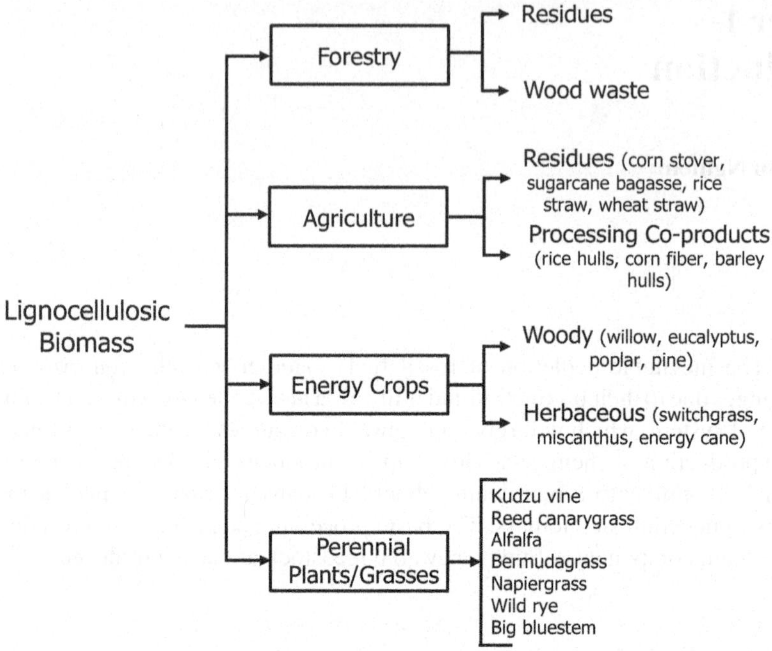

Fig. 1.1 The main sources of biomass feedstocks

available in very large quantities. The main sources of biomass feedstocks, which include forestry, agriculture, energy crops, and perennial plants and grasses, are summarized in Fig. 1.1. It is estimated that the annual global availability of forest products and residues, energy crops, and agricultural residues for the year 2050 is 1845, 4925, and 2831 million MT, respectively (Drapcho et al. 2020). The data for perennial plants and grasses are not readily available.

The three main components of biomass are cellulose, hemicellulose, and lignin, which can all be converted to ethanol via either the sugar platform or the syngas platform. In the sugar platform, cellulose and hemicellulose are hydrolyzed by either chemicals or enzymes to monomeric sugars, which subsequently can be fermented to ethanol. Prior to enzymatic hydrolysis, a process called pretreatment is needed. In the pretreatment process, part of the lignin is removed, which opens up the tight biomass structure and allows higher accessibility of cellulases and hemicellulases toward the substrates. Pretreatment, therefore, improves both the rates and yields of the fermentable sugars. Initially, in the sugar platform, lignin was considered a waste, which would be burned to provide thermal energy. Recently, the potential of lignin as a feedstock for the production of high-value chemicals and liquid fuels has been realized. Many attempts have been made to develop the processes and technologies for lignin conversion to valuable products. In the syngas platform, all three components of biomass are subjected to gasification, which produces a gaseous product known as synthesis gas, or simply, syngas. The syngas thus generated can be converted to

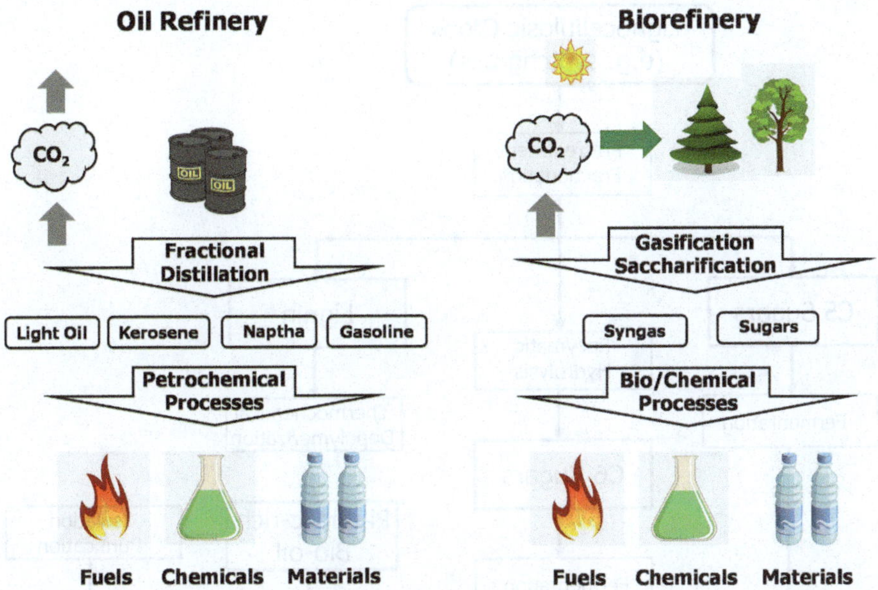

Fig. 1.2 Comparison of an oil refinery and a biorefinery

ethanol via fermentation or chemical catalysis. From a technological point of view, it can be visualized that the processes developed for biomass-based ethanol production can readily be adapted to the production of other chemicals by using suitable biocatalysts (i.e., microorganisms) or chemical catalysts. In other words, the biomass-based ethanol technology can be modified for implementation in a more complex industrial setting of a biorefinery where various chemicals and fuels are produced together. A biorefinery is very similar to an oil refinery in many aspects. A comparison is shown in Fig. 1.2. In both cases, the feedstock is first fractionated into its components, which subsequently are converted to a wide range of products in various processes. The key difference is the type of feedstock. In an oil refinery, the feedstock is crude oil. The CO_2 generated in the production and utilization, e.g., combustion, of the fuel products, enters the atmosphere. An oil refinery, therefore, is part of an open carbon cycle where the CO_2 generated simply escapes to contribute to the accumulation of greenhouse gases in the atmosphere. On the other hand, a biorefinery displays a closed carbon cycle where the CO_2 generated is absorbed and incorporated back into the new biomass feedstock during photosynthesis (Hülsey 2018).

The individual processes and range of products are different from one biorefinery to another. An example of a biorefinery is illustrated in Fig. 1.3. In this biorefinery, the biomass feedstock first is subjected to pretreatment and fractionation to separate the three main components. The C5 sugars are used for the production of xylitol via fermentation. The C6 sugars are fermented to produce ethanol and animal feed. Finally, the lignin is subjected to thermochemical depolymerization to produce phenolic-rich bio-oil, which is then used as a feedstock for the isolation of chemicals

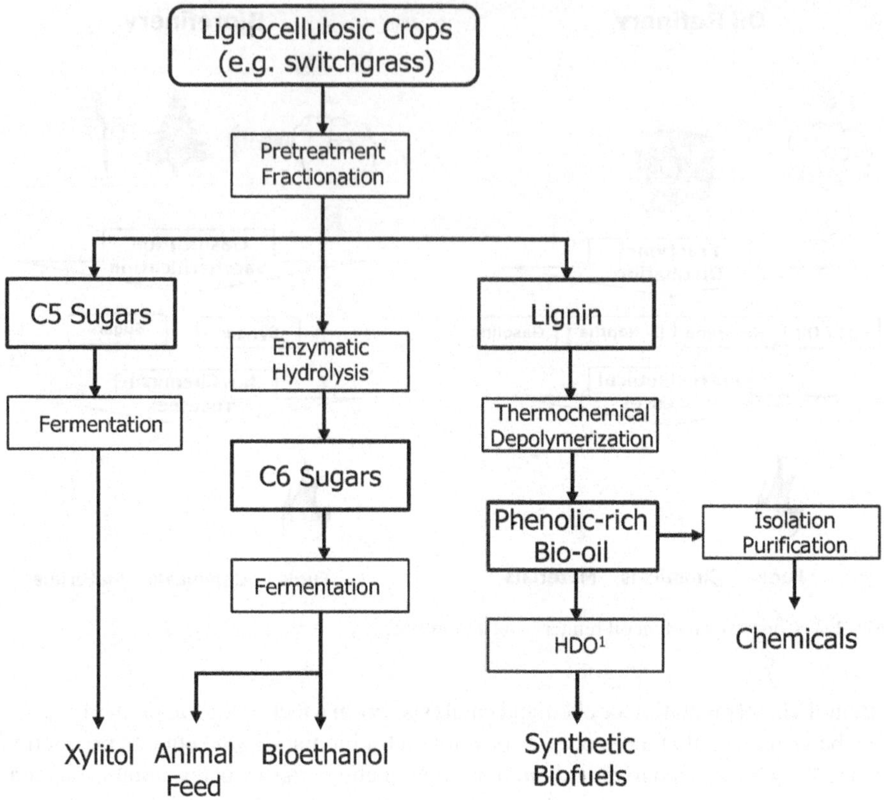

Fig. 1.3 An example of a biorefinery. *Note* (1) HDO: Hydrodeoxygenation

or production of synthetic fuels via hydrodeoxygenation. The individual processes and products in other biorefineries are different, but the general concept still is the same as shown in Fig. 1.3.

The main focus of this book is to review the most up-to-date research efforts aiming at the development of an economically feasible commercial biomass biorefinery. The strategies for biomass fractionation are discussed in Chap. 2. The biochemical conversion of cellulose is discussed in Chap. 3, which is followed by the discussion of hemicellulose biochemical conversion in Chap. 4. The biochemical conversion of lignin is discussed in Chap. 5. The thermochemical conversion of cellulose and hemicellulose is discussed in Chap. 6, followed by the thermochemical conversion of lignin in Chap. 7. The use of lignin in material applications is discussed in Chap. 8. Finally, the evaluation of various biorefinery strategies by techno-economic analysis (TEA) is discussed in Chap. 9.

References

Drapcho CM, Nghiem NP, Walker TH (2020) Biofuels engineering process technology, 2nd edn. McGraw-Hill, New York

Hülsey MJ (2018) Shell biorefinery: a comprehensive introduction. Green Energy Environ 3:318–327

USDA (2007) The future of biofuels: a global perspective. Available online https://www.ers.usda.gov/amber-waves/2007/november/the-future-of-biofuels-a-global-perspective/. Accessed 5 Dec 2021

US DOE (2021) Global ethanol production by country or region. Available online https://afdc.energy.gov/data/10331/. Accessed 5 Dec 2021

Chapter 2
Fractionation Strategies

Diep Trung Tin Le and Tae Hyun Kim

Abstract Lignocellulosic biomass has recently attracted attention as a promising option for replacing non-renewable resources such as fossil fuels with biofuels and biochemical production. In industrial biorefinery processes, the fractionation of biomass into major or intermediate components plays an important role in converting it into value-added products. However, current biomass fractionation processes still have many technical and economic barriers to be industrialized, and further research is necessary. This chapter examines the important factors that should be considered when designing an effective fractionation process that completely utilizes all biomass components. Various chemical agents that can be used for effective fractionation, process design, fractionation efficiency, and fates and characteristics of the separated and recovered components are also discussed.

2.1 Introduction

Over the past decades, the rapid development of biorefinery research due to the critical issues of fossil resource shortage and climate change has been observed. Lignocellulosic biomass, which is a renewable material, has gained increased attention as a promising alternative for fossil resources such as oils and coals.

In general, lignocellulosic biomass is categorized into several types such as agricultural residues (e.g., corn stover, sugarcane bagasse, and rice straw), energy crops (e.g., switchgrass, miscanthus, and fast-growing woods), forestry residues, and various types of cellulosic wastes (e.g., municipal solid waste and pulp mill sludges). Lignocellulosic biomass fundamentally comprises cellulose, hemicellulose, and lignin plus smaller quantities of extractives and inorganic compounds. The relative abundance of the three primary components depends on the type of biomass.

D. T. T. Le · T. H. Kim (✉)
Department of Materials Science and Chemical Engineering, Hanyang University ERICA, Ansan, Gyeonggi-do 15588, Republic of Korea
e-mail: hitaehyun@hanyang.ac.kr

D. T. T. Le
e-mail: trungtin@hanyang.ac.kr; trungtinlediep@gmail.com

N. P. Nghiem et al. (eds.), *Biomass Utilization: Conversion Strategies*,
https://doi.org/10.1007/978-3-031-05835-6_2

Biomass normally contains 35–50% cellulose, 20–35% hemicellulose, and 10–25% lignin (Cai et al. 2017).

Cellulose is the most abundant natural polymer on the earth, which consists of linear chains of β–D-glucopyranose units (>10,000) linked together via 1,4-glycosidic bonds (Fig. 2.1a) (Putro et al. 2016). Bundles of cellulose molecules are formed together as the microfibrils, which contain highly ordered regions (crystalline cellulose) along with less-ordered regions (amorphous cellulose). Cellulose has a wide

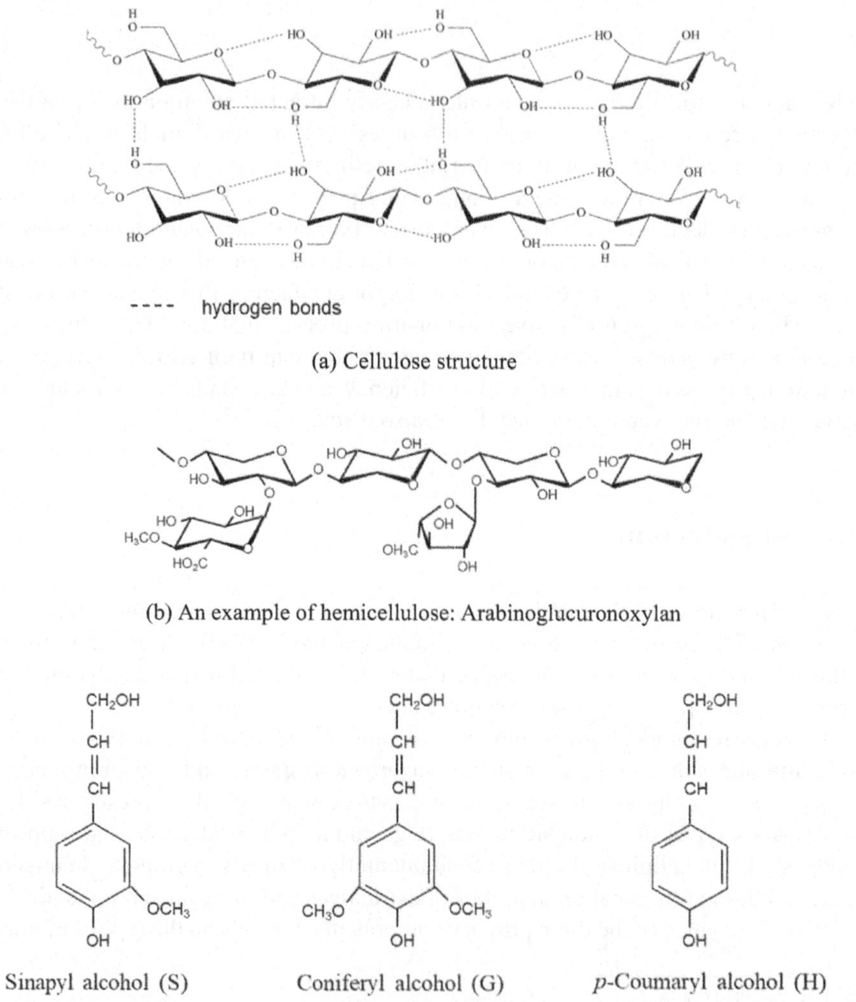

Fig. 2.1 Chemical structure of the three main constituents of lignocellulosic biomass: **a** cellulose; **b** hemicellulose; **c** lignin

range of applications in the pulp and paper industry, energy production, food industry, cosmetics, pharmaceutical industry, and others (Balat et al. 2011; Gupta et al. 2019). Hemicellulose, like cellulose, is another type of carbohydrate and consists of polysaccharides grouped together in a lower degree of polymerization (100–400). Hemicellulose mostly contains D-xylose (a five-carbon sugar) among five different types of sugar units. (Fig. 2.1b shows arabinoglucuronoxylan as an example.) It is more amorphous than cellulose, has a highly branched structure, and is more readily hydrolyzed by acids or enzymes to oligomers and monomers (Spiridon et al. 2008). Hemicellulose can be hydrolyzed and further converted into numerous value-added chemicals such as ethanol, xylitol, 1,2-butanediol, and lactic acid, which has made it a subject of increased interest in its application research. In addition, hemicellulose can also have a potential utilization in food, pharmaceutical, and paper industries as thickeners, emulsifiers, and gels (Chen et al. 2016; Farhat et al. 2017; Liu et al. 2019a, b). Lignin is a polyphenolic compound, providing mechanical strength, rigidity, and hydrophobicity to the plant cell walls while preventing cellulose and hemicellulose from microbial degradation. Lignin consists of three major monolignols, which include coniferyl alcohol, sinapyl alcohol, and *p*-coumaryl alcohol (Fig. 2.1c). Unlike the other two components of biomass, lignin is usually recognized as industrial waste and burned as a fuel. However, the full utilization of lignin as a potential feedstock has been increasingly studied. Promisingly, lignin can be fully exploited as a base material to produce aromatic compounds (benzene, toluene, and phenols) (Stewart et al. 2008).

The development and application of lignocellulosic biomass fractionation technologies are still in the early stage and will be critically needed for fundamental and applied research in the next few decades (FitzPatrick et al. 2010). One of the major barriers to biomass application is the large volume of biofuel and biochemical supply with relatively low energy content, seasonality, and discrete geographic availability of feedstock (Lipinsky 1981). Therefore, if an effective fractionation strategy is applied and target building block components can be separated prior to the further conversion processing in the chemical facilities, it will be beneficial to boost the biorefinery development and make it become more economical and practical for industrial applications.

This chapter discusses the desirable design strategy with consideration of various fractionation technologies, which are the currently widely used methods. Different solvents and reaction methods for effective extraction or separation of individual components in biomass are discussed. Moreover, the changes of lignocellulosic components in chemical structure, molecular weight distribution, and chemical–physical properties after single-step or multi-step processing are also described. Other aspects, including the outlook of desirable fractionation technologies and emerging research trends for future commercial biorefinery, are also discussed.

2.2 Fractionation vs. Pretreatment

Fractionation of biomass has been proposed as the first step of biomass processing in a biorefinery in the study of Koukios and Valkanas (1982). The primary purpose of fractionation is the separation of each component of lignocellulosic biomass (cellulose, hemicellulose, and lignin) making it feasible for further conversion to value-added products. After the fractionation, the resultant components can be utilized directly or converted into the intermediates of various products. Consequently, the high purity and recovery yield of each component are expected in the upstream process. Overall, the ideal fractionation process must meet the expectation of high purity and recovery of each component, and it has been acknowledged that an effective fractionation process that could separate lignocellulose into its three usable forms is the key to unlocking a multiple product integrated biorefinery.

On the other hand, pretreatment methods have generally been proposed and developed to enhance the enzymatic saccharification yield and rate of cellulose and hemicellulose; therefore, pretreatment is a pre-process developed for the purpose of enzymatic saccharification and/or other subsequent processes in either biological conversion or chemical conversion. In particular, pretreatment (by chemical, mechanical, or biological method) is an essential step in the biological conversion process because it improves the enzymatic saccharification reaction by allowing the enzymes to effectively react with the lignocellulosic biomass (in particular, cellulose and hemicellulose) through lignin removal.

2.2.1 Fractionation and Pretreatment in the Biorefinery Concept

The biorefinery concept has been introduced because the bio-industry in the near future can result in producing the replacement of oil refinery products with renewable feedstocks (Cherubini 2010). Biorefinery can be achieved by a wide range of fractionation technologies, which are able to separate renewable biomass into primary building blocks (carbohydrates, proteins, triglycerides), which subsequently can be converted to value-added products, such as fuels and chemicals. This concept is analogous to today's petroleum refinery, which produces multiple fuels and products from petroleum.

The primary purpose of fractionation is to produce value-added products from the constituents in biomass. Therefore, the important factors to consider in the fractionation process generally include recovery yields and purities, chemical structural modifications/changes, and economics of products and processes. On the other hand, pretreatment mainly focuses on the removal extent of components that hinder the following processes such as enzymatic saccharification and fermentation. For

Table 2.1 Comparison between biomass pretreatment and biomass fractionation

Items	Pretreatment	Fractionation
Purpose	To enhance the reaction rate and yield in the following step To enhance the enzymatic hydrolysis and bioconversion yield	To isolate or recover specific components with high purity and yield
Mechanism	Reduction of biomass recalcitrance Removal of inhibitory component	Solubilization/hydrolysis of specific chemical components
Process option	Mechanical, chemical, or biological	Mechanical, chemical, or biological
Focused products/constituents	Mainly cellulose and hemicellulose	All building blocks (C5, C6, lignin), proteins, lipids, etc.

example, the current pretreatment method primarily intends to enhance the rate and yield of subsequent enzymatic hydrolysis of the pretreated biomass by breaking down the lignin structure and disrupting the crystalline structure of cellulose (Alvira et al. 2010). A detailed comparison between biomass pretreatment and biomass fractionation is given in Table 2.1.

2.2.2 Desirable Fractionation Process Design

Unlike pretreatment, fractionation has been studied and developed to improve overall biomass utilization. The production of high-value products from the building blocks of lignocellulosic biomass can increase the economic value of biofuel and bio-based products. In general, in the concept of desirable fractionation process design, the following two points should be carefully considered and addressed (Shen et al. 2021):

(1) Low operating costs: mild reaction conditions, easy recycling, and reuse of catalysts, minimal production of waste material, simple process/operation scheme.
(2) Low capital cost: non-corrosive chemical catalysts, minimal number of equipment, and reasonable reaction conditions.

Therefore, the desirable or ideal fractionation technologies should result in

- high purity of building blocks including cellulose, hemicellulose, lignin, etc., for direct application or as a precursor for further conversion,
- the chemical modification of products beneficial to the next stage of processing,
- high recovery yield of products,
- low degradation of biomass constituents,

- low capital cost by using inexpensive construction materials and non-corrosive chemicals,
- recycle and reuse catalysts, and
- simple process scheme for easy scale-up.

However, it should be noted that currently there is no single fractionation process that meets all the above requirements. To date, many processes have been proposed for biomass fractionation, which can be simply classified into three main approaches based on target building blocks as follows:

(1) cellulose-first fractionation,
(2) hemicellulose-first fractionation, and
(3) lignin-first fractionation.

These fractionation concepts are discussed in the following section.

2.3 Fractionation Processes

2.3.1 Cellulose-First Fractionation Methods

Cellulose fractionation can be generally achieved by two approaches: (1) cellulose retention after removal of lignin and hemicellulose and (2) cellulose regeneration after complete dissolution of biomass. In both strategies, the relatively pure cellulose can be obtained and further converted into value-added products. In the cellulose retention method, solvents such as acid or alkali are used to remove lignin and hemicellulose, thus remaining in the highly pure cellulose. On the other hand, in the cellulose regeneration method, all components in biomass are completely dissolved and then cellulose is regenerated by adding an anti-solvent.

2.3.1.1 Ionic Liquids (ILs) Method

Ionic liquids (ILs) have been considered as a green and highly selective solvent for cellulose dissolution and regeneration. Ionic liquids are salts composed of an organic cation and an organic/inorganic anion. ILs are in the liquid state at room temperature and are stable up to approximately 300 °C. Cellulose is mainly dissolved in the process, and the regenerated cellulose can be formed by adding an anti-solvent such as water, methanol, or ethanol. The regenerated cellulose can be obtained as powders, tubes, beads, fibers, and films by altering the regeneration process. The cellulose regenerated from the ionic liquid fractionation process was found to have either an amorphous or crystalline structure depending on the regeneration method (Zhu et al. 2006). Some widely used ILs are 1-butyl-3-methyl-imidazolium chloride ([C_4mim]Cl), 1-butyl-3-methylimidazolium dimethyl phosphate [C4mim]-[(MeO)(MeO)PO_2], 1-ethyl-3-methylimidazolium acetate [Emim][CH_3COO] plus others (da Costa Lopes et al. 2013; van Osch et al. 2017). However, toxicity, poor biodegradability, and high chemical cost are considered the limitations in the applications of ILs to biomass fractionation.

2.3.1.2 Deep Eutectic Solvents (DES) Method

Deep eutectic solvents with the properties of both ionic liquid and organic solvent have shown a promising feature for cellulose fractionation. A DES is a homogenous mixture, formed by mixing a hydrogen bond donor (HBD) and a hydrogen bond acceptor (HBA) at certain composition, and has a melting point lower than that of its constituents. The hydrogen bonding and van der Waals force between HBD and HBA limit the recrystallization process of the individual components, therefore reducing the melting point of the DES system. DES can also be considered the next generation of ionic liquid, which is cheaper and easily recyclable. In lignocellulosic biomass dissolution, the HBD type can be used such as carbohydrate, acid, polyalcohol, or phenolic compounds, whereas the most used HBA is choline chloride (Tan et al. 2020; van Osch et al. 2017). DES can effectively reduce the cellulose crystallinity and remove lignin and hemicellulose, which can enhance the saccharification efficiency of the fractionated cellulose. More than 90% of hemicellulose from rice straw was removed as well as 70% delignification was obtained by a choline chloride–oxalic acid dihydrate DES solution. The cellulosic-rich materials were easily digestible by the enzyme at over 80% yield, while the fractionated lignin has high purity (>82%) (Hou et al. 2018). The cellulose-rich solid fraction has higher thermal degradability and higher enzyme accessibility due to the large removal of the amorphous contents (hemicellulose and lignin) (Fang et al. 2017; Yiin et al. 2017). The fractionated cellulose can be further converted to sugar for bioethanol production or used to produce cellulose nanocrystals.

2.3.1.3 Sulfite Method

Sulfite treatment includes active catalysts sulfite (SO_3^{-2}), bisulfite (HSO_3^{-1}), or a combination of two of the three catalysts sulfite (SO_3^{-2}), bisulfite (SO_3^{-2}), and sulfur dioxide (SO_2 or H_2SO_3), depending on the pH of the treatment liquor at the reaction temperature (Ingruber et al. 1985). Sulfite ions can promote the dissolution of lignin due to its nucleophilic properties, which can combine acid or alkali solvent to effectively remove lignin and hemicellulose and preserve cellulose for further conversion. During the sulfite treatment process, a considerable amount of hemicellulose is removed in the form of fermentable sugars with a smaller number of inhibitors generated (furfural and hydroxymethylfurfural (HMF)). It is worth mentioning that the sulfonation of lignin enhanced its hydrophilicity, which also limited the adverse effect on the subsequent enzymatic hydrolysis. Sulfite treatment under acidic conditions combined with disk refining for size reduction showed a promising result for cellulose fractionation. The results indicated that after acidic sulfite pretreatment to overcome recalcitrance of lignocellulose (SPORL) with 8–10% bisulfite and 1.8–3.7% sulfuric acid, large amounts of lignin and hemicellulose were dissolved. More than 90% of cellulose was retained and subsequently converted to glucose via enzymatic hydrolysis (Zhu et al. 2009).

2.3.1.4 Acid Method

Acid solvent has huge potential for sustainable lignocellulose fractionation for making high value-added products. In addition to traditional methods using sulfuric acid and hydrochloric acid, other acids such as aromatic sulfonic acid [*p*-toluenesulfonic acid (*p*-TsOH)], dicarboxylic acid (maleic acid), and phosphoric acid have been studied and applied for cellulose fractionation process (Zhu et al. 2021). In the study of Chen et al. (2017), poplar wood was fractionated by a *p*-TsOH solution of 80 wt% concentration at 80 °C for 20 min and retained more than 80% cellulose. The fractionated cellulose can be subsequently digested by enzymes to produce glucose. Maleic acid has been reported as an effective catalyst due to its easy recovery and being a much weaker acid (Su et al. 2021). It can esterify cellulose via the Fischer-Speier reaction. Recently, investigators have examined the effects of *p*-TsOH and maleic acid on switchgrass fractionation. The results indicated that maleic acid obtained more rapid and higher maximal dissolution of lignin and hemicellulose (88% and 63%, respectively) compared to *p*-TsOH (73% and 55%, respectively). The fractionated cellulose was further fibrillated by a mechanical method for making lignin-containing cellulosic nanofibrils (Su et al. 2021). In a study, it was shown that 96% of glucan in a type of biomass (e.g., corn stover, switchgrass, etc.) was retained in the solid form after fractionation with 84.0% H_3PO_4 at 50 °C for 60 min and subsequently was hydrolyzed by the enzyme (Moxley et al. 2008).

2.3.2 *Hemicellulose-First Fractionation Methods*

The structure of hemicellulose allows it to be easily fractionated by neutral hydrothermal solvents or acid solvents and further degraded in these solvents. Consequently, hemicellulose can be extracted from lignocellulosic biomass at moderate operating conditions in the form of various sugars (xylose, arabinose, xylo-oligosaccharides), furfural, and organic acids. However, the appropriate application of the fractionation process for selective degradation and recovery of hemicellulose as a target product must be carefully considered. If the desired product is a polymer material, the extraction process should be conducted under mild conditions to preserve its structural integrity and prevent its degradation. On the other hand, if small molecule compounds such as platform chemicals and liquid fuels are the targets, the simultaneous hemicellulose fractionation and degradation process should be performed (Shen et al. 2021). It should be noted that the structure of hemicellulose and its compositions in the plant cell wall are also the main factors to consider for solvent selection. It has been reported (Huang et al. 2019) that hemicellulose from hardwood and herbage basically consists of xylan, whereas that from softwood is primarily constituted of mannan. It is also mentioned that acid-based solvents are favorable for xylan, while alkali solvents are preferred for mannan. In general, both the hydrothermal process and acid process can be used effectively for hemicellulose extraction from biomass (Zhuang et al. 2016; Carvalheiro et al. 2008).

2.3.2.1 Hydrothermal Method

The hydrothermal process or hot water treatment uses water as a solvent at elevated temperatures (160–240 °C). Since there are acetyl groups in the hemicellulose structure, acetic acid can be formed during the hydrothermal stage and further results in the auto-hydrolysis of hemicellulose in water. Under severe operating conditions, hemicellulose will be significantly degraded into xylo-oligosaccharides (Zhuang et al. 2016; Moniz et al. 2013). Over 80% of hemicellulose in agricultural residues (corn stover, bamboo, and sugarcane bagasse) are separated successfully (Xiao et al. 2013; Mosier et al. 2005; Laser et al. 2002). The fermentation inhibitors (furfural, formic acid, and lactic acid) normally are produced at very low concentrations in comparison with other fractionation methods. Hemicellulose fractionation by the hydrothermal process is considered an economical and eco-friendly technology due to its use of water as a solvent without the addition of chemicals, little corrosion on reaction equipment, and low concentrations of inhibitors (Zhuang et al. 2016).

2.3.2.2 Acid Method

Inorganic acids such as sulfuric acid (H_2SO_4), hydrochloric acid (HCl), phosphoric acid (H_3PO_4), and nitric acid (HNO_3) have been widely used as solvents for hemicellulose extraction. Concentrated acid or diluted acid require different reaction conditions and have different impacts on hemicellulose fractionation efficiency. Concentrated acid extraction is usually conducted using an acid concentration of over 30% at ambient to moderate temperatures (<100 °C) for several hours. In contrast, the dilute acid method uses 0.5–5% acid concentration and is operated at higher temperatures (120–250 °C) and below 10 atm. for 5–90 min (Jung et al. 2015). The strong acidity of the solvent causes degradation of hemicellulose and results in the formation of monosaccharides and furfural. The strong solvent acidity also leads to the partial hydrolysis of cellulose and depolymerization of the oligomers formed from hemicellulose. The inorganic acid fractionation technology has some drawbacks such as corrosion of equipment materials, high acid recovery cost, and solvent toxicity, which limit its application.

Recently, organic acids have been promoted due to their highly selective dissolution and control of the depolymerization of hemicellulose under mild conditions.

The dicarboxylic maleic acid and oxalic acid are two potential alternatives to inorganic acids. During the hemicellulose extraction by maleic acid, the extent of xylose degradation is 3–10 times lower than that of sulfuric acid. These results indicate that maleic acid can be applied as a more effective fractionation solvent by providing a higher energy barrier to the degradation reaction of xylose to furfural (Jung et al. 2015). In a study using bagasse (Yan et al. 2018), 96.7% hemicellulose could be extracted in the liquid fractions and the highest xylose yield of 95.7% was obtained with 0.4 mol/L aqueous oxalic acid at 120 °C and 10 min.

2.3.3 Lignin-First Fractionation Methods

Lignin fractionation, or delignification, can also be considered a biomass pretreatment process since lignin limits the accessibility of enzymes to cellulose and hemicellulose, and hence its removal will improve both hydrolysis rates and yields of fermentable sugars. Many chemical methods of lignin removal have been studied and developed including alkali, organosolv, ionic liquids, and deep eutectic solvents. Enzymes can also be used to assist lignin extraction by hydrolyzing the carbohydrates (Xu et al. 2020). In terms of efficient utilization of lignin for potential applications, it is very important to obtain high purity of lignin with excellent chemical reactivity for further chemical or biochemical conversion. Lignin is mostly obtained in the liquor after fractionation and is precipitated to collect the pure solid form. The overall process of lignin extraction is illustrated in Fig. 2.2. Each fractionation technology as well as reaction severity can give a wide variety of lignin types, molecular weight distribution, chemical reactivity, and degree of polymerization (Jiang et al. 2020; Galkin et al. 2016). To date, three major obstacles are encountered in the one-pot lignin isolation process, which include:

(1) breaking the cell wall recalcitrance to release sugars and phenolic intermediates,
(2) reducing the oxygen contents of phenolic intermediates to produce stable fuels with high calorific value, and

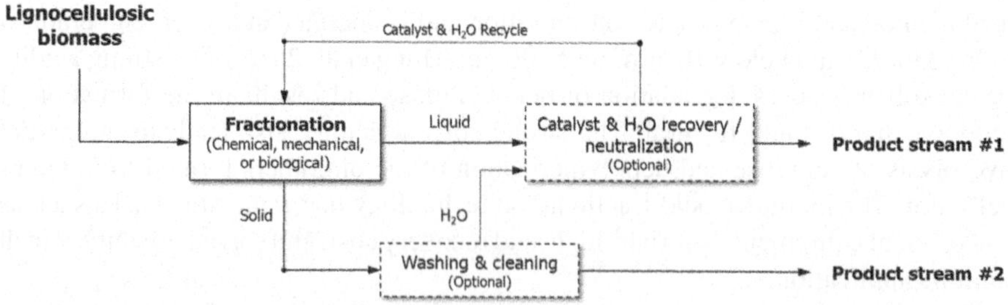

Fig. 2.2 Schematic of a simple fractionation process

Table 2.2 Various alkaline fractionation processes and their effects on lignin extraction

Description	Chemical catalysts			
	$NaOH$	NH_3	$Ca(OH)_2$	Na_2CO_3
Operating conditions				
Temperature	60–180 °C	30–210 °C	25–130 °C	60–180 °C
Pressure	Low	Low–high	Low	Low
Catalyst concentration	0.5–10%	5–30%	0.05–0.15 g/g biomass	1–30%
Lignin recovery	60–80%	50–80%	60–80%	40–60%
Corrosiveness	High	Low	Low	Medium

Source Kim et al. (2016a, b, c), Wyman et al. (2005), Yang et al. (2012), Li et al. (2011)

(3) stabilizing the active phenolic intermediates under elevated reaction conditions to prevent lignin re-condensation and enhance phenolic yields.

2.3.3.1 Alkaline Method

Alkaline fractionation technology has been studied for many years and applied on an industrial scale. The alkaline catalysts include sodium hydroxide, ammonia, calcium hydroxide, and sodium carbonate. These chemicals have been demonstrated to be efficient catalysts for the extraction of up to 80% lignin from raw biomass. The rich lignin fraction will exist in the black liquor after the fractionation stage and can be precipitated to obtain solid lignin. Alkali-based fractionation processes are carried out under mild conditions, thus eliminating the need of expensive materials and special design to deal with corrosion issues and severe operating conditions. In some of the alkaline processes, chemical catalysts such as ammonia can be recovered and reused (Yoo et al. 2011a, b). The operating conditions and other important technical details of various alkaline lignin extraction processes are summarized in Table 2.2.

2.3.3.2 Organosolv Method

In the organosolv fractionation process, the organic solvent is used to dissolve lignin from the plant cell walls to obtain high purity lignin after separation. The organosolv solvents include organic alcohols (e.g., methanol, ethanol), organic acids (e.g., formic acid, acetic acid), esters, and mixtures of solvents with or without acid/alkaline catalysts. The alcohol-based fractionation solvents are more widely used. The process temperatures range from 170 °C to 200 °C with a reaction time of 60 min using 50–80% (wt.) ethanol (Teramoto et al. 2008; Aziz et al. 1989). One of the main advantages of organosolv treatment is the extracted lignin is relatively pure, undegraded, and sulfur-free, which allows it to be used directly to produce value-added materials. At the optimal processing conditions, 20–78% of the total lignin (acid-soluble and acid-insoluble) in the raw biomass could be recovered as ethanol organosolv lignin (Sannigrahi et al. 2013). On the other hand, there are some limitations of alcohol-based organosolv process applications, which include high temperature and high pressure.

Besides the alcohol-based organosolv process, organic acid-based fractionation technology is a promising approach and has been extensively employed in lignin extraction. Formic acid, acetic acid, maleic acid, and *p*-TsOH are primarily used. Hu et al. (year) obtained promising results that showed approximately 80% of lignin was separated from corn husk and had very high purity (~99%) by the one-step formic-acid treatment at 85 °C in 5 h. Recently, investigators have examined the effects of acid hydrotrope solution of *p*-TsOH on the lignin extraction of poplar wood. About 80% of lignin dissolution was achieved at 2.5 mol/L *p*-TsOH in 60 min using a flow-through reactor at 90 °C (Wang et al. 2019).

2.3.3.3 Ionic Liquids (ILs) Method

Compared with other conventional fractionation approaches, IL fractionation technology is a novel promising method because it can selectively dissolve cellulose, hemicellulose, and lignin. Another interesting point is that the tunability of IL physical properties for lignin fractionation can be achieved by an appropriate selection of cations/anions. ILs also attract interest because of their low volatility and high thermal stability. It should be noted that for lignin isolation, the H-bonding between biomass and solvent molecules must be strong enough to break the inter- and intramolecular H-bonding in the lignocellulosic feedstocks. ILs with anions having a strong H-bond accepting ability were found to extract lignin effectively (van et al. 2017). A longer alkyl chain length of the cation in ILs enhances the lignin extraction yield since its hydrophobicity increases. Imidazolium-based ILs, such as 1-ethyl-3-methylimidazolium acetate ([C2C1Im][OAc]), 1-butyl-3-methylimidazolium chloride ([C4C1Im][Cl]), and 1-ethyl-3-methylimidazolium chloride ([C2C1Im][Cl]), have been studied and demonstrated as highly effective solvent in lignin fractionation for a variety of biomass feedstocks, including switchgrass, agave bagasse, poplar, and pine wood (Shi et al. 2013; Perez et al. 2013; Varanasi et al. 2012).

2.3.3.4 Deep Eutectic Solvent (DES) Method

As previously mentioned, DES is known as the next generation of traditional ionic liquids, having at least one hydrogen bond donor (HBD) and one hydrogen bond acceptor (HBA). Recently, many research groups used acidic DES solvents made of organic acid HBD for lignin extraction. A recent study reported that lignin dissolved by acid DES method could be precipitated by using a simple water–ethanol anti-solvent rinsing, whereas lignin isolated by near-neutral and basic DES extraction methods could not be precipitated by applying the same methodology. Adjustment of pH is needed to achieve lignin precipitation (Tan et al. 2020). It was pointed out (Tan et al. 2019) that the functional group of acid in acidic DES solvents, which has more hydroxyl group and short alkyl chain, will improve the lignin extraction efficiency. Choline chloride-lactic acid mixture is a promising DES solvent, which could extract more than 60% of lignin. Additionally, it should be noted that the extraction conditions play a critical role in the purity of the isolated lignin. It was reported that the lignin purity increased from 90% to 95.4% when the extraction time was extended from 6 to 24 h; meanwhile, the remaining carbohydrate content in the lignin sample declined from 1.12 to 0.11% (Lyu et al. 2018).

2.3.3.5 Enzyme-Assisted Method

Enzyme can be considered an effective biological agent which can assist in lignin extraction. The obtained lignin sample from mechanical methods such as milling contains a significant amount of residual carbohydrates and therefore reduces the lignin purity. Consequently, enzyme, which is mainly cellulase, is used to hydrolyze the carbohydrate residues in milled lignin samples. The assistant function of the enzyme improves the product yields of lignin extraction process, and the obtained lignin can be called cellulolytic enzyme lignin (CEL). It was shown that the lignin recovery yield from *Eucalyptus* wood could reach 95% yield after mild alkaline swelling (4% NaOH, 25 °C, 24 h) and enzymatic hydrolysis. This combined mild alkaline treatment and enzymatic hydrolysis had a minor change in the lignin structure and produced syringyl-rich lignin macromolecules (Wen et al. 2015). The use of enzymes in the lignin isolation process would reduce the use of toxic chemicals and offer economic and environmental benefits (Li et al. 2018).

2.4 Fates of Biomass Components upon Process Options and Designs

2.4.1 Fates of Components by Fractionation Methods

2.4.1.1 Chemical Methods

There are three main strategies based on chemical fractionation approaches including cellulose-first fractionation, hemicellulose-first fractionation, and lignin-first fractionation. The IL and DES are good choices for cellulose-first strategy and lignin-first strategy (van Osch et al. 2017). Lignocellulose can be selectively dissolved to various extents, and the cellulose-rich component with low lignin and hemicellulose contents can be regenerated. Hemicellulose and lignin can be fractionated via precipitation in the anti-solvent such as methanol, ethanol, or acetone, which allows for the separation and purification of lignocellulosic biomass components and further valorization. Besides, alkaline hydrogen peroxide, sulfite solvent, and acid hydrotropic have been demonstrated as promising catalysts for the delignification process in addition to preserving the cellular structure of cellulose and making it feasible for further conversion (Shen et al. 2021). For the hemicellulose-first strategy, hydrothermal methods including liquid hot-water and acid solvent treatment are mostly applied (Carvalheiro et al. 2008; Ma et al. 2012). It is proven to be effective for hemicellulose isolation. Hemicellulose exists in the liquid fraction and can be easily degraded to xylo-oligosaccharides and xylose, while the cellulose-rich solid fraction has greater surface accessibility and larger pore volume ready for the enzymatic hydrolysis. In the chemical process for lignin isolation, lignin may be subjected to re-condensation during the fractionation process. Lignin can be isolated by alkaline treatment or as a by-product of the pulping process. Novel approaches such as IL and DES can also be applied to achieve higher lignin yield and purity (Jiang et al. 2020; Xu et al. 2020).

2.4.1.2 Mechanical Methods

Mechanical fractionation is considered a crucial step to mainly result in higher carbohydrate extraction yields and the structural transformation of biomass components. The mechanical process has been demonstrated to affect the carbohydrate fractions by increasing specific surface area and porosity, reducing particle size and fiber thickness, and decreasing the crystallinity without producing toxic components in downstream (Zhu et al. 2010). The mechanical methods that can be used include grinding, chipping, and milling. These treatments lead to different size reduction with "powdering" (m to cm), coarse milling (cm to mm), intermediate micronization (cm to 100 μm), fine grinding (<100 μm), ultra-fine grinding (<30 μm), and nano-grinding (<1 μm) (Barakat et al. 2013). Achieving higher fractionation yields and retaining the integrity of the macromolecular structure of products are two primary factors to determine the mechanical process efficiency. The performance and energy consumption of a mechanical fractionation process mainly depends on the ratio of particle size distribution before and after milling, moisture content, biomass sources, and the equipment type and process parameters (feed rate, rotation speed, geometry, etc.) (Barakat et al. 2014; Silva et al. 2012). However, with the heterogeneity of lignocellulosic material, it is difficult to fully separate cellulose, hemicellulose, and lignin into pure fractions solely by a mechanical process. Therefore, it is worthy to mention that the mechanical method should be combined with the chemical method to reduce energy consumption and total cost of the fractionation process, limit the inhibiting effects caused by chemical treatments, preserve the cellulose structure, and improve the enzymatic hydrolysis process. Moreover, the smaller particle size has been proven to have a positive impact on the chemical fractionation efficiency by improving mass diffusion and heat transfer rate. The impacts of typical mechanical treatments on lignocellulosic biomass are listed in Table 2.3. A combination of various forces such as impact, compression, friction, and shear stress is applied in the mechanical fractionation process. Some illustrations of operation equipment with different stresses generated are shown in Fig. 2.3.

Table 2.3 Effects of mechanical fractionation methods on structural features of lignocellulosic biomass (modified from Barakat et al. 2014; Barakat et al. 2013)

Effects	Milling types			
	Ball milling	Disk milling	Hammer milling	Knife milling
Reduction of particle size	✓✓	✓✓	✓	✓
Increase of porosity	✓	✓	✓	✓
Increase of surface area	✓✓	✓✓	✓	✓
Reduction of DP	•	•	•	•
Reduction of crystallinity	✓✓	✓✓	✓	✓
Solubilization of cellulose	✖	✖	✖	✖
Energy consumption	High	High	High	High
Environmental toxicity	✖	✖	✖	✖
Pretreatment effect	✓✓	✓✓	✓	✓

(✓✓ = greatly positive effect, ✓ = very positive effect, • = minor positive effect, ✖ = no effect)
(DP; degree of polymerization)

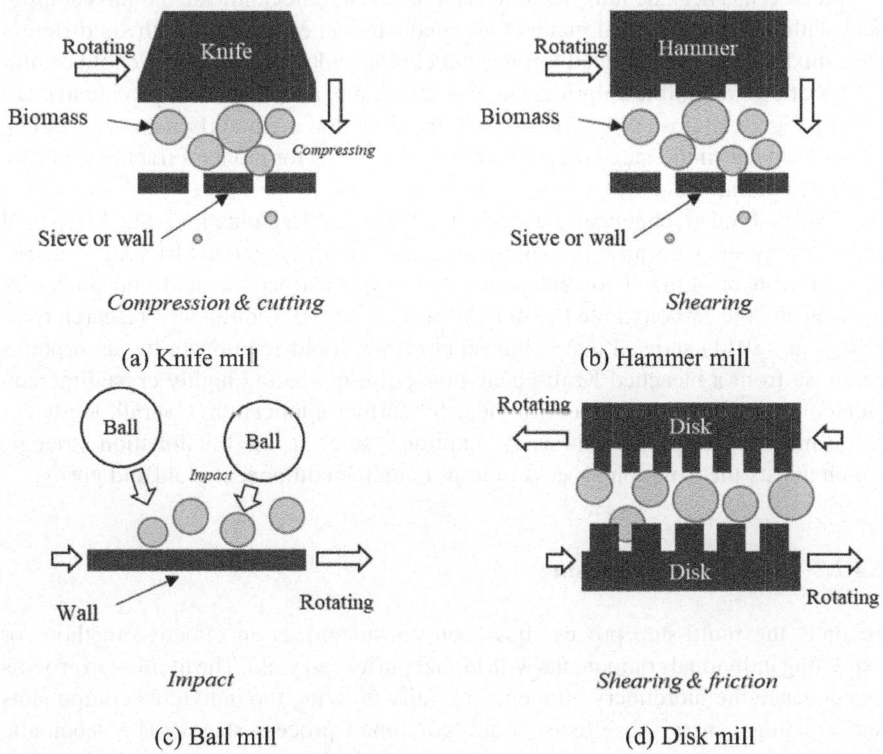

Fig. 2.3 Various types of mechanical equipment for fractionation and their milling mechanisms (modified from Barakat et al. 2014; Kim et al. 2016a, b, c)

2.4.2 *Fates of Components According to the Process Designs*

2.4.2.1 Single-Step Processing

Many studies have been conducted to isolate cellulose, hemicellulose, and lignin by one-step treatment. In general, all the treatments leave an impact on individual components to various extents. Depending on the desired purity of the products obtained from a fractionation process, an appropriate treatment with optimal conditions should be applied. Some typical effective methods used for specific biomass components are discussed in Sect. 2.4.1.1. This section discusses recent studies, which showed promising features in fractionation efficiency. Corn stover was successfully fractionated into cellulose fibers and lignin with yields of 38.9% and 25.4%, respectively, by a single-step process (Zhong et al. 2021). The optimal extraction conditions were 85.0% aqueous formic acid at 130 °C for 30 min. The obtained cellulose fibers showed great mechanical properties whereas the sugar products contained 61% xylose and 17% glucose. An environmentally friendly approach with IL and DES has been demonstrated to separate each lignocellulosic biomass component with high selectivity. Lynam et al. conducted an experiment with six different DES mixtures, which included formic acid/choline chloride, lactic acid/choline chloride, acetic acid/choline chloride, lactic acid/betaine, and lactic acid/proline to evaluate the lignocellulose solubility by DES fractionation method (Lynam et al. 2017). Several typical single-stage processes using chemicals for biomass fractionation are presented in Table 2.4.

Besides typical chemical methods for biomass fractionation, the biological approach by enzyme also is a promising and environmentally friendly process. Enzyme can be utilized for enhancing the lignin extraction yield and purity by hydrolyzing the carbohydrate fraction (Wen et al. 2015). Additionally, research from Zhu et al. (2011) showed that cellulase enzymes could fractionate the amorphous cellulose from a bleached Kraft eucalyptus pulp, producing highly crystalline and recalcitrant cellulose at up to 70% yield for further application. Overall, the enzymatic method can play a role as an additional stage in the fractionation strategy, which assists the isolation process in improving the component yield and purity.

2.4.2.2 Multi-step Processing

To date, the multi-step process has been considered as an effective method for extracting individual components with higher purity and yield. The multi-step process can enhance the biorefinery efficiency by fully utilizing the individual components and creating synergistic effects of the combined process stages. The schematic diagram of a representative two-stage fractionation process is shown in Fig. 2.4. Cellulose, hemicellulose, and lignin can be collected in solid forms after the fractionation and purification stages. The results of several studies on multi-step processing are summarized together with the main features of these methods in Table 2.5. Gener-

Table 2.4 Typical single-stage fractionation process

Fractionation method	Reaction conditions	Main features	References
DES	Acidic DES (Choline chloride/ethylene glycol = 1:2) 130 °C, 30 min, 20–27 wt% solid loading	• 72.6–86.2% glucose yield • 69.4–79.5% lignin removal	Chen et al. (2018)
SPORL	9 wt% sodium bisulfite 180 °C 30 min, L/S = 5:1	• 92.5% cellulose recovery • 32% lignin removal	Shuai et al. (2010)
p-TsOH	*p*-TsOH 80 °C, 20 min	• 90% of lignin extraction yield	Chen et al. (2017)
Acid hydrothermal	H_2SO_4 0.25–1.76% (w/v), 150–180 °C, 3–30 min	• 87.9% hemicellulose yield • Only 15.3% glucose yield	Lee et al. (2013)
Hot-water	Water, 150 °C–240 °C, L/S = 10:1	• 53% hemicellulose recovery with 72% xylan was hydrolyzed • Cellulose and lignin were not substantially affected	Moniz et al. (2013)
Aqueous ammonia	Ammonia recycled percolation, NH_3 15 wt%, 10–90 min, 170 °C, flowrate 5 ml/min, 2.3 MPa	• 70–85% lignin removal • 40–60% hemicellulose solubilized, cellulose intact	Kim et al. (2003)
Sodium hydroxide	NaOH 0.5–2.0 w/v%; 21 °C, 50 °C, 121 °C, 0.25–1;1–48; 1–96 h, L/S = 10:1	• 85.8% lignin removal • Sugar yield = 3.8 times raw biomass	Xu et al. (2010)

Note L/S: liquid/solid ratio

ally, the process can be a combination of chemical/chemical, chemical/mechanical, or chemical/biological methods. The data in Table 2.5 shows promising results on the fractionation efficiency with various degrees of component recovery and purity. These components can be further converted to value-added products such as biomaterials, biofuels, and bioenergy. However, the techno-economic analysis of these approaches should be further investigated to assess their economic feasibility and potential implementation at the industrial scale.

Figure 2.4 shows a generalized two-step fractionation process for lignocellulosic biomass. In a multi-stage process, each fractionation stage must be operated under appropriate reaction conditions according to an adopted catalyst. Subsequent unit operations after each fractionation step typically include multiple catalyst recovery/separation and product recovery units. The type of each product may have

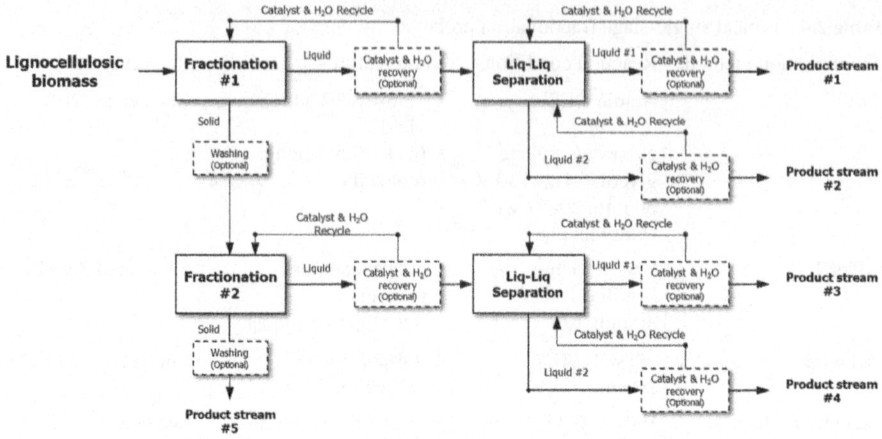

Fig. 2.4 Schematic diagram of two-stage fractionation process and various product streams

various process types depending on the type of catalyst applied and the reaction conditions. For example, if an alkaline catalyst is used in Fractionation #1, a lignin component may be produced in Product stream #1 or #2, and if an acid catalyst is used in Fractionation #1, the lignin component will be more likely to be produced in the Product stream #3, #4, or #5.

2.5 Fractionation Strategies for Future Commercial Biorefinery

2.5.1 Valorization of Biomass Constituents Produced from Fractionation Process

Utilization of the fractionated components is the major step in the biorefinery approach to produce the value-added products to enhance the economic and renewable value of the process. The natural and renewable biological resources from lignocellulosic biomass can be further applied to various industrial sectors including materials and chemicals, bioenergy, and biofuels (Shen et al. 2021). However, the valorization of the extracted components mainly depends on the purity of the isolated components, the chemical transformation that occurs during the fractionation process, the degree of polymerization, molecular weight distribution, and morphological structure modification. Therefore, an effective strategy for fractionation in addition to an appropriate valorization process must be developed and implemented to obtain the components suitable for conversion to high-value products in an economically feasible process.

Table 2.5 Multi-stage process for lignocellulosic biomass fractionation

Fractionation method	Biomass	Stage 1	Stage 2	Main features	References
Hot-water–DES	Poplar wood	Hot water (170 °C, 40 min, L/S = 30:1)	Acidic DES (CC–formic acid/acetic acid/lactic acid)	• HC recovery 54.4% (1st stage) • Lignin extraction 73–76.5% (2nd stage)	Tian et al. (2020)
Acid–Acid	Poplar wood	*p*-TsOH 2 wt%	*p*-TsOH 40 wt%	• 55.1% xylan solubilized (1st stage) • 83% lignin removal (2nd stage)	Wu et al. (2020)
Organosolv–alkaline	Wheat straw	Organosolv–50% ethanol + 1% H_2SO_4, 160 °C 50 min, L/S = 10:1	Alkaline sodium carbonate 11 wt%, 75 °C, 85 min, L/S = 4 L/kg)	• 86.68% cellulose recovery (2nd stage) • Lignin recovery 62.5% • 6.5% w/v ethanol titer obtained	Yuan et al. (2018)
Aqueous $AlCl_3$–DES	Poplar	$AlCl_3$ 0.005–0.02 M, 130–150 °C, 1 h, L/S = 10:1	DES (EG:CC:$AlCl_3$ = 1:2:0.1) 80 °C, 1 h	• HC fractionation (1st stage) with 5.4 g/L concentration • Glucose yield increased from 15.7–96.3%	Ma et al. (2021)
Hot water with BmimCl–alkaline solvent	Corn cob	Hot water, 190 °C, 30 min, L/S = 10:1	NaOH 2 wt%, 90 °C 1.5 h, L/S = 30 BmimCl 130 °C,0.5–5 h, L/S = 25/1	• 86.1% cellulose purity with 86.9% cellulose recovery • 76.7% lignin recovery and 97.3% HC recovery	Araujo et al. (2019)

(continued)

Table 2.5 (continued)

Fractionation method	Biomass	Stage 1	Stage 2	Main features	References
Hot water–aqueous ammonia	Corn stover	Hot water 190–210 °C, 30 min	15 wt% ammonia 170–190 °C 30 min	• Hemicellulose removal (1st stage) and lignin removal (2nd stage) • 83–86% hemicellulose recovery, 75–81% lignin removal, 78–85% cellulose retained	Kim et al. (2006)
Aqueous ammonia–liquid hot water	Corn stover	15 wt% ammonia 60 °C 8 h, L/S = 10:1	Hot water 190–210 °C, 10 min	• 47% lignin recovery and 68% lignin purity • 65% xylan recovery and 78% xylan purity	Yoo et al. (2011a, b)
Acetic acid–aqueous ammonia	Empty fruit bunches	Acetic acid 3.0–7.0 wt%, 170–190 °C, 10–20 min, L/S = 10:1	Ammonium hydroxide 5–20 wt%, 140–220 °C, 5–25 min	• 53.6% HC dissolution and 59.5% lignin removal • 65.3% glucan in solid	Kim et al. (2016a, b, c)
ILs– Alkaline treatment	Sugarcane bagasse	BmimCl, 110 °C under N_2 atmosphere, L/S = 50:1	Alkaline NaOH 3%, 50 °C, 45 min, L/S = 25:1	• 47.17% cellulose recovery, 33.9% HC recovery, 54.6% lignin recovery	Lan et al. (2011)
Sulfite solution–disk milling	Spruce and red pine	0–12 wt% bisulfate, 0–7.36 wt% sulfuric acid, 180 °C, 30 min	Disk refiner (8 in and 12 in) with 0.25 mm disk gap	• 90% cellulose recovery and converted by enzyme hydrolysis	Zhu et al. (2009)
Enzyme—DES	Bamboo shoot shell	*Galactomyces* sp. CCZU11-1, 30 °C, 3 days	CC:Oxalic acid (1:2), 120 °C, 1.5 h	• 77% HC removal and 20% lignin removal (1st stage) • 90% sugar hydrolysis by enzyme	Dai et al. (2017)

Note CC: Choline chloride, HC: hemicellulose, BmimCl: 1-butyl-3-methylimidazolium chloride, EG: ethylene glycol

2.5.1.1 Cellulose Valorization

Cellulose can be converted into various building blocks and precursors such as glucose, alcohols (ethanol, ethylene glycol, and alditol), organic acids (gluconic acid, lactic acid, and levulinic acid), and 5-hydroxymethylfural (5-HMF). Cellulosic ethanol production has been studied for many years and applied in the industrial sector as a biofuel process. Typically, fractionated cellulose is hydrolyzed to glucose by enzymes and subsequently fermented to ethanol by selected microbial strains (Liu et al. 2019a, b). To date, several processes have been developed for cellulosic ethanol production including separate hydrolysis and fermentation (SHF), simultaneous saccharification and fermentation (SSF), simultaneous saccharification and co-fermentation (SSCF), and separate hydrolysis and co-fermentation process (SHCF). In each process, different types of enzymes and strains are used under optimal conditions (time, temperature, pH, solid loading) for cellulose hydrolysis and glucose fermentation. Commercial enzyme products, which contain both cellulases and β-glucosidases, have been developed. The microorganism of choice for use in industrial processes is *Saccharomyces cerevisiae*. In addition to ethanol, other products have been investigated. It was reported that the fractionated cellulose obtained with maleic acid could be directly fibrillated through micro-fluidization to produce lignocellulose nanofibrils (LCNF) (Cai et al. 2020). In comparison with the cellulose nanofibrils obtained by directly hydrolyzed bleached pulp, the LCNF had a similar diameter from several to ten nanometers. A high concentration of 2,3-butanediol (90.2 g/L) was produced from the cellulose obtained by DES fractionation. The cellulose was effectively saccharified at 89.6% glucose yield and subsequently fermented to produce 2,3-butanediol (Chen et al. 2018).

2.5.1.2 Hemicellulose Valorization

Xylose, which is the major monomer unit obtained from hemicellulose in the liquid fraction during the fractionation process, can be dehydrated to furfural via acid catalysis. Various solvents such as water, γ-valerolactone (GVL), 2-methyltetrahydrofuran (2-MTHF), and dimethyl sulfoxide (DMSO) can be used for the conversion of xylose to furfural (Luo et al. 2019). It was reported that the xylan-based sugar products of hemicellulose in the DES fractionation liquor could reach the concentration of 96.8 g/L xylose equivalents. This xylan-rich liquor was successfully converted into furfural at 160 °C with an 84.6% yield (Chen et al. 2018).

2.5.1.3 Lignin Valorization

Lignin is the most abundant aromatic natural compound, which has high potential in commercial applications. Lignin typically is separated in the pulp and paper industry and is being burned to supply heat and energy for the process. Moreover, lignin is an important source of aromatic compounds, and it can be considered as a sustainable platform for the chemicals and materials manufacturing industries. Several chemicals can be produced from isolated lignin such as vanillin, aromatic hydrocarbons (benzene, toluene, xylene), and phenol (Schutyser et al. 2018). Lignin from the sulfite pulping process can be used to produce vanillin on an industrial scale. The process for vanillin production is mainly based on the alkaline oxidative depolymerization of lignin in which treating the lignin-rich liquor with oxidants (O_2) at alkaline pH and high temperature (130–200 °C) and pressures to enable the depolymerization of lignin (Fache et al. 2016). In addition, the production of low-cost and high-quality carbon fiber from lignin precursors has been widely reported. Lignin is initially converted into a fiber by a spinning process (wet spinning, melt spinning, or dry spinning), then stabilized by oxidation at elevated temperatures in air or oxygen for enhancing the integrity and stability of the lignin fibers via crosslinking. Finally, the lignin fibers go through the carbonization stage at high temperatures (800–1400 °C) in an inert atmosphere (Qu et al. 2021; Baker and Rials 2013). Thermoplastic elastomers, polymeric foams, membranes, and resins can also be prepared from lignin (Ragauskas et al. 2014).

2.5.2 *Effective Strategies for Enhanced Values in Large-Scale Production*

The three potential benefits, including economics, environment, and society, that biomass utilization to produce fuels, chemicals, and materials can bring were mentioned in a recent study (Yamakawa et al. 2018). In terms of economic impact, the great possibility of large investment in the biorefinery industry increases, resulting in regional growth and competitiveness as well as reducing the dependence on fossil fuel consumption. From the environmental perspective, renewable energy can cut down waste and greenhouse gas emissions in addition to their adverse impacts on soil, water, and atmosphere. Moreover, using renewable feedstocks from lignocellulosic biomass for manufacturing processes can be considered a sustainable production for complete utilization of carbon without raising the atmospheric carbon dioxide concentration. Plants using carbon for growth in photosynthesis will be effectively converted to valuable bio-based products. Subsequently, the carbon component will be released at the end of the product life cycle and will be re-absorbed. The net result is that there will be no carbon dioxide increase. Finally, in the social aspects, the development of a bio-based economy promotes regional and rural growth and creates more jobs in both the farming and industrial sectors. Therefore, it is critical to develop an effective fractionation strategy for the large-scale production of high-value products discussed previously.

A biomass fractionation process suitable for commercial implementation should be beneficial for both economic and environmental aspects. Such a process should

(1) allow easy recovery for complete utilization of all fractionated components for the production of value-added products,

(2) generate fractionated components with high yield and purity for further conversion,
(3) allow recycling and reuse of the fractionation catalysts for savings in production costs,
(4) use low or non-corrosive chemicals to reduce capital costs,
(5) reduce energy consumption.

References

Alvira P, Tomás-Pejó E, Ballesteros M, Negro MJ (2010) Pretreatment technologies for an efficient bioethanol production process based on enzymatic hydrolysis: a review. Biores Technol 101(13):4851–4861

Araujo D, Vilarinho M, Machado A (2019) Effect of combined dilute-alkaline and green pretreatments on corncob fractionation: pretreated biomass characterization and regenerated cellulose film production. Ind Crops Prod 141:111785

Baker DA, Rials TG (2013) Recent advances in low-cost carbon fiber manufacture from lignin. J Appl Polym Sci 130(2):713–728

Balat M (2011) Production of bioethanol from lignocellulosic materials via the biochemical pathway: a review. Energy Convers Manage 52(2):858–875

Barakat A, de Vries H, Rouau X (2013) Dry fractionation process as an important step in current and future lignocellulose biorefineries: a review. Biores Technol 134:362–373

Barakat A, Mayer-Laigle C, Solhy A, Arancon RA, De Vries H, Luque R (2014) Mechanical pretreatments of lignocellulosic biomass: towards facile and environmentally sound technologies for biofuels production. RSC Adv 4(89):48109–48127

Cai J, He Y, Yu X, Banks SW, Yang Y, Zhang X, ... Bridgwater AV (2017) Review of physicochemical properties and analytical characterization of lignocellulosic biomass. Renew Sustain Energy Rev 76:309–322

Cai C, Hirth K, Gleisner R, Lou H, Qiu X, Zhu JY (2020) Maleic acid as a dicarboxylic acid hydrotrope for sustainable fractionation of wood at atmospheric pressure and ≤100 C: mode and utility of lignin esterification. Green Chem 22(5):1605–1617

Carvalheiro F, Duarte LC, Gírio FM (2008) Hemicellulose biorefineries: a review on biomass pretreatments. J Sci Indus Res, 849–864

Chen L, Dou J, Ma Q, Li N, Wu R, Bian H, ... Zhu JJ (2017) Rapid and near-complete dissolution of wood lignin at ≤80 C by a recyclable acid hydrotrope. Sci Adv 3(9):e1701735

Chen GG, Qi XM, Guan Y, Peng F, Yao CL, Sun RC (2016) High strength hemicellulose-based nanocomposite film for food packaging applications. ACS Sustai Chem Eng 4(4):1985–1993

Chen Z, Bai X, Wan C (2018) High-solid lignocellulose processing enabled by natural deep eutectic solvent for lignin extraction and industrially relevant production of renewable chemicals. ACS Sustain Chem Eng 6(9):12205–12216

Cherubini F (2010) The biorefinery concept: using biomass instead of oil for producing energy and chemicals. Energy Convers Manage 51(7):1412–1421

da Costa Lopes AM, João KG, Morais ARC, Bogel-Łukasik E, Bogel-Łukasik R (2013) Ionic liquids as a tool for lignocellulosic biomass fractionation. Sustain Chem Process 1(1):1–31

Dai Y, Zhang HS, Huan B, He Y (2017) Enhancing the enzymatic saccharification of bamboo shoot shell by sequential biological pretreatment with Galactomyces sp. CCZU11-1 and deep eutectic solvent extraction. Bioprocess Biosyst Eng 40(9):1427–1436

Fache M, Boutevin B, Caillol S (2016) Vanillin production from lignin and its use as a renewable chemical. ACS Sustain Chem Eng 4(1):35–46

Fang C, Thomsen MH, Frankær CG, Brudecki GP, Schmidt JE, AlNashef IM (2017) Reviving pretreatment effectiveness of deep eutectic solvents on lignocellulosic date palm residues by prior recalcitrance reduction. Ind Eng Chem Res 56(12):3167–3174

Farhat W, Venditti R, Quick A, Taha M, Mignard N, Becquart F, Ayoub A (2017) Hemicellulose extraction and characterization for applications in paper coatings and adhesives. Ind Crops Prod 107:370–377

FitzPatrick M, Champagne P, Cunningham MF, Whitney RA (2010) A biorefinery processing perspective: treatment of lignocellulosic materials for the production of value-added products. Biores Technol 101(23):8915–8922

Galkin MV, Samec JS (2016) Lignin valorization through catalytic lignocellulose fractionation: a fundamental platform for the future biorefinery. Chemsuschem 9(13):1544–1558

Gupta PK, Raghunath SS, Prasanna DV, Venkat P, Shree V, Chithananthan C, ... Geetha K (2019) An update on overview of cellulose, its structure and applications. Cellulose, 846–1297

Hou XD, Lin KP, Li AL, Yang LM, Fu MH (2018) Effect of constituents molar ratios of deep eutectic solvents on rice straw fractionation efficiency and the micro-mechanism investigation. Ind Crops Prod 120:322–329

Huang C, Wang X, Liang C, Jiang X, Yang G, Xu J, Yong Q (2019) A sustainable process for procuring biologically active fractions of high-purity xylooligosaccharides and water-soluble lignin from Moso bamboo prehydrolyzate. Biotechnol Biofuels 12(1):1–13

Ingruber O (1985) Sulfite science, part I: sulfite pulping cooking liquor and the four bases. Sulfite science and technology. Edited by Ingruber o, Kocurek m, Wong A 4:3–23

Jiang X, de Assis CA, Kollman M, Sun R, Jameel H, Chang HM, Gonzalez R (2020) Lignin fractionation from laboratory to commercialization: chemistry, scalability and techno-economic analysis. Green Chem 22(21):7448–7459

Jung YH, Kim KH (2015) Acidic pretreatment. In: Pretreatment of biomass. Elsevier , pp 27–50

Kim TH, Kim JS, Sunwoo C, Lee YY (2003) Pretreatment of corn stover by aqueous ammonia. Biores Technol 90(1):39–47

Kim TH, Lee YY (2006) Fractionation of corn stover by hot-water and aqueous ammonia treatment. Biores Technol 97(2):224–232

Kim JS, Lee YY, Kim TH (2016a) A review on alkaline pretreatment technology for bioconversion of lignocellulosic biomass. Biores Technol 199:42–48

Kim DY, Kim YS, Kim TH, Oh KK (2016b) Two-stage, acetic acid-aqueous ammonia, fractionation of empty fruit bunches for increased lignocellulosic biomass utilization. Biores Technol 199:121–127

Kim SM, Dien BS, Singh V (2016c) Promise of combined hydrothermal/chemical and mechanical refining for pretreatment of woody and herbaceous biomass. Biotechnol Biofuels 9(1):1–15

Koukios EG, Valkanas GN (1982) Process for chemical separation of the three main components of lignocellulosic biomass. Ind Eng Chem Prod Res Dev 21(2):309–314

Lan W, Liu CF, Sun RC (2011) Fractionation of bagasse into cellulose, hemicelluloses, and lignin with ionic liquid treatment followed by alkaline extraction. J Agric Food Chem 59(16):8691–8701

Laser M, Schulman D, Allen SG, Lichwa J, Antal MJ Jr, Lynd LR (2002) A comparison of liquid hot-water and steam pretreatments of sugar cane bagasse for bioconversion to ethanol. Biores Technol 81(1):33–44

Lee JY, Ryu HJ, Oh KK (2013) Acid-catalyzed hydrothermal severity on the fractionation of agricultural residues for xylose-rich hydrolyzates. Bioresour Technol 132:84–90; Li MF, Yang S, Sun RC (2016) Recent advances in alcohol and organic acid fractionation of lignocellulosic biomass. Bioresour Technol 200:971-980

Li T, Lyu G, Saeed HA, Liu Y, Wu Y, Yang G, Lucia LA (2018) Analytical pyrolysis characteristics of enzymatic/mild acidolysis lignin (EMAL). BioResources 13(2):4484–4496

Li X, Kim TH (2011) Low-liquid pretreatment of corn stover with aqueous ammonia. Biores Technol 102(7):4779–4786

Lipinsky ES (1981) Chemicals from biomass: petrochemical substitution options. Science 212(4502):1465–1471

Liu X, Lin Q, Yan Y, Peng F, Sun R, Ren J (2019a) Hemicellulose from plant biomass in medical and pharmaceutical application: a critical review. Curr Med Chem 26(14):2430–2455
Liu CG, Xiao Y, Xia XX, Zhao XQ, Peng L, Srinophakun P, Bai FW (2019b) Cellulosic ethanol production: progress, challenges and strategies for solutions. Biotechnol Adv 37(3):491–504
Luo Y, Li Z, Li X, Liu X, Fan J, Clark JH, Hu C (2019) The production of furfural directly from hemicellulose in lignocellulosic biomass: a review. Catal Today 319:14–24
Lynam JG, Kumar N, Wong MJ (2017) Deep eutectic solvents' ability to solubilize lignin, cellulose, and hemicellulose; thermal stability; and density. Biores Technol 238:684–689
Lyu G, Li T, Ji X, Yang G, Liu Y, Lucia LA, Chen J (2018) Characterization of lignin extracted from willow by deep eutectic solvent treatments. Polymers 10(8):869
Ma CY, Xu LH, Zhang C, Guo KN, Yuan TQ, Wen JL (2021) A synergistic hydrothermal-deep eutectic solvent (DES) pretreatment for rapid fractionation and targeted valorization of hemicelluloses and cellulose from poplar wood. Biores Technol 341:125828
Ma MG, Jia N, Zhu JF, Li SM, Peng F, Sun RC (2012) Isolation and characterization of hemicelluloses extracted by hydrothermal pretreatment. Biores Technol 114:677–683
Moniz P, Pereira H, Quilhó T, Carvalheiro F (2013) Characterisation and hydrothermal processing of corn straw towards the selective fractionation of hemicelluloses. Ind Crops Prod 50:145–153
Mosier N, Hendrickson R, Ho N, Sedlak M, Ladisch MR (2005) Optimization of pH controlled liquid hot-water pretreatment of corn stover. Biores Technol 96(18):1986–1993
Moxley G, Zhu Z, Zhang YHP (2008) Efficient sugar release by the cellulose solvent-based lignocellulose fractionation technology and enzymatic cellulose hydrolysis. J Agric Food Chem 56(17):7885–7890
Perez-Pimienta JA, Lopez-Ortega MG, Varanasi P, Stavila V, Cheng G, Singh S, Simmons BA (2013) Comparison of the impact of ionic liquid pretreatment on recalcitrance of agave bagasse and switchgrass. Biores Technol 127:18–24
Putro JN, Soetaredjo FE, Lin SY, Ju YH, Ismadji S (2016) Pretreatment and conversion of lignocellulose biomass into valuable chemicals. RSC Adv 6(52):46834–46852
Qu W, Yang J, Sun X, Bai X, Jin H, Zhang M (2021) Towards producing high-quality lignin-based carbon fibers: A review of crucial factors affecting lignin properties and conversion techniques. Int J Biol Macromol 189:768–784
Ragauskas AJ, Beckham GT, Biddy MJ, Chandra R, Chen F, Davis MF, ... Wyman CE (2014) Lignin valorization: improving lignin processing in the biorefinery. Science 344(6185)
Sannigrahi P, Ragauskas AJ (2013) Fundamentals of biomass pretreatment by fractionation. Aqueous pretreatment of plant biomass for biological and chemical conversion to fuels and chemicals, 201–222
Schutyser W, Renders AT, Van den Bosch S, Koelewijn SF, Beckham GT, Sels BF (2018) Chemicals from lignin: an interplay of lignocellulose fractionation, depolymerisation, and upgrading. Chem Soc Rev 47(3):852–908
Shen X, Sun R (2021) Recent advances in lignocellulose prior-fractionation for biomaterials, biochemicals, and bioenergy. Carbohy Poly, 117884
Shi J, Gladden JM, Sathitsuksanoh N, Kambam P, Sandoval L, Mitra D, ... Singh S (2013) One-pot ionic liquid pretreatment and saccharification of switchgrass. Green Chem 15(9):2579–2589
Shuai L, Yang Q, Zhu JY, Lu FC, Weimer PJ, Ralph J, Pan XJ (2010) Comparative study of SPORL and dilute-acid pretreatments of spruce for cellulosic ethanol production. Biores Technol 101(9):3106–3114
Silva GGD, Couturier M, Berrin JG, Buléon A, Rouau X (2012) Effects of grinding processes on enzymatic degradation of wheat straw. Biores Technol 103(1):192–200
Spiridon I, Popa VI (2008) Hemicelluloses: major sources, properties and applications. In: Monomers, polymers and composites from renewable resources. Elsevier, pp 289–304
Stewart D (2008) Lignin as a base material for materials applications: chemistry, application and economics. Ind Crops Prod 27(2):202–207

Su C, Hirth K, Liu Z, Cao Y, Zhu JY (2021) Acid hydrotropic fractionation of switchgrass at atmospheric pressure using maleic acid in comparison with p-TsOH: Advantages of lignin esterification. Ind Crops Prod 159:113017
Tan YT, Chua ASM, Ngoh GC (2020) Deep eutectic solvent for lignocellulosic biomass fractionation and the subsequent conversion to bio-based products–a review. Biores Technol 297:122522
Tan YT, Ngoh GC, Chua ASM (2019) Effect of functional groups in acid constituent of deep eutectic solvent for extraction of reactive lignin. Biores Technol 281:359–366
Teramoto Y, Lee SH, Endo T (2008) Pretreatment of woody and herbaceous biomass for enzymatic saccharification using sulfuric acid-free ethanol cooking. Biores Technol 99(18):8856–8863
Tian D, Guo Y, Hu J, Yang G, Zhang J, Luo L, ... Shen F (2020) Acidic deep eutectic solvents pretreatment for selective lignocellulosic biomass fractionation with enhanced cellulose reactivity. Int J Biolog Macromolecules 142:288–297
van Osch DJ, Kollau LJ, van den Bruinhorst A, Asikainen S, Rocha MA, Kroon MC (2017) Ionic liquids and deep eutectic solvents for lignocellulosic biomass fractionation. Phys Chem Chem Phys 19(4):2636–2665
Varanasi P, Singh P, Arora R, Adams PD, Auer M, Simmons BA, Singh S (2012) Understanding changes in lignin of Panicum virgatum and Eucalyptus globulus as a function of ionic liquid pretreatment. Biores Technol 126:156–161
Wang Z, Qiu S, Hirth K, Cheng J, Wen J, Li N, ... Zhu JY (2019) Preserving both lignin and cellulose chemical structures: flow-through acid hydrotropic fractionation at atmospheric pressure for complete wood valorization. ACS Sustain Chem Eng 7(12):10808–10820
Wen JL, Sun SL, Yuan TQ, Sun RC (2015) Structural elucidation of whole lignin from Eucalyptus based on preswelling and enzymatic hydrolysis. Green Chem 17(3):1589–1596
Wu X, Zhang T, Liu N, Zhao Y, Tian G, Wang Z (2020) Sequential extraction of hemicelluloses and lignin for wood fractionation using acid hydrotrope at mild conditions. Ind Crops Prod 145:112086
Wyman CE, Dale BE, Elander RT, Holtzapple M, Ladisch MR, Lee YY (2005) Coordinated development of leading biomass pretreatment technologies. Biores Technol 96(18):1959–1966
Xiao X, Bian J, Peng XP, Xu H, Xiao B, Sun RC (2013) Autohydrolysis of bamboo (Dendrocalamus giganteus Munro) culm for the production of xylo-oligosaccharides. Biores Technol 138:63–70
Xu J, Cheng JJ, Sharma-Shivappa RR, Burns JC (2010) Sodium hydroxide pretreatment of switchgrass for ethanol production. Energy Fuels 24(3):2113–2119
Xu J, Li C, Dai L, Xu C, Zhong Y, Yu F, Si C (2020) Biomass fractionation and lignin fractionation towards lignin valorization. Chemsuschem 13(17):4284–4295
Yamakawa CK, Qin F, Mussatto SI (2018) Advances and opportunities in biomass conversion technologies and biorefineries for the development of a bio-based economy. Biomass Bioenerg 119:54–60
Yan Y, Zhang C, Lin Q, Wang X, Cheng B, Li H, Ren J (2018) Microwave-assisted oxalic acid pretreatment for the enhancing of enzyme hydrolysis in the production of xylose and arabinose from bagasse. Molecules 23(4):862
Yang L, Cao J, Jin Y, Chang HM, Jameel H, Phillips R, Li Z (2012) Effects of sodium carbonate pretreatment on the chemical compositions and enzymatic saccharification of rice straw. Biores Technol 124:283–291
Yiin CL, Quitain AT, Yusup S, Uemura Y, Sasaki M, Kida T (2017) Choline chloride (ChCl) and monosodium glutamate (MSG)-based green solvents from optimized cactus malic acid for biomass delignification. Biores Technol 244:941–948
Yoo CG, Nghiem NP, Hicks KB, Kim TH (2011a) Pretreatment of corn stover using low-moisture anhydrous ammonia (LMAA) process. Biores Technol 102(21):10028–10034
Yoo CG, Lee CW, Kim TH (2011b) Two-stage fractionation of corn stover using aqueous ammonia and hot-water. Appl Bioch Biotechnol 164(6):729–740; Zhang YHP, Ding SY, Mielenz JR, Cui JB, Elander RT, Laser M, ... Lynd LR (2007) Fractionating recalcitrant lignocellulose at modest reaction conditions. Biotechnol Bioeng 97(2):214–223

Yuan Z, Wen Y, Kapu NS, Beatson R (2018) Evaluation of an organosolv-based biorefinery process to fractionate wheat straw into ethanol and co-products. Ind Crops Prod 121:294–302
Zhong X, Yuan R, Zhang B, Wang B, Chu Y, Wang Z (2021) Full fractionation of cellulose, hemicellulose, and lignin in pith-leaf containing corn stover by one-step treatment using aqueous formic acid. Ind Crops Prod 172:113962
Zhu J, Chen L, Cai C (2021) Acid hydrotropic fractionation of lignocelluloses for sustainable biorefinery: advantages, opportunities, and research needs. ChemSusChem J 14(15):3031–3046
Zhu JY, Pan XJ, Wang GS, Gleisner R (2009) Sulfite pretreatment (SPORL) for robust enzymatic saccharification of spruce and red pine. Biores Technol 100(8):2411–2418
Zhu S, Wu Y, Chen Q, Yu Z, Wang C, Jin S, ... Wu G (2006) Dissolution of cellulose with ionic liquids and its application: a mini-review. Green Chem 8(4):325–327
Zhu JY, Pan X, Zalesny RS (2010) Pretreatment of woody biomass for biofuel production: energy efficiency, technologies, and recalcitrance. Appl Microbiol Biotechnol 87(3):847–857
Zhu JY, Sabo R, Luo X (2011) Integrated production of nano-fibrillated cellulose and cellulosic biofuel (ethanol) by enzymatic fractionation of wood fibers. Green Chem 13(5):1339–1344
Zhuang X, Wang W, Yu Q, Qi W, Wang Q, Tan X, ... Yuan Z (2016) Liquid hot-water pretreatment of lignocellulosic biomass for bioethanol production accompanying with high valuable products. Bioresour Technol 199:68–75

Chapter 3
Biochemical Conversion of Cellulose

Daehwan Kim, Youngmi Kim, and Sun Min Kim

Abstract The recovery of cellulose/hemicellulose fractions from renewable lignocellulose and their integrated sugar platform processes is crucial for the production of textiles, chemicals, polymers, biofuels, and economic value-added molecules. In order to perform an effective sugar recovery, hydrothermal pretreatment is considered a valuable strategy to disrupt a complex crystalline structure of lignocellulose and improve the hydrolysis of cellulose by solubilizing hemicellulose/lignin components, and decreasing the feedstock recalcitrance. However, certain factors such as types of substrates, chemical compositions, degree of cellulose/hemicellulose polymerizations (DP), chemical properties, and pretreatment conditions (temperature, retention time, catalyst) influence the generation of undesirable inhibitory compounds, which can affect enzyme and microbial activities. This work aims to address the current biorefinery technologies for cellulose conversion, primary parameters for efficient enzymatic catalysis, lignocellulose-derived soluble inhibitors, and their performances on enzymes. Furthermore, the advanced strategies to avoid and/or minimize the negative effects of potential inhibitors are summarized and discussed.

3.1 Introduction

Cellulose and hemicellulose are the major components in plant biomass. They play a central role in the production of renewable biofuels and biomaterials as an alternative feedstock to replace fossil-based fuels (e.g., petroleum, coal, and natural gas) and

D. Kim (✉)
Department of Biology, Hood College, Frederick, MD, USA
e-mail: kimd@hood.edu

Y. Kim
Department of Agricultural Engineering Technology, University of Wisconsin – River Falls, River Falls, WI, USA
e-mail: youngmi.kim@uwrf.edu

S. M. Kim
Cargill Starches, Sweeteners and Texturizers North America, Dayton, OH, USA
e-mail: sun_min_kim@cargill.com

N. P. Nghiem et al. (eds.), *Biomass Utilization: Conversion Strategies*,
https://doi.org/10.1007/978-3-031-05835-6_3

chemicals. Currently, traditional fossil-based fuels support approximately 80% of the worldwide industrial/technological energy supply and need (Cárdenas et al. 2019). Lignocellulosic biomass can replace fossil-based energy resources to resolve the issues associated with these energy systems, such as limited energy resources, rapid climate change, and other environmental issues caused by the heavy dependence on fossil-based fuels and chemicals. Lignocellulosic materials, such as agricultural and forestry wastes, field-grown grasses, crop residues, municipal wastes, and other plant residues are regarded as competitive renewable energy sources due to their plentiful availability and their large-scale feasibility at low cost.

Lignocellulosic ethanol, for example, has been considered a potential renewable transportation biofuel for several decades. The U.S. and Brazil are the leading producers of biofuels (mainly ethanol) in the world. The U.S. produced 15.8 billion gallons of ethanol fuel and Brazil produced around 7.1 billion gallons in 2017 (https://ethanolrfa.org/markets-and-statistics/annual-ethanol-production). However, lignocellulosic bioethanol has not been widely commercialized because of the difficulties associated with the complex structure of lignocellulose, technical challenges, and cost efficiency. The physical structure of lignocellulose is defined by its cellulose, hemicellulose, and lignin constituents, which are constructed from carbon dioxide, water, and sunlight through photosynthesis. In particular, the lignin acts as a glue to bind with cellulose and hemicellulose, providing physical strength and inflexibility to the plant cell wall as well as protecting plant cells from foreign pathogens (Kim 2018; Kim and Ku 2018). Furthermore, the chemically stable aromatic structure of lignin, crystallinity of cell walls carbohydrates, and the chemical linkages, such as hydrogen, ester, and ether bonds that exist between the cellulose–hemicellulose and cellulose–lignin make the plant cell wall recalcitrant to chemical or biological decomposition. Thus, a pretreatment process is required to disrupt the highly sophisticated and complex structure of the plant cell wall and enhance the enzyme accessibility of cellulose for producing fermentable sugars.

Various pretreatment approaches have been applied to enhance cellulose conversion at low enzyme loadings. However, pretreatments often generate undesirable molecules that are inhibitory to cellulolytic enzymes and microbes during the subsequent fermentation process. For example, soluble sugars and phenolic compounds released during hydrothermal pretreatments are known to inhibit or deactivate cellulase enzymes, resulting in the decrease of final sugar titers. Additionally, inhibitory compounds, such as furan derivatives and phenolics, also have negative effects on cell viability during microbial fermentation. The presence of these inhibitory compounds increases the processing costs as they necessitate the use of more enzymes for hydrolysis and/or a pre-conditioning step to remove the inhibitory molecules after pretreatment. The complex nature of lignocellulose, the need for pretreatment steps to enhance enzymatic accessibility of cellulose, and the formation of inhibitory compounds during the pretreatment are among the key obstacles to practical utilization of lignocellulose as an alternative feedstock for fuels and chemicals. This chapter reviews the recent progress in enzymatic hydrolysis of cellulose, discusses the types of cellulase inhibitors and potential mitigation strategies, and provides the recent advances in the microbial fermentation of lignocellulose-derived sugars.

3.2 Biorefinery Technologies for Lignocellulosic Biomass Conversion

Efficient conversion of cellulosic biomass to glucose requires several steps. One of the essential steps in the initial process is pretreatment, which aims to increase enzyme accessibility to cellulose while minimizing sugar losses. The enhanced enzyme accessibility of lignocellulose could be achieved by decrystallizing cellulose, removing hemicellulose, reducing lignin recalcitrance, increasing surface area and porosity, and decreasing particle size. However, at harsh pretreatment conditions, sugars are degraded and produce toxic compounds to enzymes and microorganisms. Considering various types of feedstocks including soft wood, hard wood, crops, industrial hemp, corn stover, sugarcane bagasse, switchgrass, poplar, and corn fiber, it is hard to select the best pretreatment method (Galbe and Wallberg 2019). An ideal pretreatment should apply to a wide range of lignocellulose feedstocks, preserve more cellulose while making it susceptible to cellulolytic enzymes, and minimize the formation of degradation products that inhibit the subsequent steps (Kim 2018).

After pretreatment, saccharification and fermentation are carried out either separately (separate hydrolysis and fermentation, SHF) or simultaneously (simultaneous saccharification and fermentation, SSF). SHF allows both hydrolysis and fermentation to be performed at their optimal conditions, but end-products such as glucose and cellobiose could inhibit enzyme actions, limiting cellulose conversion during the saccharification step. SSF is performed in a single vessel where both hydrolysis and fermentation occur at the same time. This can potentially lower the capital costs and reduce end-product inhibition as sugars are fermented as they are being formed. Since cellulases are most active at temperatures higher than the optimal temperature for fermentation, SSF may require higher cellulase loadings than SHF to compensate for the suboptimal enzyme activity. Typically, higher ethanol yields have been observed by SSF than SHF, although final ethanol yields depend on pretreatment methods and types of biomass (Sukhang et al. 2020). Steam-exploded corn stover had 13% higher ethanol yields in SSF than SHF due to the reduction of glucose inhibition in the enzymatic hydrolysis during SSF, the detoxifying effect of fermentation, and the positive effect of lower concentrations of some of the inhibitors on the fermentation (Öhgren et al. 2007). For dilute acid and liquid hot water pretreated *Arundo donax* biomass, ethanol concentration (g/L) at the end of SSF and SHF were comparable. Considering overall process time, SSF had higher ethanol productivity (g/L h) compared to SHF for both hot water and dilute acid pretreated samples (Loaces et al. 2017). Regardless of detoxification methods applied, steam-exploded poplar wood had higher ethanol productivities in SSF (0.64–1.08 g/L/h) than SHF (0.11–0.37 g/L/h) (Cantarella et al. 2004).

Consolidated bioprocessing (CBP) has the potential for low-cost production of ethanol or bio-based chemicals. Enzyme production, saccharification, and fermentation are integrated within a single step using microorganisms that produce their own cellulolytic and hemicellulolytic enzymes as well as ferment hexose and pentose sugars. This can be achieved by either a "native cellulolytic" strategy, engineering native cellulolytic microorganisms to produce the desired product, or a "recombinant cellulolytic" strategy, genetically modifying non-cellulolytic microorganisms that possess high fermenting ability to have cellulolytic activities as well (Jouzani and Taherzadeh 2015). According to the recent techno-economic analysis of SHF vs. CBP of sugar cane bagasse pretreated with sodium hydroxide, 66 MM gal/y and 76.2 MM gal/y ethanol can be produced from CBP and SHF, respectively, from 3,500 U.S. tons of feed per day (Raftery and Karim 2017). However, due to zero enzyme cost in CBP, the ethanol selling price was estimated to be lower in CBP ($1.31 per gal) compared to SHF ($1.64 per gal). The ethanol selling price of CBP was projected to be 20% lower than SHF, but assumptions for the analysis were based on lab-scale experiments due to a lack of scale-up studies of CBP. Much progress has been made in the area of CBP, but many challenges remain. Microorganisms for CBP need to produce sufficient levels of cellulolytic enzymes without sacrificing production yields, co-ferment hexose and pentose sugars, and have a high tolerance to toxic compounds (Haan et al. 2014).

3.3 Primary Factors for Effective Enzymatic Hydrolysis

3.3.1 Types of Enzymes

Cellulases are enzymes that hydrolyze the β-1,4-D-glycosidic bonds of cellulose to glucose, cellobiose, and cello-oligosaccharides. Three types of enzymes act together synergistically to hydrolyze cellulose to glucose. Endoglucanases (EGs; EC 3.2.1.4) randomly cleave inter-bonds of amorphous cellulose. Exoglucanases, also known as cellobiohydrolases (CBHs), produce cellobiose either from reducing (CBH I; EC 3.2.1.176) or non-reducing (CBH II; EC 3.2.1.91) ends of oligosaccharide chains generated by endoglucanases. β-glucosidases (BGs), also called cellobiases (EC 3.2.1.21), hydrolyze cellobiose or short oligosaccharides to glucose. BGs increase glucose production rates by decreasing cellobiose inhibition (Shokrkar et al. 2018).

Lytic polysaccharide monooxygenases (LPMOs) were discovered recently (Levasseur et al. 2013). It was found that LPMOs improve cellulose hydrolysis by acting synergistically with cellulases. In the Carbohydrate Active Enzymes (CAZy) database, LPMOs are classified in the "Auxiliary Activities" family (Levasseur et al. 2013). LPMOs are copper-dependent enzymes that cleave polysaccharides including chitin, cellulose, xyloglucan, mixed-linkage glucan, glucomannan, and starch by an oxidative process involving O_2/H_2O_2 and an electron donor (Caldararu et al. 2019). LPMOs bind to the flat, solid, well-ordered surface of crystalline nano-fibrils in cellulose and break inaccessible crystalline polysaccharides into shorter and thinner insoluble fragments, promoting an overall faster and more complete surface degradation (Fig. 3.1). By cleaving glycosidic linkages of crystalline nano-fibrils in cellulose by hydroxylation at the C1 or C4 carbon, more new chain ends can be generated. As a result, substrates become more prone to attack by the classical EGs and

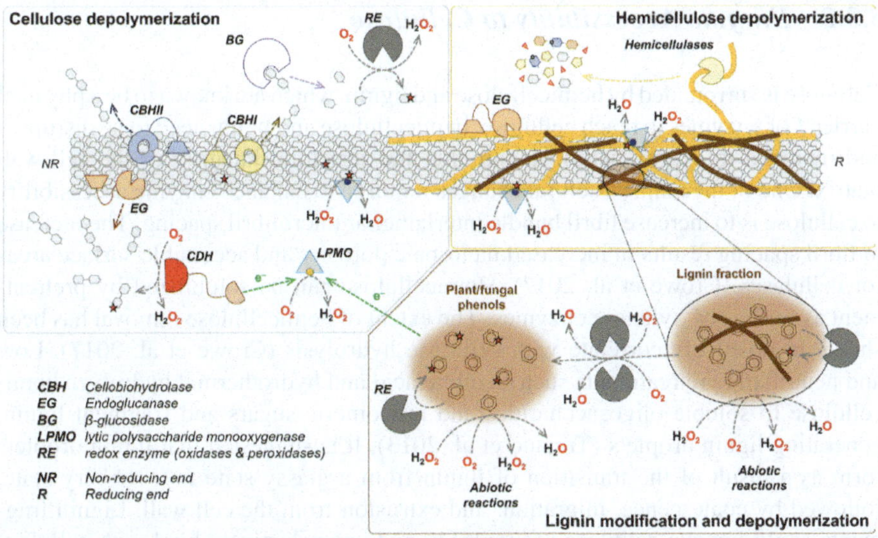

Fig. 3.1 Enzymatic depolymerization of plant cell wall polysaccharides: a cellulose fibril (gray) covered with hemicellulose (orange) and lignin (brown). Reproduced from Østby et al. "Enzymatic processing of lignocellulosic biomass: principles, recent advances, and perspectives." Published in *J Ind Microbiol Biotechnol,* 2020 (Østby et al. 2020))

CBHs (Eibinger et al. 2014). LPMOs are monooxygenases; however, H_2O_2 is the preferred co-substrate over O_2. LPMOs can function in the absence of O_2, but high H_2O_2 concentration leads to inactivation (Bissaro et al. 2017; Kuusk et al. 2018). By controlling H_2O_2 supply in anaerobic hydrolysis such as maintaining low H_2O_2 concentration or reducing H_2O_2 supply over time, LPMO activity can be enhanced resulting in improved saccharification yields of pretreated biomass (Bissaro et al. 2017; Müller et al. 2018). Various electron donors activate LPMOs such as ascorbic acid, glutathione, gallic acid, resveratrol, catechin, caffeic acid, sinapic acid, and also hydroquinone. Moreover, the cell wall matrix, insoluble high molecular weight lignin, also serves as an electron donor (Westereng et al. 2015). Lignosulfonates produced from lignin during sulfite pretreatment also have the potential to act as electron donors to activate LPMOs, which could reduce costs associated with the addition of small molecule reductants (Chylenski et al. 2017b). Metal ions can also enhance LPMO activities. Glycoside hydrolase 61 (GH61) proteins that belong to the LPMO family lack measurable hydrolytic activity in the absence of metal ions, but the presence of divalent metal ions can increase their activity (Jung et al. 2015). Especially, the presence of cobalt (Co^{2+}) enhanced GH61 activity 52-fold higher for carboxymethyl cellulose substrate.

3.3.2 Enzyme Accessibility to Cellulose

Cellulose is surrounded by hemicellulose and lignin, which are known to be a physical barrier. For enzymes to reach cellulose, hemicellulose and lignin need to be disrupted and removed to expose cellulose and alter the structural integrity of the cell wall matrix. One of the main structural changes required to enhance enzyme accessibility to cellulose is to increase fibril bundle interlamellar microfibril spacing. The increase in fibril spacing results in increased nanoscale porosity and accessible surface areas for cellulases (Crowe et al. 2017). Hemicellulose can be solubilized by pretreatment as well as by xylanase enzymes. The extent of hemicellulose removal has been shown to positively correlate with cellulose hydrolysis (Crowe et al. 2017). Low and neutral pH pretreatments such as dilute acid and hydrothermal hydrolyze hemicellulose to soluble oligosaccharides and monomeric sugars and fragment lignin, generating lignin droplets (Trajano et al. 2013). It is suggested that lignin droplets form as a result of the transition of lignin from a glassy state to a rubbery state, followed by coalescence, migration, and extrusion from the cell wall. Lignin fragments could impede cellulose accessibility and unproductively bind with cellulase enzymes. Especially lignin droplets on the cellulose surface inhibit the procession of enzymes along the surface and inner layers (Li et al. 2014a, b).

Although degrees of hemicellulose and lignin removal vary by pretreatment methods and types of biomass, pretreatment, in general, aims to destroy the spatial structure, expose the cellulose surface, and increase the pore volume of the plant cell wall structure. At a severe acidic pretreatment condition, lignin is solubilized and lignin structure is rearranged, which increases pore volume further (Hsu et al. 2010). High-pH pretreatments disrupt lignin but have less impact on hemicellulose (Galbe and Wallberg 2019). Alkali pretreatment could cleave ferulate and *p*-coumarate cross-linkages between hemicellulose and lignin and could remove aromatic ring structures of lignin, increasing pore volume (Alam et al. 2019). Similar work done by Yoo et al. (2016) elucidated that the modification of ferulate and *p*-coumarate esters could also improve the cellulolytic saccharification. The pore volume of liquid hot water, H_2SO_4, and NaOH pretreated *Miscanthus* were compared (Alam et al. 2019). At the optimal conditions for each pretreatment method, the NaOH pretreated sample had the most pore volume, as confirmed by multiple assays including Simons stain, Congo red dye adsorption, cellulase enzyme adsorption, and N_2 adsorption. Pore volume had a high correlation coefficient (0.9) with hexose yields and bioethanol yields. For *Populus*, H_2SO_4 was more effective to increase pore area and cellulose accessibility compared to liquid hot water and NaOH pretreatment (Meng et al. 2015). Dilute acid pretreated samples had 4.3 times more total pore area with smaller pore diameters than NaOH-pretreated samples. Hemicellulose removal was more important for *Populus* to increase cellulose accessibility than lignin removal, but lignin restricts xylan accessibility, which in turn controls the access of cellulase to cellulose. Similar results were observed for sugarcane bagasse that removal of hemicellulose had a stronger positive correlation with enzymatic digestibility than delignification (Lv et al. 2013). Pore volume and specific surface area of corn stover were increased significantly after ultrafine grinding, alkaline hydrogen peroxide, dilute acid, and ammonia fiber expansion pretreatments (Li et al. 2019). Especially pore volume and specific surface area were increased by 13 and 25 times, respectively, by hydrogen peroxide. Both pore volume and specific surface area had a positive correlation with glucose yields.

3.3.3 Solids (Cellulose Substrate) Concentration

Along with low enzyme dosage, high titer from high solids hydrolysis is one of the key approaches to make cellulose conversion economically feasible. Generally, high-solids hydrolysis is considered above 15% (w/w) solids loading, but to be economically feasible, initial solids loading should be above 20% (w/w) (Chen and Liu 2017). It is challenging to conduct high-solids hydrolysis due to mass transfer limitation, high viscosity, and high inhibitor concentration. Inhibitors such as phenolics, acetic acid, furan compounds, and xylooligosaccharides can be removed by washing after pretreatment (Liu and Kong 2016). Washing three times can remove almost all inhibitors (Nogueira et al. 2018). However, a large amount of water is used in the post-washing step consisting of 10–20% of total water consumption (Tan et al. 2020).

Various types of bioreactors such as stirred tank and membrane bioreactors and operation modes such as batch, fed-batch, and continuous modes have been investigated to perform high-solids hydrolysis (Pito et al. 2018). The most commonly used bioreactors are stirred tank bioreactors, in which adequate mass and heat transfer can be achieved by different types of agitation systems, such as Rushton turbine, Elephant ear impeller, Helical ribbon impeller, Peg mixer, and Anchor impeller. Enzymatic hydrolysis and separation can be done in membrane bioreactors, allowing inhibitor removal and possible enzyme recovery and reuse. Fed-batch mode is promising in high-solids loading hydrolysis, which provides better mixing and fast adjustment of pH and temperature (Hernández-Beltrán and Hernández-Escoto 2018). By lowering viscosity throughout hydrolysis, high sugar conversion can be achieved in fed-batch mode. High-solids loading (30%, w/v) in fed-batch mode had comparable hydrolysis conversion efficiency to 10% solids loading in batch mode (Snatos-Rocha et al. 2018). Fed-batch hydrolysis can increase solids loading up to 50% (w/v) when combined with two-stage enzymatic hydrolysis and PEG 4000 addition (Cheng et al. 2020). Feeding substrate and enzyme mixture concomitantly resulted in better sugar conversion than adding whole enzyme mixtures at the beginning. Especially enzyme activity diminished gradually as enzymatic hydrolysis proceeded when adding the whole enzyme mixture at once (Gong et al. 2020). In addition, feeding substrate more frequently and in smaller amounts prevents accumulation of biomass, and hence reduces the power required for agitation (Snatos-Rocha et al. 2018). When enzymes and substrates were added simultaneously in fed-batch mode, 52% higher energy efficiency was obtained than in batch mode (Corrêa et al. 2016).

3.4 Inhibitors of Cellulase Enzymes

3.4.1 *Cellulose–Hemicellulose–Lignin-Derived Soluble Inhibitors*

In general, pretreatment of lignocellulose releases various compounds in the aqueous phase that are inhibitory to hydrolysis and fermentation (Kim et al. 2013, 2011). These inhibitory compounds, both naturally occurring and process derived, interfere with the subsequent enzymatic and fermentation steps of the pretreated biomass. The generation of soluble inhibitory compounds during pretreatment is considered one of the most significant barriers to the enzymatic hydrolysis of lignocellulose (Ko et al. 2015a, b). These soluble inhibitory compounds can be categorized into two groups: sugars and non-sugar compounds. The sugar-based inhibitory compounds include sugar monomers and oligomers (e.g., xylo-oligomers) derived from mostly hemicellulose and extractive fractions of the biomass. The non-sugar inhibitors include furan derivatives (e.g., furfural and 5-hydroxymethylfurfural (HMF)) and other mostly phenolic compounds formed from the degradation of sugar monomers, organic acids (acetic, levulinic, formic acids), low and high molecular weight phenolics, and aromatic compounds released from lignin–carbohydrate complexes (Palmqvist 2000a, b). Figure 3.2 summarizes the pathways of inhibitor formation from lignocellulosic biomass.

The nature and levels of the soluble enzyme and fermentation inhibitors generated during pretreatment depend on many factors: types of lignocellulose feedstocks, catalysts used in the pretreatment, pretreatment severity (temperature, duration, acidity), and solids loading (Kim et al. 2015). Regardless of the feedstock and pretreatment

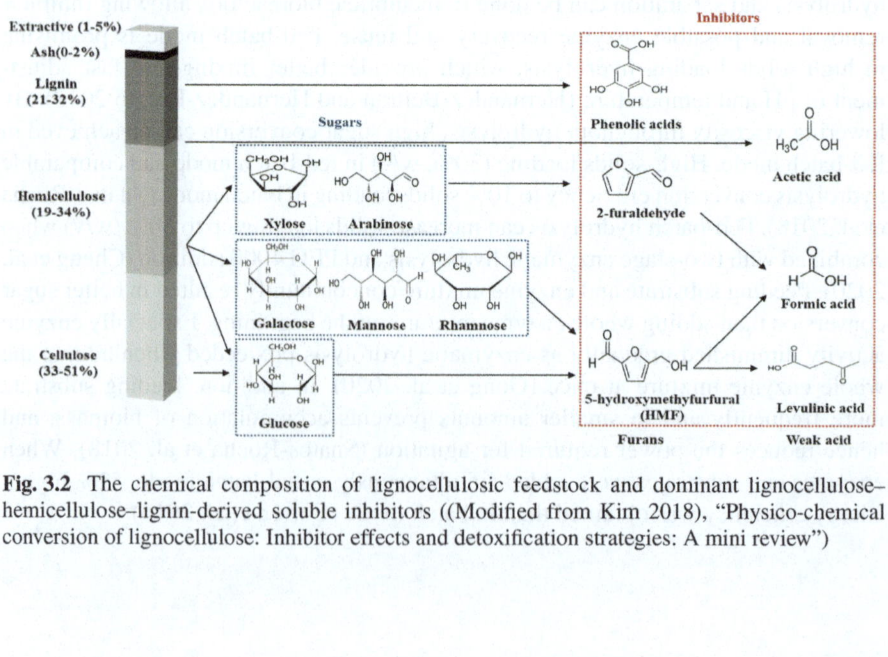

Fig. 3.2 The chemical composition of lignocellulosic feedstock and dominant lignocellulose–hemicellulose–lignin-derived soluble inhibitors ((Modified from Kim 2018), "Physico-chemical conversion of lignocellulose: Inhibitor effects and detoxification strategies: A mini review")

method applied, processing the lignocellulosic feedstock at a high concentration of solids (>15% dry solids) is a requirement to reduce the water consumption and downstream processing costs of lignocellulosic conversion. However, the levels of soluble inhibitors also increase with the solids loading in the pretreatment process. Thus, the increased inhibitory effects of the water-soluble fraction of the pretreated slurry may offset the benefit of high-solids loading pretreatment in the subsequent enzymatic and fermentation processes unless the potential adverse effects of those soluble inhibitors are avoided by applying a proper mitigation strategy. The enhanced accessibility of enzymes to high-severity pretreated cellulose could also be partially canceled by the increased presence of inhibitors released during the high-severity pretreatment. In addition, the inhibitory effects of soluble inhibitors are often synergistic, suggesting that the compounded inhibitory effects of the various soluble inhibitors are more significant than the combined incremental inhibition of each inhibitor (Alam et al. 2019). In addition, the inhibitory impact of soluble inhibitors will be even more amplified at low cellulase loadings, which is required to make the lignocellulosic bioprocessing economically attractive (Jönsson and Martín 2016). Thus, developing an efficient mitigation strategy to overcome the inhibitory impacts of soluble inhibitors is a critical consideration in improving the economic feasibility of enzyme-based lignocellulose processing.

3.4.1.1 Monosaccharides, Oligosaccharides, and Furan Derivatives

The hemicellulose fraction of lignocellulosic material releases mono- and oligomeric pentose and hexose sugars during pretreatment. The resulting oligomers have a broad distribution of degrees of polymerization (Moniz et al. 2016) and can be hydrolyzed to monomeric sugars, which can be decomposed at more severe pretreatment conditions, in other words, at high reaction severities. Depending on the combined severity of the reaction and the acidity of the water resulting from the released acetic acid, these hemicellulose derived C5 and C6 mono- and oligomeric sugars may further degrade to furfural and HMF through acid-catalyzed dehydration reactions. Subsequent rehydration of HMF leads to the formation of levulinic and formic acid. Under harsh reaction conditions, these furan derivates can further proceed to form condensation products called humins. Recent studies have shown that xylose and xylose oligomers interfere with the binding of cellulase enzymes with cellulose by obstructing the active sites of the cellulase enzymes (Cao et al. 2013; Qing et al. 2010). Interestingly, the inhibitory effect of xylose oligomers varies with cellulase enzyme components. Studies have indicated that, while the presence of xylose and xylose oligomers reduces overall cellulase activity, β-glucosidase was not impacted by xylose and xylo-oligomers (Qing et al. 2010). It has been demonstrated in many studies that sugar oligomers are generally more inhibitory than monomeric sugars at the equivalent concentrations (Kim et al. 2011; Qing et al. 2010). The extent of inhibition caused by sugar oligomers on cellulase enzymes varies. While some studies claim that sugar oligomers are strong cellulase inhibitors (Fang et al. 2018; Kumar and Wyman 2014; Qing et al. 2010), other studies have found the inhibition by sugar oligomers to be minimal or negligible (Kim et al. 2011; Zhai et al. 2016). The study by Fang et al. (2018) utilized the pre-hydrolyzate of softwood (loblolly pine), which contains a high level of mannose oligomers, to study its impact on cellulase activities. The authors found that among the soluble inhibitors in the softwood pretreatment liquid, oligomeric sugars were the most inhibitory to the commercial and recombinant cellulase enzymes they tested. On the other hand, studies have also suggested the minimal or limited impact of sugar oligomers on cellulose hydrolysis. Kim et al. (2011) showed that less than a 10% yield increase was observed using Solka-Floc® when all the xylo-oligomers in the pretreatment liquid were hydrolyzed to xylose. The glucose yield ceased to increase even in the absence of xylo-oligomers in the same study. These results suggest that the anticipated beneficial effect of removing sugar-oligomers on cellulose conversion is limited due to the presence of more potent enzyme inhibitors in the pretreatment liquid. Similarly, Zhai et al. (2016) demonstrated that the oligomeric sugars derived from both softwood and hardwood exhibited little inhibition in cellulose hydrolysis by enzymes and identified sugar monomers and phenolics as the major contributors to cellulase inhibition.

3.4.1.2 Lignin-Derived Soluble Phenolics

Soluble phenolic compounds are released from the lignin and phenolic ester groups within the hemicellulose fraction of lignocellulose during physicochemical pretreatment (Jönsson and Martín 2016). Hydrothermal pretreatment, for example, partially solubilizes lignin-derived phenolic compounds due to partial depolymerization of lignin and lignin–carbohydrates complexes, generating a water-soluble fraction containing the soluble lignin residues. The amounts and compositions of phenolic compounds released during hydrothermal pretreatments also greatly vary depending on the raw feedstocks, type and severity of pretreatment, and solids concentration during pretreatment (Liu et al. 2020). The most common phenolic compounds found in the water-soluble fraction of the pretreated lignocellulose include simple phenolic acids (syringic acid, ferulic acid, vanillic acid, syringaldehyde, *trans*-cinnamic acid, coumaric acid, *p*-hydroxybenzoic acid, gallic acid, and caffeic acid) and polymeric phenolic residues (anthocyanins, lignans, tannic acid, ellagic acid, epicatechin, etc.) (González-bautista et al. 2017; Jönsson and Martín 2016). Among these phenolic compounds, ferulic, vanillic, and syringic acids; vanillin; and syringaldehyde are the most prevalent in the liquid obtained from hydrothermally pretreated hardwoods. On the other hand, softwood releases ferulic, vanillic, *p*-hydroxybenzoic, and syringic acids, as well as coniferyaldehyde as major phenolics during autohydrolysis (Domínguez et al. 2017). The presence of these soluble phenolics in cellulose hydrolysis and microbial fermentation can be problematic as they may negatively impact the conversion efficiencies of cellulose and fermentation yields (Ximenes et al. 2010, 2011; Kim et al. 2017).

Inhibitory levels of the phenolic compounds vary depending on their structure and composition. Studies indicated that non-productive lignin–enzyme interactions are caused by phenolic hydroxyl groups of lignin compounds, which results in reduced cellulose and xylan hydrolysis (Boukari et al. 2011). The study by Li et al. (2014a, b) used vanillin as a model phenolic compound to test the effects of different functional groups (aldehyde, methoxyl, or phenolic hydroxyl group) in vanillin on cellulase inhibition. Their results showed that the presence of aldehyde and phenolic hydroxyl groups in vanillin strongly correlated with cellulase inhibition, while the effect of the methoxyl group on cellulase inhibition was negligible. However, another study by Zhai et al. (2018) has provided a contradictory result regarding the effect of the hydroxyl group on cellulase inhibition. It has shown that the inhibitory effect of the hydroxyl group in phenolic compounds derived from steam-pretreated hardwood was negligible. The same study has indicated that the carbonyl group of phenolics is the major contributor to cellulase inhibition. It has been suggested that phenolic carbonyl compounds increase phenolic hydrophobicity, resulting in increased unproductive interactions between phenolics and enzymes (Zhai and Hu 2018). Similarly, the carboxylic acid groups of phenolics and lignin exhibited less inhibitory effects than phenolics with aldehyde or alcohol groups due to the decreased lignin–cellulase binding (Nakagame et al. 2011a; Qin et al. 2016; Zhai et al. 2018). The study by Mathibe et al. (2020) used several model lignin derivatives to study their effects on endo-1,4-β-xylanase and found that the presence of carbonyl groups in phenolics exhibited more substantial inhibitory effects than hydroxyl groups on endo-1,4-β-xylanase, resulting in over 50% reduction in activity at concentrations as low as 0.5 mg/ml.

The inhibitory effect can be even more pronounced with oligomeric phenolics, such as tannic acid (Rasmussen et al. 2017; Tejirian and Xu 2011; Ximenes et al. 2011), than with simple monomeric phenolic compounds. Tannic acid, for example, deactivates β-glucosidase in *Aspergillus niger* at a ratio of 1.5 mg tannic acid/mg protein, resulting in an 80% loss of cellobiase activity. In contrast, simple phenolic acids, such as gallic, cinnamic, ferulic, *p*-coumaric, and vanillin, resulted in less than 30% deactivation at the same mg inhibitor to mg protein ratio (Ximenes et al. 2011). Rasmussen et al. (2017) have identified 26 oligo-phenol cellulase inhibitors formed by pentose degradation and condensation reactions with phenolic compounds during hydrothermal pretreatment of wheat straw. The study also has shown that oligo-phenolics were significantly more inhibitory to cellulases than xylo-oligomers. On the other hand, Zhai et al. (2018) have shown that smaller phenolics (<1 kDa) exhibit about twice more potent inhibition than relatively high molecular weight lignin.

3.4.2 Enzyme Interactions with Lignin

Non-productive binding between the residual lignin in the pretreated lignocellulose and cellulase enzymes is recognized as the primary cause of poor cellulose conversion (Kim 2018; Ázar et al. 2020; Bordignon et al. 2022). The residual lignin responsible for causing unproductive binding with cellulase components includes both undissolved structural lignin and lignin fraction that has been solubilized, condensed, and precipitated upon pretreatment (Liu and Kong 2016). The non-productive binding of cellulases with residual lignin also limits enzyme recyclability (Liu and Kong 2016). The impact of non-productive binding tends to be emphasized at a low enzyme loading, which is a necessary condition to make cellulose conversion economically competitive. Because of the inherent structural and chemical complexity associated with lignin and how it interacts with enzymes, alleviating the inhibitory effects of residual lignin remains a significant obstacle in achieving sustainable biochemical conversion of lignocellulose. While the impact of lignin on cellulase enzymes is widely studied and extensively investigated, the relationships between the inhibition and the properties of lignin are still not well understood due to the complex nature of lignin and the significant variability in the physicochemical characteristics and contents among the lignin sources. Recent studies suggest that lignin content alone does not predict the extent of inhibition, and the inhibition effects vary depending on the sources of lignin and the pretreatment methods applied (Qin et al. 2014). For example, lignin isolated from softwood is more inhibitory than lignin from herbaceous materials (Wang et al. 2020a, b). Hao et al. (2019) showed that dilute acid pretreatment generates more inhibitory lignin than alkaline pretreatment. Studies found that the concentration, structure, molecular weight, and hydrophobicity of residual lignin change with pretreatment, and the inhibitory effect of the residual lignin on cellulases increases with the pretreatment severity (Kellock et al. 2019). While residual lignin may act as a physical barrier to cellulases by blocking their access to cellulose (Djajadi and Meyer 2018), the interactions of lignin and enzymes leading to unproductive binding and loss of active cellulase enzymes have been identified as the major culprit of the inhibition caused by residual lignin (Hao et al. 2019; Kellock et al. 2019). Studies have found that the governing interactions that result in the unspecific cellulase adsorption to lignin include hydrophobic and electrostatic interactions and hydrogen bonding between the residual lignin and cellulases (Michelin et al. 2016; Nakagame et al. 2011a; Qin et al. 2014; Strobel et al. 2016, 2015; Yarbrough et al. 2015). More than one type of interaction is likely involved in the non-productive binding between lignin and cellulases (Guo et al. 2014; Lu et al. 2016).

3.5 Strategies to Improve Enzymatic Hydrolysis of Cellulose

3.5.1 Enzyme Cocktail Optimization

Several research studies have been conducted to improve hydrolysis efficiency and reduce operating costs by optimizing enzyme cocktails, engineering microorganisms for improved enzyme properties and stability, and recycling enzymes (Adsul et al. 2020; Binod et al. 2019; Bordignon et al. 2022). Enzyme cocktail optimization to achieve synergistic effects of different classes of enzymes, auxiliary enzymes, and supplements is required to lower enzyme loading and shorten hydrolysis time as well as increase sugar yields. Individual enzymes have specific modes of action that would not be sufficient to hydrolyze cellulose completely. An enzyme cocktail comprising various classes of enzymes enables one enzyme to act on the products of another enzyme (Lopes et al. 2018). The types of enzymes and their ratio are important in enzyme cocktail optimization. Recent work (Agrawal et al. 2018) optimized cellulase enzyme cocktails through a central composite design. A Celluclast® 1.5 L, β-glucosidase, and xylanase cocktail in a protein ratio of roughly 1:2:2 resulted in the highest hydrolysis yields (>95%) of cellulose at an enzyme dose of 49 mg per g biomass. The optimized enzyme cocktail enabled a 25% reduction in enzyme dosage. Sugar yields were improved by replacing cellulase with xylanase and LPMOs. The relative amounts of xylanase and LPMOs that were needed to replace cellulase were different at various solid loadings (Hu et al. 2015). At 20% solid loading, 1.3–2 times more xylanase was needed, but 3.4–6.7 times fewer LPMOs were needed compared to 2% solid loading to achieve the maximum sugar yields. A similar study (Berlin et al. 2007) demonstrated that an optimized enzyme mixture using accessory enzymes allowed for about a twofold reduction in the total protein required to achieve 99% glucose yield and 88% xylose yield.

Synergistic effects between glycoside hydrolases including exo–exo synergism, processive endo–endo synergism, endo–exo synergism, and intramolecular synergy between catalytic domains and cellulose-binding domains are required for efficient degradation of cellulose (Thoresen et al. 2021). However, cellulase interaction is not always positive, so enzyme compatibility is important when optimizing enzyme cocktails. For example, CBH can be obtained from various microorganisms such as *Hypocrea jecorina*, *Clostridium stercorarium*, *Trichoderma longibrachiatum*, and *Clostridium thermocellum*. CBH II enzymes from different microorganisms could be exchangeable without sacrificing activity, while CBH I enzymes from different microorganisms did not exhibit the same exchangeability. When designing enzyme cocktails, the compatibility of endo- and exo-cellulase needs to be considered (Thoresen et al. 2021).

One of the challenges for enzyme cocktail preparation is that specific tailoring is necessary according to the types of biomass and pretreatment methods. For example, xylanase supplementation was more effective for pretreatments that remove less xylan, such as AFEX. Acetyl group removal during pretreatment increased xylanase efficiency and enhanced the accessibility of cellulose and/or cellulase effectiveness (Kumar and Wyman 2009). When the enzyme cocktail was optimized by adjusting the ratio of EG II, CBH I, and BG I, an enzyme mixture comprising more than 90% CBH I was required for hydrolysis of dilute acid pretreated corncob residue and sulfuric acid pretreated wheat straw at 20% solid loading, probably due to the strong adsorption of CBH I on lignin. A high proportion of EG II (65%) in the enzyme mixture was needed for hydrolysis of ammonium sulfite pretreated wheat straw, followed by CBH I (21.2%) and BG I (13.8%) (Du et al. 2020). In another study, pectinases and laccase had a positive impact on enzymatic hydrolysis of dilute acid pretreated wheat straw while showing a negative impact on the hydrolysis of steam pretreated wheat straw, which might be due to competition between enzymes for cellulase binding sites and slight inhibition of β-glucosidase activity (Agrawal et al. 2018). Even when the same pretreatment conditions are applied, different types of biomass require their enzyme optimization. In one example, steam-pretreated poplar and corn stover required different enzyme ratios. At 20% solids loading, the highest sugar yields were observed with enzyme cocktail of cellulase (Celluclast® 1.5 L), xylanase, and LPMOs in the proportion of 87%, 10%, and 3%, respectively, for steam-pretreated poplar, and 65%, 33%, and 2%, respectively, for steam-pretreated corn stover (Hu et al. 2015). Similarly, sulfite-cooked sugarcane bagasse and Norway spruce required different enzyme cocktails to maximize cellulose conversion. Relatively similar enzyme doses were required to hydrolyze sugarcane bagasse, at 28%, 30%, and 30% of total protein for *Tr*Cel7A (CBH I), *Tr*Cel7B (EG I), and *Tr*Cel6A (CBH II), respectively. However, acid sulfite-cooked Norway spruce required a higher proportion of endoglucanase *Tr*Cel7B, which comprised nearly half (48%) of the total protein loading (Chylenski et al. 2017a).

3.5.2 Conditioning/Removal of Soluble Inhibitory Compounds

Mitigating the soluble inhibitors in lignocellulose bioprocessing is critical to reducing the enzyme cost, one of the significant cost contributors of lignocellulose bioconversion. Physical separation of the water-soluble fraction of pretreated biomass slurry followed by hot water washing of the solids fraction of the pretreated biomass can prevent the inhibitory effects of soluble inhibitors on the subsequent enzymatic processing of the pretreated cellulose (Kim et al. 2013). The separated liquid fraction of the pretreated slurry, which mainly contains sugars and sugar oligomers derived from hemicellulose, could be processed separately from the cellulose-rich solids fraction. Although the separation and hot water washing entail increased water and energy consumption in the overall process, this approach enables a separate processing route that can be optimized for hemicellulose utilization and more diversified end-products. The potential applications of the hemicellulose recovered in the pretreatment liquid have been widely studied. For example, hemicellulose obtained from lignocellulose has been studied extensively as an ingredient for functional foods due to its various health benefits such as prebiotic, antimicrobial, and anti-inflammatory effects (Gullón et al. 2014; Moniz et al. 2016). Also, xylo-oligomers and pentose sugars in the pretreatment liquid can be separated, purified, and processed into various functional biomaterials, such as nutraceuticals, films, and hydrogel drug delivery vehicles (Peng and She 2014; Sun et al. 2015). The soluble lignin-derived phenolics are considered renewable and promising sources of bioactive compounds that exhibit antioxidant, antimicrobial, and pharmacological activities (Domínguez et al. 2017). Alternatively, the pretreatment liquid can be detoxified to mitigate the effects of the soluble inhibitors and combined with the pretreated cellulose in the subsequent hydrolysis and fermentation process, in which both hexose and pentose sugars are utilized to produce a final product. These approaches ensure the maximum utilization of sugars derived from lignocellulose in producing a single identified product in the co-fermentation process.

3.5.3 Chemical/Biological Detoxification with Additives

Various chemical/biological detoxification methods have been applied to mitigate the effects of phenolic inhibitors, resulting in improved cellulose and fermentation efficiencies (Kim et al. 2016). Examples of such approaches can be categorized into chemical, physical, and biological methods. Chemical methods include overliming (Zhang et al. 2018) and sulfite treatment (Larsson et al. 2001a, b). Examples of physical removal of phenolics include vacuum evaporation (Larsson et al. 2001a, b) and removal by adsorbents, such as polymeric resins or activated carbon (Zhai et al. 2016). Recently, fly ash has been suggested as an efficient, no-cost adsorbent to remove the soluble phenolics in pretreatment liquids (Freitas and Farinas 2017). However, chemical and physical detoxification approaches often result in the loss of fermentable sugars and the generation of waste streams, which could add to the overall cost of lignocellulose conversion. Biological methods utilize microorganisms or enzymes to detoxify the soluble phenolic inhibitors. Several microorganisms, such as *Cupriavidus basilensis, Coniochaeta ligniaria*, and *Kurthia huakuii* have been identified as promising candidates to metabolize the soluble phenolic inhibitors (Kim 2018; Nichols et al. 2020; Saha et al. 2015). Bioabatement using *Coniochaeta ligniaria* NRRL3061, for example, removed soluble cellulase inhibitors in hydrothermally pretreated corn stover liquid, resulting in a 20–40% increase in sugar yields even at low enzyme loadings (Cao et al. 2015; Kim et al. 2016). The effect was comparable to overliming, while the approach did not require any chemicals and minimized the generation of waste streams.

Lignin blocking additives that can prevent the non-specific binding between cellulases and lignin have also been widely studied. Such agents include surfactants (such as Tween, PEG, Triton X100, and polyvinylpyrrolidone) to increase cellulose accessibility to cellulases by lowering the surface tension of lignocellulose and/or improving the thermal stability of cellulases (Chen et al. 2018; Wang et al. 2020a, b); amphiphilic lignin derivatives to enhance the activity, stability, and recyclability of cellulases (Cheng et al. 2017; Lin et al. 2019); and ionic polymers (such as lignosulfonates and polyacrylamide) to increase cellulase binding to cellulose while preventing the unproductive binding of cellulases to lignin (Wang et al. 2015). Proteins and peptides, such as bovine serum albumin (BSA), peptone, yeast extract, expansin, and artificial peptides, have been extensively studied to reduce the interactions between residual lignin and enzymes (Brondi et al. 2019; Florencio et al. 2016; Kim et al. 2021). Among the lignin blocking agents, BSA has proven to be effective in various types of pretreated lignocellulose to reduce the non-productive binding between cellulases and lignin. However, its high cost prohibits application on a commercial scale. Thus, low-cost alternative proteins that can perform similarly to BSA have also been studied. Such potential substitutes include soybean proteins, whey, and bacterial expansin, all of which have shown a noticeable improvement in glucose yields compared to the control conditions without the additives (Luo et al. 2019). In addition to the effectiveness of the mitigation method applied, the cost and availability of additives would be essential in selecting the best strategy. Future research should focus on developing effective mitigation strategies with minimal costs to the overall processing.

3.5.4 Surface Hydrophobicity Modification of Lignin and Cellulases

The adsorption affinity of cellulases towards lignin is inversely proportional to lignin content and hydrophobicity (Heiss-blanquet et al. 2011; Yang and Pan 2015). Hydrophobic interactions have been recognized as the dominant force, especially between the carbohydrate-binding module (CBM) of cellulases (endoglucanases, cellobiohydrolases) and lignin (Kawakubo et al. 2010). The amino acid sequence of the CBM determines its hydrophobicity and affinity toward lignin. Costaouëc et al. (2013) compared the lignin adsorption of EGI and CBHI and found that Cel7B (EGI) CBM exhibited a higher level of hydrophobicity and affinity for lignin than Cel7A (CBHI). Strobel et al. (2015) have demonstrated that adding negatively charged amino acids to CBM through site-directed mutagenesis improved cellulose hydrolysis. The study also showed that removing positively charged residues from the linker reduced the adsorption specificity toward lignin. It has been suggested that the hydrophobic interaction between CBM and lignin occurs at the three tyrosine amino acid residues of CBM that dominate the binding with cellulose (Guo et al. 2014; Rahikainen et al. 2013). The results suggest that mutation of the CBMs of cellulase components could reduce lignin binding and improve cellulose hydrolysis.

A recent study by Wang et al. (2020a, b) investigated the adsorption/desorption characteristics of different cellulase components on enzymatic residual lignin (ERL) pretreated by dilute acid with different severities. The results indicate that the hydrophobicity of residual lignin increases with the severity of dilute acid pretreatment due to the increased formation of condensed aromatics and the elimination of hydrophilic groups at higher severities. The same study found that the inhibitory effect was more severe for endoglucanase (Cel5A) than cellobiohydrolase (Cel7A) or β-glucosidase (Cel3A). The binding of each cellulase component by lignin, as measured by assaying the cellulase activities in supernatants, increased with pretreatment severity, while desorption was lower for ERL pretreated at higher severities (Wang et al. 2020a, b).

Hydrophobicity of lignin varies depending on the sources of lignin and pretreatment method and affects the loss of cellulases via hydrophobic attraction (Nakagame et al. 2011a; Notley and Norgren 2010). Considering that lignin is more hydrophobic than cellulose, the increase in lignin content in the pretreated lignocellulose (especially for hydrothermal pretreatments that mainly solubilize hemicellulose) could increase the hydrophobicity of the pretreated biomass, which in turn leads to the loss of active cellulase through unproductive binding with lignin. The pH-dependent behavior of non-productive binding also suggests that electrostatic attractions between lignin and cellulase may lead to the loss of enzymes during hydrolysis (Lan et al. 2013; Nakagame et al. 2011a; Rahikainen et al. 2013; Yarbrough et al. 2015). For example, at pH 4.8, the positively charged enzymes Cel6A and Cel5A adsorb more strongly to the net negatively charged lignin than Cel7A and Cel 7B, which possess a net negative charge at the same pH (Nakagame et al. 2011a). Studies have also shown that charge modification of the amino acids in CBMs results in a change in cellulase affinity toward lignin (Yarbrough et al. 2015). Similarly, surface charge modification of lignin in the pretreated biomass, or carrying out the cellulose hydrolysis at an elevated pH (pH 5.2–6.2), has proven to be an efficient way to reduce the non-productive binding between lignin and cellulases (Lan et al. 2013; Lu et al. 2017). Lastly, hydrogen bonding between cellulase components and dissolved lignin components is identified as another culprit of the unproductive binding of enzymes to lignin (Mhlongo et al. 2015; Michelin et al. 2016; Tejirian and Xu 2011). It has been suggested that phenolic and aliphatic hydroxyl groups are involved in the adsorption of cellulases to lignin through hydrogen bonding (Nakagame et al. 2011b; Pan 2008; Sewalt et al. 1997). However, studies indicate that hydrogen bonding is not a dominating attractive force between cellulase and lignin (Qin et al. 2014; Strobel et al. 2015).

Modification of cellulolytic enzymes has been evaluated to change their hydrophobicity or surface charge, either through succinylation or acetylation (to reduce the lignin–cellulase hydrophobic interactions) or through the construction of linker peptide variants (to reduce electrostatic attractions between lignin and cellulases) (Strobel et al. 2015). The binding capacities of cellulase enzymes also can be modified through lignin modification. Attempts have been made to increase the hydrophilicity of residual lignin through carboxylation, hydroxypropylation, or sulfonation (Ying et al. 2018). A recent study by Ying et al. (2018) measured the effects of three different chemical modifications (sulfonation, oxidation, carboxylation) of alkali lignin on cellulase adsorption and found that they all improved the wettability of lignin, phenolic hydroxyl contents, and negative charges, leading to reduced cellulase adsorption to the modified lignin. Elevating the pH of hydrolysis to increase the lignin surface charge and electrostatic repulsion between cellulases and lignin has also been suggested as a potential strategy to reduce lignin-induced cellulase inhibition (Lu et al. 2017). Laccase treatment has also been proven to improve cellulase efficiency by reducing the binding of cellulase components to lignin through delignification and lignin modification (Rico et al. 2014).

3.6 Microbial Fermentation of Lignocellulose for Biofuels and Biochemicals

3.6.1 Biofuels—Ethanol

Bioethanol is commonly produced from food crops (e.g., corn, sugarcane, and sugar beet), which have high contents of carbohydrates to be converted into fermentable sugars for ethanol production ($C_6H_{12}O_6 \longrightarrow 2\,C_2H_5OH + 2\,CO_2$). The high octane number of 108 in bioethanol is also attractive for an internal combustion engine vehicle to meet anti-knocking requirements. However, the feedstock trend has shifted from food crops to lignocellulosic biomass (inedible waste materials) to encourage sustainable energy resources and achieve the environmentally friendly substitution of gasoline and diesel at a low cost. For example, the recently reported prices of corn and sugarcane were $185.9/ton and $60.9/ton, respectively, while their residuals of corn stover and sugarcane bagasse were $58.5/ton and $36.38/ton, respectively (Bonomi et al. 2016; Silva and Marques 2017). The recent annual world ethanol production is represented in Table 3.1. Since early 2010, some start-up/large companies for bioethanol production have been founded to support the domestic fuel demands with export to other countries. However, most of the bioethanol production companies were bankrupted or sold to other companies due to the business disparity between cellulosic feedstock and cellulosic ethanol price, and technical process challenges in pretreatment and subsequent operations https://ethanolrfa.org/statistics/annual-ethanol-production/). A recent economic analysis determined that the cost for bioethanol production from food crops (corn, sugar beet, or sugarcane) ranged from $0.81 to $2.35/gallon ethanol, while those from lignocellulosic biomass were estimated between $1.91 and $3.48/gallon ethanol, requiring 2.36 times or higher costs for lignocellulosic bioethanol (Rosales-Calderon and Arantes 2019).

Even though lignocellulosic biomass is a cost-competitive feedstock compared to starch and sugar-rich crops, some challenges of feedstock (e.g., lignin-rich lignocellulosic materials), processing technologies, and the relations between the cost of biomass, ethanol, and gasoline prices remain barriers to economically feasible bioethanol production. In the past few decades, a significant number of efforts have led to the discovery and characterization of sugar-fermenting bacteria and yeast. They can consume and metabolize the sugar compounds from enzymatically hydrolyzed biomass (hydrolysate) and directly generate ethanol as a by-product. It is essential to maximize the monosaccharide recovery rather than disaccharide or polysaccharide since most of the fermentation bacteria and yeast prefer to digest monomers (mainly glucose) in the hydrolysate as their energy resource. *Zymomonas mobilis* and *Saccharomyces cerevisiae* are frequently utilized as fermentation strains with the advantages of simple preparation, rapid cell growth, high ethanol productivity, low cost, and biocompatibility (Gandla et al. 2018; Orrego et al. 2018a, b), but they can naturally only metabolize hexose sugars (e.g., glucose, sucrose, fructose, and mannose) and not pentose sugars (e.g., xylose and arabinose). *Candida shehatae*, *Pachysolen tannophilus*, and *Scheffersomyces stipitis*, have been introduced in fermentation for

Table 3.1 A: Annual world fuel ethanol production (2018–2020, million gallons) and B: the major ethanol-producing plants in the world. Data sources: Renewable Fuels Association (https://ethanolrfa.org/statistics/annual-ethanol-production/)

A. Annual world fuel ethanol production (million gallons)				
Region	2018	2019	2020	% of world production
United States	16,091	15,778	13,926	53
Brazil	7,990	8,590	7,930	30
European Union	1,450	1,370	1,250	5
China	770	1,000	880	3
Canada	460	520	428	2
India	430	510	515	2
Thailand	390	430	400	2
Argentina	290	280	230	1
Others	529	522	500	2
Total	28,400	29,000	26,059	100
B. The start-up cellulosic ethanol plants in the worlds				
Year	Name	Region	Million gallons/year	Note
2012	Beta renewables	Crescentino, Italy	40	Sold in 2018
2015	DuPont	Nevada, USA	30	Sold in 2017 to Verbio Vereinigte BioEnergie AG (German company)
2015	Abengoa	Kansas, USA	25	Bankrupt in 2016
2014	GranBio	Brazil	20	Suspended operations in 2016 and resumed in 2019
2014	Raizen	Brazil	40	Use in sugarcane residue
2014	POET-DSM Advanced Biofuels	Iowa, USA	20–25	50/50 joint venture company of Royal DSM (Netherlands) and POET, LLC (USA)
2017	Enviral	Slovakia	16.7	–

their ability to consume both C5 and C6 sugars; however, they are not preferred due to low production efficiency and their sensitivity to potential inhibitors, acid molecules, and the final ethanol product (Demiray et al. 2018).

In order to achieve a higher product yield, engineered microorganisms are employed, which are manipulated in xylose-utilizing genes (e.g., xylulokinase, xylose reductase, and xylitol dehydrogenase) and can simultaneously co-ferment xylose and glucose into ethanol (Kwak and Jin 2017). Genetically modified strains would be vulnerable to toxic compounds; for example, an evolutionarily adapted fermenting strain was found to be incompatible when different pretreatment, feedstock, and/or other hydrolysates were applied for the fermentation (Hoang Nguyen Tran et al. 2020). This is similar to engineered xylose-fermenting strains, where successful utilization of both C5 and C6 sugars is observed in the absence of inhibitors, but intolerance to organic acids, furan aldehydes, aliphatic compounds, and lignin-derived molecules resulted in similar or lower productivity compared to wild-type strains. Other approaches focused on improving selected microbes for inhibitor tolerance, minimizing the generation of inhibitory compounds before enzymatic saccharification, and screening adequate biomass which is susceptible to be degraded, for increasing ethanol productivity (Kim 2018; Kumari and Singh 2018). Previous work reported that a thermostable and inhibitor-resistant strain (Fm13, *S. cerevisiae*) showed high ethanol production (>90% theoretical yield) and was compatible with other hydrolysates obtained from steam explosion pretreated sugarcane bagasse, showing 7.7 times higher ethanol production yield than those from the *S. cerevisiae* benchmark strain control tests (Favaro et al. 2013). Similar research with a fungus *Trametes versicolor* and bacterium *Escherichia coli* has demonstrated these microbes as potential alternative ethanol-fermenting strains for spruce wood and sugarcane bagasse hydrolysates, respectively (Jönsson 2001; Larsson et al. 2001a, b; Wang et al. 2013). To address the current technological drawbacks and limitations, multiple research efforts are underway to develop high-yielding, co-fermenting strains that are tolerant to inhibitors and suitable for diverse hydrolysates.

3.6.2 Hydrogen

Hydrogen gas (H_2) is one of the most attractive clean energy resources that can replace the current fossil fuels and fossil-dependent energy without concerns of environmental pollution while providing a higher energy generation (> 2.5 times, 122 kJ/g) than hydrocarbon resources (Cárdenas et al. 2019, 2020). Since hydrogen gas does not have carbon molecules (zero carbon), the energy is released when two hydrogen molecules react with molecular oxygen, forming water molecules as a by-product:

$$2H_2(\text{gas}) + O_2(\text{gas}) \rightarrow 2H_2O + \text{energy} \tag{3.1}$$

With this unique and simple reaction ability with oxygen gas to generate energy, hydrogen has not only been utilized for internal combustion engines in commercial vehicles such as passenger cars and trucks but also for fuel cells that can convert chemical energy into electric energy. In addition, the relatively high energy efficiency of hydrogen fuel can be applied for hydrogen-powered aircraft/rocket propulsion, hydrogen fuel–cell propulsion systems, and as a primary backup power source for fuel cells. However, the current hydrogen fuel is traditionally generated from fossil fuels via steam-methane reforming, partial oxidation of crude oil and methane, and gasification of coal. These fossil-based processing methods are responsible for approximately >90% yields of total hydrogen production (Brentner et al. 2010). Due to the inadequacy of alternative technologies and lack of cost-effective resources, a small proportion of hydrogen is formed by bio-catalytic approaches (e.g., photo-decomposition of organic compounds, electrolysis of water, and gasification of agricultural feedstock). Although the current petrochemical reforming processes widely generate valuable chemicals, such as acetic acid, butyric acid, and hydrogen, biological methods receive scientific attention as substitutable for fossil-derived chemicals. The fundamental concept of the biological process is to transform organic wastes into hydrogen gas using microbes or disintegrate wastewater molecules in an eco-friendly manner to release hydrogen molecules. Examples of microorganisms include photo-metabolic active microalgae, cyanobacteria, and microbial strains that can fix nitrogen and convert organic substrates in anaerobic conditions. The common biological methods are (i) direct/indirect photo-dissociation (photo-lysis), (ii) photo fermentation, and (iii) dark fermentation. Photo-decomposition uses the photon energy from a diverse range of visible and invisible lights, such as ultraviolet, X-rays, and gamma rays, in photochemical reactions that result in the generation of hydrogen. Hydrogen-producing microbes are capable of capturing light energy and converting carbohydrates or organic materials into hydrogen and carbon dioxide under anaerobic conditions. Nitrogen degrading enzymes (e.g., nitrogenase) secreted from microbes are involved in a key process of nitrogen fixation during photo fermentation to fix atmospheric nitrogen gas (N_2) and the conversion of N_2 to ammonia NH_3, which can contribute toward the generation of hydrogen (Mishra et al. 2019).

$$C_6H_{12}O_6 + 6H_2O + \text{sun light} \rightarrow 12H_2 + 6CO_2 \quad (3.2)$$

Unlike photo fermentation, dark fermentation occurs in the absence of light, and its hydrogen yield is relatively lower than other approaches in the presence of light sources. Despite low hydrogen productivity, dark fermentation is considered a promising biological process since various organic wastes and agricultural feedstock can be applied as energy carriers with less energy demand, cost-effectiveness, and a simple process and control. Carbohydrates in organic substrates are hydrolyzed to monomeric fermentable sugars that subsequently undergo acidogenesis and acetogenesis during dark fermentation. In biological digestion pathways, acetic acid, volatile fatty acids, carbon dioxide, and hydrogen are formed as by-products. Experimental factors such as substrate type, sugar content, pH, temperature, and carbon/nitrogen ratio can determine the hydrogen yield. In the methanogenesis pathway, following acetogenesis, hydrogen is metabolized to form methane and carbon dioxide; therefore, limiting methane generation or hydrogen consumption is important to increase the hydrogen yield. Several studies observed that stable hydrogen production was achievable when a high C/N ratio (around 53.4) organic substrate was anaerobically cultivated, which inhibited the methanogenesis process. This batch fermentation test continued producing hydrogen for 144 days with periods of four stages and different C/N ratios that resulted in changes in the production of hydrogen, carbon dioxide, and methane. Hydrogen yield was improved with the C/N levels, which indicates possible changes in the metabolic pathways in dominant hydrogen-producing microbes and their activity (Hernández et al. 2014; Jo et al. 2007; Saint-Amans et al. 2001). The theoretical maximum yield of hydrogen from a six-carbon monosaccharide is 4 mol H_2/mol monomer, while actual hydrogen yields are lower than this due to the formation of other by-products such as alcohols, butyric acid, malic acid, lactic acid, and propionic acid (Hay et al. 2013). A five-carbon sugar (xylose), a sugar alcohol (mannitol), and a simple polyol compound (glycerol) also can be anaerobically fermented into hydrogen with theoretical yields of 3.33 mol H_2/mol xylose, 5 mol H_2/mol mannitol, and 3 mol H_2/mol glycerol, respectively. Since carbohydrate monomers are the primary substrates in producing hydrogen, several pretreatments are applied to agricultural wastes prior to dark fermentation to enhance hydrogen efficiency. Comparative hydrogen production from different carbon sources (pure sugars, lignocellulosic sugars, or organic wastes) and further fermentation conditions are described in Table 3.2. Since the high concentration of monomeric sugars is the primary factor driving fermentation pathways, further study is required with respect to C/N rich organic materials, cellulose–hemicellulose containing agricultural wastes, culture operations, mixing conditions, changes in metabolism, cell/product inhibition, and fermentation types.

3.6.3 Other Applications for Biochemical Production

Methanol (CH_3OH) is the most preferable industrial alcohol with several advantages, including a high octane rating, cost-effective large-scale production from natural gas and/or petroleum-based resources, and utility as a valuable raw material for petrochemical applications such as formaldehyde, acetic acid, methyl *tert*-butyl ether (MTBE), and alcohol for the esterification process (Khoshtinat et al. 2010). Methanol is competitive with other hydrocarbon fuels in price, reactivity, and recoverability (Siddiquee and Rohani 2011). To minimize the incremental costs for improving the efficiency and environmental impact of internal combustion engines in the vehicle, lignocellulosic feedstocks or waste materials have been utilized for bio-methanol production. Various agricultural wastes are used for bio-methanol generation through different pretreatment processes (Table 3.3A). Although bio-methane is a promising biofuel and a substrate for the synthesis of biochemical hydrocarbon molecules, fewer investigations have been conducted on the optimization and scale-up process of bio-methane production.

Butanol (C_4H_9OH) is a simple four-carbon alcohol that is widely used as a solvent in chemicals, cosmetics, textile processes, hydraulic/brake fluids, painting, and other coating applications with ambient-cured enamels. It is also largely applied as a chemical intermediate for butyl acrylate and *tert*-butoxide production, and as a primary component in the manufacture of antibiotics, hormones, and pharmaceuticals (Jouzani and Taherzadeh 2015). Compared to ethanol, butanol

Table 3.2 Comparison of hydrogen yields from pure sugars, lignocellulosic biomass, and organic wastes with and without pretreatment

Carbon source	Strain	Pretreatment	H_2 yield (mol/mol)	References
Glucose	*Ethanoligenens harbinense* YUAN-3	–	2.62	Zhang et al. (2015)
Glucose	*Clostridium butyricum* INET1	–	2.24	Yin and Wang (2017)
Glucose	*Escherichia coli* BW25113	–	1.82	Mathews et al. (2010)
Glycerol	*Halanaerobium saccharolyticum*	–	0.58	Kivistö et al. (2010)
Galactose	*E. aerogenes*	–	1.26	Ren et al. (2009)
Xylose	*Clostridium bei-jerinckii*	–	1.77	An et al. (2014)
Rice bran	*Enterobacter ludwigii*	Acid	545 mL/L H_2	Tandon et al. (2018)
Wheat straw	*Escherichia coli* WDHL	Dilute acid	269 mL H_2/g substrate	Lopez-hidalgo et al. (2017)
Rice straw	*Clostridium lentocellum*	–	3 mmol H_2/g substrate	Zhang et al. (2019)
Wood fibers	*C. thermocellum* ATCC27405	–	1.6	Xiong et al. (2018)
Wastepaper	*Ruminococcus albus*	–	2.29 (282.7 L/Kg dry biomass)	Ntaikou et al. (2010)
Coffee mucilage + organic wastes	Mesophilic mixed culture[a]	–	0.248 (1.295 L H_2/L substrate)	Cárdenas et al. (2020, 2019)

[a] Dark fermentation was tested without an inoculum of pure strain

has a longer hydrocarbon chain and is more characteristically close to gasoline, making it attractive as an alternative fuel. Since economical butanol production is feasible when the final yield is over 25% of raw feedstock (Kumar and Gayen 2012), lignocellulose and waste materials are considered potential substrates. Similarly, lignocellulosic materials have been utilized in pretreatment, hydrolysis, and fermentation for bio-butanol production; however, this approach is limited with a lack of technology and cost feasibility. The most common bio-butanol production is performed through acetone–butanol–ethanol (ABE) microbial fermentation; however, the formation of inhibitory compounds during the lignocellulosic biomass pretreatment limits effective enzymatic saccharification and microbial performance. To overcome these issues, inhibitor-resistant cellulolytic enzymes and/or genetically engineered microbes have been introduced to the consolidated bioprocessing

Table 3.3 Utilization of lignocellulosic biomass for bio-methanol and bio-butanol production

(A) Bio-methanol

Biomass	Pretreatment	Yield	References
Plant feedstock	-	0.59 kg MeOH/kg dry solids	Arteaga-pérez et al. (2016)
Forest residue	Catalytic gasification	62%–66% MeOH/kg dry feedstock	Carvalho et al. (2017)

(B) Bio-butanol

Biomass	Pretreatment	Microbe	Yield	References
Sugarcane bagasse/rice straw	H_2O_2 and NaOH at 120 °C	Mixed bacterial culture	1.95 g/L (bagasse), 2.93 g/L (rice straw)	Cheng et al. (2012)
Corn cobs	Alkali	*A. cellulovorans* 743B and *C. beijerinckii* NCIMB 8092	8.3 g/L	Wen et al. (2014)
Corn stover	Deep eutectic solvents	*Clostridium saccharobutylicum*	5.63 g/L	Guo-Chao et al. (2016)
Sucrose + sugarcane juice	–	Engineered *Clostridium tyrobutyricum*	16 g/L	Zhang et al. (2017)

(CBP) method (Sheikh et al 2014). The CBP system comprises three major biological processes (cellulolytic enzyme production, hydrolysis of pretreated biomass, and microbial fermentation of both pentose and hexose sugars) in a single bioreactor, which is able to minimize processing steps, costs, and management. Since there is no natural microbe to satisfy all the desirable properties of the CBP system, various kinds of metabolically engineered microbes could be capable of meeting the current challenges. Some bio-butanol research using lignocellulosic feedstocks is summarized in Table 3.3B.

3.6.4 Future Prospects

The fractionation of reducing sugars from cellulose/hemicellulose in agricultural crop-based residues has been fulfilled for sugar platform processes, production of various biochemicals, and further applications to foods, textiles, and liquid biofuels. The use of waste crop solids has gained attention and inspired substantial research due to the cheap and plentiful availability of these resources. During the past few decades, significant advances have been made in bioprocess research for better utilization of lignocellulosic feedstocks at a low cost; however, choosing the best method remains a challenge because all methods have their virtues and demerits. Also, one approach is not generally applicable for all lignocellulosic biomass since they have different properties of cellulose crystallinity, degree of polymerization (DP), chemical composition, and potential inhibitors that depend on pretreatment type and experimental conditions (Kim 2018). Considering these challenges, recent research has concentrated more on the genetic/metabolic engineering of biomass, enzymes, and microbes to enhance the yields of sugars and final products from feedstocks and subsequent hydrolysates, respectively. Further advances in the knowledge and manipulation of strains would be a feasible strategy for reaching industrial viability.

Acknowledgements The authors gratefully express their thanks to Dr. Ryan E. Cobb (Corteva Agriscience, Indianapolis-IN, USA) for the internal review of this manuscript. Daehwan Kim also acknowledges support from the Maryland E-Nnovation Initiative Fund (MEIF) administered by the Maryland Department of Commerce, and the Maryland Energy Innovation Institute (MEII)—Energy seed grant.

References

Adsul M, Sandhu SK, Singhania RR, Gupta R, Puri SK (2020) Designing a cellulolytic enzyme cocktail for the efficient and economical conversion of lignocellulosic biomass to biofuels. Enzyme Microb Technol 133:109442

Agrawal R, Semwal S, Kumar R, Mathur A, Gupta RP (2018) Synergistic enzyme cocktail to enhance hydrolysis of steam exploded wheat straw at pilot scale. Front. Energy Res 6:1–11

Alam A, Zhang R, Liu P, Huang J, Wang Y, Hu Z, Madadi M, Sun D, Hu R, Ragauskas AJ, Tu Y, Peng L (2019) A finalized determinant for complete lignocellulose enzymatic saccharification potential to maximize bioethanol production in bioenergy Miscanthus. Biotechnol Biofuels, 1–22

An D, Li Q, Wang X, Yang H, Guo L (2014) Characterization on hydrogen production performance of a newly isolated Clostridium beijerinckii YA001 using xylose. Int J Hydrog Energy 39:19928–19936

Arteaga-Pérez LE, Gómez-Cápiro O, Karelovic A, Jiménez R (2016) A modelling approach to the techno-economics of Biomass-to-SNG/Methanol systems: standalone vs integrated topologies. Chem Eng Sci 286:663–678

Ázar RISL, Bordignon SE, Laufer C, Specht J, Ferrier D, Kim D (2020) Effect of lignin content on cellulolytic saccharification of liquid hot water pretreated sugarcane bagasse. Molecules 25:623

Berlin A, Maximenko V, Gilkes N, Saddler J (2007) Optimization of enzyme complexes for lignocellulose hydrolysis. Biotechnol Bioeng 97:287–296

Binod P, Gnansounou E, Sindhu R, Pandey A (2019) Enzymes for second generation biofuels : recent developments and future perspectives. Bioresour Technol Rep 5:317–325

Bissaro B, Røhr ÅK, Müller G, Chylenski P, Skaugen M, Horn SJ, Vaaje-Kolstad G, Eijsink VGH (2017) Oxidative cleavage of polysaccharides by monocopper enzymes depends on H_2O_2. Nat Publ Gr 13:1123–1128

Bordignon SE, Ximenes E, Olavo Micali P, Carreira Nunes, CC, Kim D, Boscolo M, Gomes E, Filho, EXF, da Silva R, Ladisch, MR (2022) Combined sugarcane pretreatment for the generation of ethanol and value-added products. Front Energy Res 10:834966

Bonomi A, Junqueira TL, Chagas MF, Gouveia VLR, Watanabe MDB, Cavalett O (2016) Techno-economic and environmental assessment of second generation ethanol: short and long term prospects. Chem Eng Trans 50:439–444

Boukari I, Donohue MO, Rémond C, Chabbert B (2011) Enzymatic probing a family GH11 endo-β-1,4-xylanase inhibition mechanism by phenolic compounds: role of functional phenolic groups. J Mol Catal B Enzym 72:130–138

Brentner LB, Jordan PA, Zimmerman JB (2010) Challenges in developing biohydrogen as a sustainable energy source: implications for a research agenda. Environ Sci Technol 44:2243–2254

Brondi MG, Vasconcellos VM, Giordano RC, Farinas CS (2019) Alternative low-cost additives to improve the saccharification of lignocellulosic biomass. Appl Biochem Biotechnol 187:461–473

Cantarella M, Cantarella L, Gallifuoco A, Spera A, Alfani F (2004) Comparison of different detoxification methods for steam-exploded poplar wood as a substrate for the bioproduction of ethanol in SHF and SSF 39. Process Biochem 39:1533–1542

Cao G, Ximenes E, Nichols NN, Frazer SE, Kim D, Cotta MA, Ladisch M (2015) Bioabatement with hemicellulase supplementation to reduce enzymatic hydrolysis inhibitors. Bioresour Technol 190:412–415

Cao G, Ximenes E, Nichols NN, Zhang L, Ladisch M (2013) Biological abatement of cellulase inhibitors. Bioresour Technol 146:604–610

Caldararu O, Oksanen E, Ryde U, Hedegård ED (2019) Mechanism of hydrogen peroxide formation by lytic polysaccharide monooxygenase. Chem Sci 10:576–586

Cárdenas ELM, Zapata-Zapata AD, Kim D (2020) Modeling dark fermentation of coffee mucilage wastes for hydrogen production: artificial neural network model vs. fuzzy logic model. Energies 13:1–13

Cárdenas ELM, Zapata-Zapata AD, Kim D (2019) Hydrogen production from coffee mucilage in dark fermentation with organic wastes. Energies 12:71

Carvalho L, Furusjö E, Kirtania K, Wetterlund E, Lundgren J, Anheden M, Wolf J (2017) Techno-economic assessment of catalytic gasification of biomass powders for methanol production. Bioresour Technol 237:167–177

Chen H, Liu Z (2017) Enzymatic hydrolysis of lignocellulosic biomass from low to high solids loading. Eng Life Sci 17:489–499

Chen Y, Zhou Y, Qin Y, Liu D, Zhao X (2018) Evaluation of the action of tween 20 non-ionic surfactant during enzymatic hydrolysis of lignocellulose: pretreatment, hydrolysis conditions and lignin structure. Bioresour Technol 269:329–338

Cheng C, Che P, Chen B, Lee W, Lin C, Chang J (2012) Biobutanol production from agricultural waste by an acclimated mixed bacterial microflora. Appl Energy 100:3–9

Cheng MH, Kadhum HJ, Murthy GS, Dien BS, Singh V (2020) High solids loading biorefinery for the production of cellulosic sugars from bioenergy sorghum. Bioresour Technol 318:124051

Cheng N, Koda K, Tamai Y, Yamamoto Y, Takasuka TE, Uraki Y (2017) Optimization of simultaneous saccharification and fermentation conditions with amphipathic lignin derivatives for concentrated bioethanol production. Bioresour Technol 232:126–132

Chylenski P, Forsberg Z, Ståhlberg J, Várnai A, Lersch M, Bengtsson O, Sæbø S, Jarle S, Eijsink VGH (2017a) Development of minimal enzyme cocktails for hydrolysis of sulfite-pulped lignocellulosic biomass. J Biotechnol 246:16–23

Chylenski P, Petrović DM, Müller G, Dahlström M, Bengtsson O, Lersch M, Siika M, Horn SJ, Eijsink VGH (2017b) Enzymatic degradation of sulfite—pulped softwoods and the role of LPMOs. Biotechnol Biofuels, 1–13

Corrêa LJ Jose, Badino AC, Cruz AJG (2016) Power consumption evaluation of different fed-batch strategies for enzymatic hydrolysis of sugarcane bagasse. Bioprocess Biosyst Eng, 825–833

Costaouëc T, Pakarinen A, Várnai A, Puranen T, Viikari L (2013) The role of carbohydrate binding module (CBM) at high substrate consistency : comparison of *Trichoderma reesei* and *Thermoascus aurantiacus* Cel7A (CBHI) and Cel5A (EGII). Bioresour Technol 143:196–203

Crowe JD, Zarger RA, Hodge DB (2017) Relating nanoscale accessibility within plant cell walls to improved enzyme hydrolysis yields in corn stover subjected to diverse pretreatments. J Agric Food Chem 65:8652–8662

da Silva HJT, Marques PV (2017) Evolution of production costs in brazilian sugar-energy sector. China-USA Bus Rev 16(3):93–107

den Haan R, Rensburg EV, Rose SH, Görgens JF, Zyl, W.H.v., (2014) Progress and challenges in the engineering of non-cellulolytic microorganisms for consolidated bioprocessing. Curr Opi Biotechnol 33:32–38

Demiray E, Karatay SE, Dönmez G (2018) Evaluation of pomegranate peel in ethanol production by *Saccharomyces cerevisiae* and *Pichia stipitis*. Energy 159:988–994
Djajadi DT, Meyer AS (2018) Cellulases adsorb reversibly on biomass lignin. Biotechol. Bioeng. 115:2869–2880
Domínguez H, Moure A, Garrote G (2017) Effect of hydrothermal pretreatment on lignin and antioxidant activity. Hydrothermal processing in biorefineries: production of bioethanol and high added-value compounds of second and third generation biomass. Springer International Publishing, 5–42
dos Santos-Rocha MSR, Pratto B, Jacob L, Colli A, Maria R, Garcia R, José A, Cruz G (2018) Assessment of different biomass feeding strategies for improving the enzymatic hydrolysis of sugarcane straw. Ind Crop Prod 125:293–302
Du J, Liang J, Gao X, Liu G, Qu Y (2020) Optimization of an artificial cellulase cocktail for high-solids enzymatic hydrolysis of cellulosic materials with different pretreatment methods. Bioresour Technol 295:122272
Eibinger M, Ganner T, Bubner P, Ros S, Kracher D, Haltrich D, Ludwig R, Plank H, Nidetzky B (2014) Cellulose surface degradation by a lytic polysaccharide monooxygenase and its effect on cellulase hydrolytic. J Biol Chem 289:35929–35938
Fang H, Kandhola G, Rajan K, Djioleu A (2018) Effects of oligosaccharides isolated from pinewood hot water pre-hydrolyzates on recombinant cellulases. Front Bioeng Biotechnol 6:1–11
Favaro L, Basaglia M, Trento A, Rensburg E Van, García-Aparicio M, Zyl WH Van, Casella S (2013) Exploring grape marc as trove for new thermotolerant and inhibitor-tolerant *Saccharomyces cerevisiae* strains for second-generation bioethanol production. Biotechnol Biofuels, 1–14
Florencio C, Badino AC, Farinas CS (2016) Soybean protein as a cost-effective lignin-blocking additive for the saccharification of sugarcane bagasse. Bioresour Technol 221:172–180
Freitas JV, Farinas CS (2017) Sugarcane bagasse fly ash as a no-cost adsorbent for removal of phenolic inhibitors and improvement of biomass saccharification. ACS Sustain Chem Eng 5:11727–11736
Galbe M, Wallberg O (2019) Pretreatment for biorefineries : a review of common methods for efficient utilisation of lignocellulosic materials. Biotechnol Biofuels, 1–26
Gandla ML, Martín C, Jönsson LJ (2018) Analytical enzymatic saccharification of lignocellulosic biomass for conversion to biofuels and bio-based chemicals. Energies 11:2936
Gong Z, Wang X, Yuan W, Wang Y, Zhou W, Wang G, Liu Y (2020) Fed-batch enzymatic hydrolysis of alkaline organosolv-pretreated corn stover facilitating high concentrations and yields of fermentable sugars for microbial lipid production. Biotechnol Biofuels 13:1–15
González-Bautista E, Santana-Morales JC, Ríos-Fránquez FJ, Poggi-Varaldo HM, Ramos-Valdivia AC, Cristiani-Urbina E, Ponce-Noyola T (2017) Phenolic compounds inhibit cellulase and xylanase activities of *Cellulomonas flavigena* PR-22 during saccharification of sugarcane bagasse. Fuels 196:32–35
Gullón P, Gullón B, Cardelle-Cobas A, Luis J, Pintado M, Maria A (2014) Effects of hemicellulose-derived saccharides on behavior of *Lactobacilli* under simulated gastrointestinal conditions. Food Res Int 64:880–888
Xu G-C, Ding J-C, Han R-Z, Dong J-J, Ni Y (2016) Enhacing cellulose accessibility of corn stover by deep eutectic solvent pretreatment for butanol fermentation. Bioresour Technol 203:364–369
Guo F, Shi W, Sun W, Li X, Wang F, Zhao J, Qu Y (2014) Differences in the adsorption of enzymes onto lignins from diverse types of lignocellulosic biomass and the underlying mechanism. Biotechnol Biofuels 7:1–10
Hao X, Li Y, Wang J, Qin Y, Zhang J (2019) Adsorption and desorption of cellulases on/from lignin-rich residues from corn stover. Ind Crop Prod 139:111559
Hay JXW, Wu TY, Juan JC, Jahim JM (2013) Biohydrogen production through photo fermentation or dark fermentation using waste as a substrate: overview, economics, and future prospects of hydrogen usage. Biofuels. Bioprod Biorefining 7:334–352

Heiss-blanquet S, Zheng D, Lopes N, Lapierre C, Baumberger S (2011) Effect of pretreatment and enzymatic hydrolysis of wheat straw on cell wall composition, hydrophobicity and cellulase adsorption. Bioresour Technol 102:5938–5946

Hernández-Beltrán JU, Hernández-Escoto H (2018) Enzymatic hydrolysis of biomass at high-solids loadings through fed-batch operation. Biomass Bioenergy 119:191–197

Hernández MA, Rodríguez Susa M, Andres Y (2014) Use of coffee mucilage as a new substrate for hydrogen production in anaerobic co-digestion with swine manure. Bioresour Technol 168:112–118

Tran PHN, P, Ko JK, Gong G, Um Y, Lee SM (2020) Improved simultaneous co-fermentation of glucose and xylose by *Saccharomyces cerevisiae* for efficient lignocellulosic biorefinery. Biotechnol Biofuels 13:1–14

Hsu T, Guo G, Chen W, Hwang W (2010) Bioresource technology effect of dilute acid pretreatment of rice straw on structural properties and enzymatic hydrolysis. Bioresour Technol 101:4907–4913

Hu J, Chandra R, Arantes V, Gourlay K, Dyk JSV, Saddler JN (2015) Bioresource technology the addition of accessory enzymes enhances the hydrolytic performance of cellulase enzymes at high solid loadings. Bioresour Technol 186:149–153

Jo JH, Jeon CO, Lee DS, Park JM (2007) Process stability and microbial community structure in anaerobic hydrogen-producing microflora from food waste containing kimchi. J Biotechnol 131:300–308

Jönsson LJ, Martín C (2016) Pretreatment of lignocellulose: formation of inhibitory by-products and strategies for minimizing their effects. Bioresour Technol 199:103–112

Jouzani GS, Taherzadeh MJ (2015) Advances in consolidated bioprocessing systems for bioethanol and butanol production from biomass: a comprehensive review. Biofuel Res J 2:152–195

Jung S, Song Y, Myeong H, Bae H (2015) Enhanced lignocellulosic biomass hydrolysis by oxidative lytic polysaccharide monooxygenases (LPMOs) GH61 from Gloeophyllum trabeum. Enzyme Microb Technol 77:38–45

Kawakubo T, Karita S, Araki Y, Watanabe S, Oyadomari M, Takada R, Tanaka F, Abe K, Watanabe T, Honda Y, Watanabe T (2010) Analysis of exposed cellulose surfaces in pretreated wood biomass using carbohydrate-binding module (CBM)-cyan fluorescent protein (CFP). Biotechnol Bioeng 105:499–508

Kellock M, Maaheimo H, Marjamaa K, Rahikainen J, Zhang H, Holopainen-mantila U, Ralph J, Tamminen T, Felby C, Kruus K (2019) Effect of hydrothermal pretreatment severity on lignin inhibition in enzymatic hydrolysis. Bioresour Technol 280:303–312

Khoshtinat M, Amin NAS, Noshadi I (2010) A review of methanol production from methane oxidation via non-thermal plasma reactor. World Acad Sci Eng Technol 62:354–358

Kim D (2018) Physico-chemical conversion of lignocellulose: Inhibitor effects and detoxification strategies: a mini review. Molecules 23:309

Kim D, Ku S (2018) Bacillus cellulase molecular cloning, expression, and surface display on the outer membrane of *Escherichia coli*. Molecules 23:503

Kim D, Orrego D, Ximenes EA, Ladisch MR (2017) Cellulose conversion of corn pericarp without pretreatment. Bioresour Technol 245:511–517

Kim D, Ximenes EA, Nichols NN, Cao G, Frazer SE, Ladisch MR (2016) Maleic acid treatment of biologically detoxified corn stover liquor. Bioresour Technol 216:437–445

Kim D, Yoo CG, Schwarz J, Dhekney S, Kozak R, Laufer C, Ferrier D, Mackay S, Ashcraft M, Williams R, Kim S (2021) Effect of lignin-blocking agent on enzyme hydrolysis of acid pretreated hemp waste. RSC Adv 11:22025–22033

Kim Y, Kreke T, Hendrickson R, Parenti J, Ladisch MR (2013) Fractionation of cellulase and fermentation inhibitors from steam pretreated mixed hardwood. Bioresour Technol 135:30–38

Kim Y, Kreke T, Ko JK, Ladisch MR (2015) Hydrolysis-determining substrate characteristics in liquid hot water pretreated hardwood. Biotechnol Bioeng 112:677–687

Kim Y, Ximenes E, Mosier NS, Ladisch MR (2011) Soluble inhibitors/deactivators of cellulase enzymes from lignocellulosic biomass. Enzyme Microb Technol 48:408–415

Kivistö A, Santala V, Karp M (2010) Hydrogen production from glycerol using halophilic fermentative bacteria. J Eur Ceram Soc 101:8671–8677

Ko JK, Kim Y, Ximenes E, Ladisch MR (2015a) Effect of liquid hot water pretreatment severity on properties of hardwood lignin and enzymatic hydrolysis of cellulose. Biotechnol Bioeng 112:252–262

Ko JK, Ximenes E, Kim Y, Ladisch MR (2015b) Adsorption of enzyme onto lignins of liquid hot water pretreated hardwoods. Biotechnol Bioeng 112:447–456

Kumar M, Gayen K (2012) Developments in biobutanol production: new insights. Appl Energy 88:1999–2012

Kumar R, Wyman CE (2014) Strong cellulase inhibition by mannan polysaccharides in cellulose conversion to sugars. Biotechnol Bioeng 111:1341–1353

Kumar R, Wyman CE (2009) Effect of xylanase supplementation of cellulase on digestion of corn stover solids prepared by leading pretreatment technologies. Bioresour Technol 100:4203–4213

Kumari D, Singh R (2018) Pretreatment of lignocellulosic wastes for biofuel production: a critical review. Renew Sustain Energy Rev 90:877–891

Kuusk S, Bissaro B, Kuusk P, Forsberg Z, Eijsink VGH, Sørlie M (2018) Kinetics of H_2O_2-driven degradation of chitin by a bacterial lytic polysaccharide monooxygenase. J Biol Chem 293:523–531

Kwak S, Jin YS (2017) Production of fuels and chemicals from xylose by engineered *Saccharomyces cerevisiae*: a review and perspective. Microb Cell Fact 16:1–15

Lan TQ, Lou H, Zhu JY (2013) Enzymatic saccharification of lignocelluloses should be conducted at elevated pH 5.2-6.2. Bioenergy Res 6:476–485

Larsson N, Nilvebrant N-O, Jönsson LJ (2001a) Effect of overexpression of *Saccharomyces cerevisiae* Pad1p on the resistance to phenylacrylic acids and lignocellulose hydrolysates under aerobic and oxygen-limited conditions 57:167–174

Larsson S, Cassland P, Jönsson LJ (2001b) Development of a *Saccharomyces cerevisiae* strain with enhanced resistance to phenolic fermentation inhibitors in lignocellulose hydrolysates by heterologous expression of laccase. Appl Environ Microbiol 67:1163–1170

Levasseur A, Drula E, Lombard V, Coutinho PM, Henrissat B (2013) Expansion of the enzymatic repertoire of the CAZy database to integrate auxiliary redox enzymes. Biotechnol Biofuels, 1–14

Li H, Pu Y, Kumar R, Ragauskas AJ, Wyman CE (2014a) Investigation of lignin deposition on cellulose during hydrothermal pretreatment, its effect on cellulose hydrolysis, and underlying mechanisms. Biotechnol Bioeng 111:485–492

Li J, Zhang H, Lu M, Han L (2019) Comparison and intrinsic correlation analysis based on composition, microstructure and enzymatic hydrolysis of corn stover after different types of pretreatments. Bioresour Technol 293:122016

Li Y, Qi B, Wan Y (2014b) Inhibitory effect of vanillin on cellulase activity in hydrolysis of cellulosic biomass. Bioresour Technol 167:324–330

Lin X, Wu L, Huang S, Qin Y, Qiu X, Lou H (2019) Effect of lignin-based amphiphilic polymers on the cellulase adsorption and enzymatic hydrolysis kinetics of cellulose. Carbohydr Polym 207:52–58

Liu H, Chen X, Zhang Y, Lu M, Lyu H, Han L, Xiao W (2020) Alcoholysis of ball-milled corn stover: the rnhanced conversion of carbohydrates into biobased chemicals over combination catalysts of [Bmim-SO_3H][HSO_4] and $Al_2(SO_4)_3$. Energy Fuels 34:7085–7093

Liu H, Kong TH (2016) Toward a fundamental understanding of cellulase-lignin interactions in the whole slurry enzymatic saccharification process. Biofuels. Bioprod Biorefining 10:648–663

Loaces I, Schein S, Noya F (2017) Ethanol production by *Escherichia coli* from Arundo donax biomass under SSF, SHF or CBP process configurations and in situ production of a multifunctional glucanase and xylanase. Bioresour Technol 224:307–313

Lopes AM, Filho EXF, Moreira LRS (2018) An update on enzymatic cocktails for lignocellulose breakdown. J Appl Microbiol 125:632–645

Lopez-Hidalgo AM, Sánchez A, León-Rodríguez AD (2017) Simultaneous production of bioethanol and biohydrogen by *Escherichia coli* WDHL using wheat straw hydrolysate as substrate. Fuel 188:19–27

Lu X, Wang C, Li X, Zhao J (2017) Temperature and pH in fl uence adsorption of cellobiohydrolase onto lignin by changing the protein properties. Bioresour Technol 245:819–825

Lu X, Zheng X, Li X, Zhao J (2016) Adsorption and mechanism of cellulase enzymes onto lignin isolated from corn stover pretreated with liquid hot water. Biotechnol Biofuels 9:118

Luo X, Liu J, Zheng P, Li M, Zhou Y, Huang L, Chen L, Shuai L (2019) Promoting enzymatic hydrolysis of lignocellulosic biomass by inexpensive soy protein. Biotechnol Biofuels 12:51

Lv S, Yu Q, Zhuang X, Yuan Z, Wang W, Wang Q, Qi W (2013) The Influence of hemicellulose and lignin removal on the enzymatic digestibility from sugarcane bagasse. Bioengerg Res 6:1128–1134

Mathews J, Li Q, Wang G (2010) Characterization of hydrogen production by engineered Escherichia coli strains using rich defined media. Biotechnol Bioprocess Eng 15:686–695

Mathibe BN, Malgas S, Radosavljevic L, Kumar V, Shukla P (2020) Lignocellulosic pretreatment—mediated phenolic by—products generation and their effect on the inhibition of an endo-1,4-β- xylanase from *Thermomyces lanuginosus* VAPS 24. 3 Biotech 10:1–11

Meng X, Wells T, Sun Q, Huang F, Ragauskas A (2015) Insights into the effect of dilute acid, hot water or alkaline pretreatment on the cellulose accessible surface area and the overall porosity of *Populus*. Green Chem 17:4239–4246

Mhlongo SI, Haan RD, Viljoen-bloom M, Zyl WHV (2015) Lignocellulosic hydrolysate inhibitors selectively inhibit/deactivate cellulase performance. Enzyme Microb Technol 81:16–22

Michelin M, Ximenes E, Polizeli MLTM, M., Ladisch, M.R., (2016) Effect of phenolic compounds from pretreated sugarcane bagasse on cellulolytic and hemicellulolytic activities. Bioresour Technol 199:275–278

Mishra P, Krishnan S, Rana S, Singh L, Sakinah M, Ab Wahid Z (2019) Outlook of fermentative hydrogen production techniques: an overview of dark, photo and integrated dark-photo fermentative approach to biomass. Energy Strateg Rev 24:27–37

Moniz P, Ling A, Duarte LC, Kolida S, Rastall RA, Pereira H, Carvalheiro F (2016) Assessment of the bifidogenic effect of substituted xylo-oligosaccharides obtained from corn straw. Carbohydr Polym 136:466–473

Müller G, Chylenski P, Bissaro B, Eijsink VGH, Horn SJ (2018) The impact of hydrogen peroxide supply on LPMO activity and overall saccharification efficiency of a commercial cellulase cocktail. Biotechnol Biofuels, 1–17

Nakagame S, Chandra RP, Kadla JF, Saddler JN (2011a) The isolation, characterization and effect of lignin isolated from steam pretreated Douglas-fir on the enzymatic hydrolysis of cellulose. Bioresour Technol 102:4507–4517

Nakagame S, Chandra RP, Kadla JF, Saddler JN (2011b) Enhancing the enzymatic hydrolysis of lignocellulosic biomass by increasing the carboxylic acid content of the associated lignin. Biotechnol. Bioengin. 108:538–548

Nichols NN, Mertens JA, Dien BS, Hector RE, Frazer SE (2020) Recycle of fermentation process water through mitigation of inhibitors in dilute-acid corn stover hydrolysate. Bioresour Technol Rep 9:100349

Nogueira C, Eduardo C, Padilha DA, Laura A, Leitão DS, Rocha PM, Macedo GRD, Silvino E (2018) Enhancing enzymatic hydrolysis of green coconut fiber—pretreatment assisted by tween 80 and water effect on the post-washing. Ind Crop Prod 112:734–740

Notley SM, Norgren M (2010) Surface energy and wettability of spin-coated thin films of lignin isolated from wood. Langmuir 11:5484–5490

Ntaikou I, Antonopoulou G, Lyberatos G (2010) Biohydrogen production from biomass and wastes via dark fermentation: a review. Waste Biomass Valorization 1:21–39

Öhgren K, Bura R, Lesnicki G, Saddler J, Zacchi G (2007) A comparison between simultaneous saccharification and fermentation and separate hydrolysis and fermentation using steam-pretreated corn stover. Process Biochem 42:834–839

Orrego D, Zapata-Zapata AD, Kim D (2018a) Ethanol production from coffee mucilage fermentation by S. cerevisiae immobilized in calcium-alginate beads. Bioresour Technol Rep 3:200–204

Orrego D, Zapata-Zapata AD, Kim D (2018b) Optimization and scale-up coffee mucilage fermentation for ethanol production. Energies 11:786

Østby H, Hansen LD, Horn SJ, Eijsink VGH, Várnai A (2020) Enzymatic processing of lignocellulosic biomass: principles, recent advances and perspectives. J Ind Microbiol Biotechnol 47:623–657

Palmqvist E (2000a) Fermentation of lignocellulosic hydrolysates. I: inhibition and detoxifi cation. Bioresour Technol 74:17–24

Palmqvist E (2000b) Fermentation of lignocellulosic hydrolysates. II: inhibitors and mechanisms of inhibition. Bioresour Technol 74:25–33

Pan X (2008) Role of functional groups in lignin inhibition of enzymatic hydrolysis of cellulose to glucose. J Biobased Mater Bioenergy 2:25–32

Peng P, She D (2014) Isolation, structural characterization, and potential applications of hemicelluloses from bamboo : a review. Carbohydr Polym 112:701–720

Qin C, Clarke K, Li K (2014) Interactive forces between lignin and cellulase as determined by atomic force microscopy. Biotechnol Biofuel 7:1–9

Qin L, Li WC, Liu L, Zhu JQ, Li X, Li BZ, Yuan YJ (2016) Biotechnology for Biofuels Inhibition of lignin—derived phenolic compounds to cellulase. Biotechnol Biofuels 9:70

Qing Q, Yang B, Wyman CE (2010) Xylooligomers are strong inhibitors of cellulose hydrolysis by enzymes. Bioresour Technol 101:9624–9630

Raftery JP, Karim MN (2017) Economic viability of consolidated bioprocessingutilizing multiple biomasss substrates for commercial-scale cellulosic bioethanol production. Biomass Bioenerg 103:35–46

Rahikainen JL, Martin-Sampedro R, Heikkinen H, Rovio S, Marjamaa K, Tamminen T, Rojas OJ, Kruus K (2013) Inhibitory effect of lignin during cellulose bioconversion: the effect of lignin chemistry on non-productive enzyme adsorption. Bioresour Technol 133:270–278

Rasmussen H, Tanner D, Sørensen HR, Meyer AS (2017) New degradation compounds from lignocellulosic biomass pretreatment: routes for formation of potent oligophenolic enzyme inhibitors. Creen Chem 19:464–473

Ren Y, Wang J, Liu Z, Ren Y, Li G (2009) Hydrogen production from the monomeric sugars hydrolyzed from hemicellulose by *Enterobacter aerogenes*. Renew Energy 34:2774–2779

Rico A, Rencoret J, Río JC, Martínez AT, Gutiérrez A (2014) Pretreatment with laccase and a phenolic mediator degrades lignin and enhances saccharification of Eucalyptus feedstock. Biotechnol Biofuels 7:1–14

Rosales-Calderón O, Arantes V (2019) A review on commercial scale high-value products that can be produced alongside cellulosic ethanol. Biotechnol Biofuels 12:240

Saha BC, Nichols NN, Qureshi N, Kennedy GJ, Iten LB, Cotta MA (2015) Pilot scale conversion of wheat straw to ethanol via simultaneous saccharification and fermentation. Bioresour Technol 175:17–22

Saint-Amans S, Girbal L, Andrade J, Ahrens K, Soucaille P (2001) Regulation of carbon and electron flow in *Clostridium butyricum* VPI 3266 grown on glucose-glycerol mixture. J Bacteriol Res 183:1748–1754

Sewalt VJH, Glasser WG, Beauchemin KA (1997) Lignin Impact on fiber degradation. 3. Reversal of inhibition of enzymatic hydrolysis by chemical modification of lignin and by additives. J Agric Food Chem 45:1823–1828

Sheikh MI, Kim C, Park H, Nam H, Lee S, Jo S, Lee J, Kim W (2014) A synergistic effect of pretreatment on cell wall structural changes in barley straw (*Hordeum vulgare L.*) for efficient bioethanol production. J Sci Food Agric 95:843–850

Shokrkar H, Ebrahimi S, Engineering C (2018) In the field synergism of cellulases and amylolytic enzymes in the hydrolysis of microalgal carbohydrates, Biofuels. Bioprod Biorefining 12:749–755

Siddiquee MN, Rohani S (2011) Lipid extraction and biodiesel production from municipal sewage sludges: a review. Renew Sustain Energy Rev 15:1067–1072

Pito MS, Rodríguez-Jasso RM, Michelin M, Flores-Gallegos AC (2018) Bioreactor design for enzymatic hydrolysis of biomass under the biore fi nery concept. Chem Eng J 347:119–136

Strobel KL, Pfeiffer KA, Blanch HW, Clark DS (2016) Engineering Cel7a carbohydrate binding module and linker for reduced lignin inhibition. Biotechnol Bioeng 113:1369–1374

Strobel KL, Pfeiffer KA, Blanch HW, Clark DS (2015) Structural insights into the affinity of Cel7A carbohydrate-binding module for lignin. J Biol Chem 290:22818–22826

Sukhang S, Choojit S, Reungpeerakul T (2020) Bioethanol production from oil palm empty fruit bunch with SSF and SHF processes using *Kluyveromyces marxianus* yeast. Cellulose 2:301–314

Sun J, Zhang Z, Xiao F, Jin X (2015) Production of xylooligosaccharides from corncobs using ultrasound-assisted enzymatic hydrolysis. Food Sci Biotechnol 24:2077–2081

Tan L, Zhong J, Jin Y, Sun Z, Tang Y, Kida K (2020) Production of bioethanol from unwashed-pretreated rapeseed straw at high solid loading. Bioresour Technol 303:122949

Tandon M, Thakur V, Tiwari KL, Jadhav SK (2018) *Enterobacter ludwigii* strain IF2SW-B4 isolated for bio-hydrogen production from rice bran and de-oiled rice bran. Environ Technol Innov 10:345–354

Tejirian A, Xu F (2011) Inhibition of enzymatic cellulolysis by phenolic compounds. Enzyme Microb Technol 48:239–247

Thoresen M, Malgas S, Mafa MS, Pletschke BI (2021) Revisiting the phenomenon of cellulase action: not all endo-and exo-cellulase interactions are synergistic. Catalysts 11:1–13

Trajano HL, Engle NL, Foston M, Ragauskas AJ, Tschaplinski TJ, Wyman CE (2013) The fate of lignin during hydrothermal pretreatment. Biotechnol Biofuels 6:1–16

Wang J, Wang J, Lu Z, Zhang J (2020a) Adsorption and desorption of cellulase on/from enzymatic residual lignin after alkali pretreatment. Ind Crop Prod 155:112811

Wang Q, Liu S, Yang G, Chen J, Ni Y (2015) Cationic polyacrylamide enhancing cellulase treatment efficiency of hardwood kraft-based dissolving pulp. Bioresour Technol 183:42–46

Wang X, Yomano LP, Lee JY, York SW, Zheng H, Mullinnix MT, Shanmugam KT, Ingram LO (2013) Engineering furfural tolerance in *Escherichia coli* improves the fermentation of lignocellulosic sugars into renewable chemicals. PNAS 110:4021–4026

Wang Y, Meng X, Jeong K, Li S, Leem G, Kim KH, Pu Y, Ragauskas AJ, Yoo CG (2020b) Investigation of a lignin-based deep eutectic solvent using p-hydroxybenzoic acid for efficient woody biomass conversion. ACS Sustain. Chem. Eng. 8:12542–12553

Wen Z, Wu M, Lin Y, Yang L, Lin J, Cen P (2014) Artificial symbiosis for acetone-butanol-ethanol (ABE) fermentation from alkali extracted deshelled corn cobs by co-culture of *Clostridium beijerinckii* and *Clostridium cellulovorans*. Microb. Cell Factories, 1–11

Westereng,B, Cannella D, Agger JW, Jørgensen H, Andersen ML, Eijsink VGH, Felby C (2015) Enzymatic cellulose oxidation is linked to lignin by long-range electron transfer. Nat. Publ. Gr, 1–9

Ximenes E, Kim Y, Mosier N, Dien B, Ladisch M (2011) Deactivation of cellulases by phenols. Enzyme Microb Technol 48:54–60

Ximenes E, Kim Y, Mosier N, Dien B, Ladisch M (2010) Inhibition of cellulases by phenols. Enzyme Microb Technol 46:170–176

Xiong W, Reyes LH, Michener WE, Maness PC, Chou KJ (2018) Engineering cellulolytic bacterium *Clostridium thermocellum* to co-ferment cellulose- and hemicellulose-derived sugars simultaneously. Biotechnol Bioeng 115:1755–1763

Yang Q, Pan X (2015) Correlation between lignin physicochemical properties and inhibition to enzymatic hydrolysis of cellulose. Biotechnol Bioengin 113:1213–1224

Yarbrough JM, Mittal A, Mansfield E, Ii LET, Hobdey SE, Sammond DW, Bomble YJ, Crowley MF, Decker SR, Himmel ME, Vinzant TB (2015) New perspective on glycoside hydrolase binding to lignin from pretreated corn stover. Biotechnol Biofuels 8:214

Yin Y, Wang J (2017) Isolation and characterization of a novel strain *Clostridium butyricum* INET1 for fermentative hydrogen production. Int J Hydrogen Energy 42:12173–12180

Ying W, Shi Z, Yang H, Xu G, Zheng Z, Yang J (2018) Effect of alkaline lignin modification on cellulose—lignin interactions and enzymatic saccharification yield. Biotechnol Biofuels 11:1–13

Yoo CG, Kim H, Lu, F, Azarpira A, Pan X, Oh KK, Kim JS, Ralph J, Kim TH (2016) Understanding the physicochemical characteristics and the improved enzymatic saccharification of Corn Stover pretreated with aqueous and gaseous ammonia. BioEnergy Res 9(1):67–76

Zhai R, Hu J (2018) Extent of enzyme inhibition by phenolics derived from pretreated biomass Is significantly influenced by the size and carbonyl group content of the phenolics. ACS Sustainable Chem Eng 6:3823–3829

Zhai R, Hu J, Saddler JN (2018) The inhibition of hemicellulosic sugars on cellulose hydrolysis are highly dependant on the cellulase productive binding, processivity, and substrate surface charges. Bioresour Technol 258:79–87

Zhai R, Hu J, Saddler JN (2016) What are the major components in steam pretreated lignocellulosic biomass that inhibit the efficacy of cellulase enzyme mixtures? ACS Sustainable Chem. Eng 4:3429–3436

Zhang J, Yu L, Xu M, Yang ST, Yan Q, Lin M, Tang IC (2017) Metabolic engineering of Clostridium tyrobutyricum for n-butanol production from sugarcane juice. Appl Microbiol Biotechnol 101:4327–4337

Zhang K, Ren NQ, Wang AJ (2015) Fermentative hydrogen production from corn stover hydrolyzate by two typical seed sludges: Effect of temperature. Int J Hydrogen Energy 40:3838–3848

Zhang L, Li Y, Liu X, Ren N, Ding J (2019) Lignocellulosic hydrogen production using dark fermentation by *Clostridium lentocellum* strain Cel10 newly isolated from *Ailuropoda melanoleuca* excrement. RSC Adv 9:11179–11185

Zhang Y, Xia C, Lu M, Tu M (2018) Effect of overliming and activated carbon detoxification on inhibitors removal and butanol fermentation of poplar prehydrolysates. Biotechnol Biofuels, 1–14

Chapter 4
Biochemical Conversion of Hemicellulose

Ryan J. Stoklosa

Abstract Renewable carbon from lignocellulosic biomass offers the opportunity to displace petroleum products to foster greater bioeconomy development and sustainability initiatives. Improvements continue to be made in biorefinery processes for the complete utilization of all plant cell wall biopolymers. To achieve this goal, fermentable sugars from hemicellulose should be effectively utilized alongside glucose from cellulose. The polysaccharide xylan is the most abundant form of hemicellulose present in common bioenergy feedstocks such as hardwoods, cereal grasses, and agricultural industry residues. Once deconstructed, the xylan polysaccharide produces the five-carbon sugar xylose. This pentose sugar can be biochemically converted to a wide array of products ranging from bioethanol, organic acids, or higher value chemicals. This chapter explores biochemical conversion strategies for utilizing xylose obtained from lignocellulose biomass to generate biofuels or other value-added chemicals.

4.1 Hemicellulose Fractionation for Biochemical Conversion

Lignocellulosic biomass provides the largest source of renewable carbon that can be utilized to generate bio-based fuels or chemicals through biochemical processing (Rogers et al. 2017). As more countries attempt to shift away from fossil-based resources such as petroleum, more priority has been given to cultivating different varieties of lignocellulosic biomass in order to provide new feedstock sources for biofuels (Fulton et al. 2015), increasing crop value for rural development (Kim and Dale 2015),

Mention of trade names or commercial products in this publication is solely for providing specific information and does not imply recommendation or endorsement by the U.S. Department of Agriculture. USDA is an equal opportunity provider and employer.

R. J. Stoklosa (✉)
Sustainable Biofuels and Co-Products Research Unit, Eastern Regional Research Center, USDA, ARS, Wyndmoor, PA 19038, USA
e-mail: ryan.stoklosa@usda.gov

N. P. Nghiem et al. (eds.), *Biomass Utilization: Conversion Strategies*,
https://doi.org/10.1007/978-3-031-05835-6_4

and offsetting greenhouse gas emissions through sustainable land management and biochemical conversion practices (Dale et al. 2014). Although biomass cultivation will vary from region to region, the plant cell wall components of cellulose, hemicellulose, and lignin remain ubiquitous between lignocellulosic feedstocks but vary in quantity. A remaining challenge is to develop biochemical conversion processes for lignocellulosic feedstocks that can utilize and convert the entire gamut of plant cell wall biopolymers.

Being the largest component in lignocellulose, the polysaccharide cellulose is of primary interest for biochemical conversion. Most chemical pretreatments for lignocellulosic feedstocks employ chemistries and processing parameters that operate by preserving cellulose while altering or removing the hemicellulose and lignin fraction from the plant cell wall (Nan et al. 2018; Shen et al. 2019; Wang et al. 2018). The pretreatment process makes the cellulosic portion of the biomass amenable to deconstruction downstream through enzymatic hydrolysis to obtain glucose for fermentation to bio-based ethanol. These processes all occur within the setting of a biorefinery, which is similar to current petrochemical refineries except plant biomass is the feedstock used to produce liquid transportation fuels instead of petroleum. Because profit margins are narrow for bioethanol production, biorefineries must also consider generating value-added co-products from residual hemicellulose or lignin to improve revenue and process economics (Humbird et al. 2011). Lignin's aromatic heterogeneity allows it to be valorized to a wide array of products ranging from chemicals (Khwanjaisakun et al. 2020), bio-based resins or polymers (Fang et al. 2020), or hydrocarbon fuels (Saraeian et al. 2020); however, this same heterogeneity can also present challenges for upgrading (Liu et al. 2019). Opposite lignin, hemicellulose can be more easily upgraded using a biochemical conversion strategy as long as the polysaccharide can be recovered at high yield without substantial degradation during chemical pretreatment (Stoklosa and Hodge 2015).

While cellulose is a homopolysaccharide with a crystalline structure, hemicellulose is different by being an amorphous polysaccharide that is heterogenous and more easily extracted from the plant cell wall using chemical pretreatment. Hemicellulose composition can vary greatly among different biomass feedstocks, but, in general, it accounts for 20–30% of the dry weight in the plant cell wall (Ebringerová 2005). One of the most common forms of hemicellulose is xylan, which is a polysaccharide of $\beta(1\text{–}4)$ linked units of xylose that is predominant in feedstocks such as grasses, hardwoods, and other agro-industrial residues (Deutschmann and Dekker 2012). Different chemical moieties ranging from arabinose, acetic acid, and glucuronic acid can be substituted at varying degrees along the xylan polysaccharide chain (Naran et al. 2009; Carpita and Whittern 1986). Other forms of hemicellulose include glucomannans and galactoglucomannans which are much more predominant in softwood species ranging from pine to spruce (Ebringerová 2005). Although hemicellulose composition varies among biomass species, these fractions share similar traits of being easily extractable under certain pretreatment chemistries. Historically, the pulp and paper industry has exploited this feature of hemicellulose by designing alkaline pulping processes that can remove large quantities of hemicellulose while retaining the cellulose portion for downstream paper production (Stoklosa and Hodge 2014).

The issue with a typical pulping step, such as the Kraft process, is the process severity that degrades extracted hemicellulose to saccharic acids (Helmerius et al. 2010). More recent research has focused on novel pretreatment processing of biomass that can fractionate out the hemicellulose portion from the plant cell wall without significant degradation.

Pretreatments that function by preserving hemicellulose and limiting polysaccharide degradation can range from organic solvents, ionic liquids, or less severe alkaline chemistries. These pretreatments function by either retaining hemicellulose within the plant cell wall or by extracting hemicellulose and converting the polysaccharide to soluble oligosaccharides. Ethanol organosolv pretreatment on hybrid poplar can solubilize up to 60% of xylan after 50 min of pretreatment into a stable oligomeric form while only removing 2.8% of the original cellulose (Bär et al. 2018). This allows the pretreatment process to generate separate streams of fermentable sugars in the form of hexoses from the cellulosic portion and dissolved pentoses from xylan to be biochemically converted downstream. A similar processing approach has been utilized with ionic liquid (IL) pretreatment on lignocellulosic biomass. ILs are salts comprised of an organic cation with an organic or inorganic anion that maintains a liquid phase at standard temperate and pressure conditions (Raj et al. 2018). ILs function in a pretreatment by dissolving the entire plant cell wall biopolymer structure due to strong interaction with hydrogen bonds (Usmani et al. 2020). Individual components of the dissolved biomass can be recovered using distinct precipitation techniques such as acetone or hot water to recover cellulose while organic solvents with water are able to precipitate hemicellulose and lignin fractions (Usmani et al. 2020). The recovered cellulosic fraction has typically shown reduced crystallinity in the polysaccharide which leads to greater monomeric sugar yields from enzymatic hydrolysis (Raj et al. 2018). As with other pretreatments, the isolation and recovery of hemicellulose can be problematic when using ILs. For example, the IL triethylammonium hydrogen sulfate was able to provide up to 80% glucose yields from cellulose and greater than 85% delignification during pretreatment, but the dissolved hemicellulose was discovered to be solubilized mostly into monomeric sugars that were then converted into furfural due to dehydration during the pretreatment (Brandt-Talbot et al. 2017). Other types of ILs that are known as deep eutectic solvents (DES) have shown to be more favorable for hemicellulose recovery. An alkaline DES consisting of choline chloride mixed with different alkaline hydrogen bond donors exhibited high xylan solubility while maintaining a branched polysaccharide structure in solution without degradation (Yang et al. 2021). Other DES such as choline chloride mixed with urea and copper chloride could provide high xylan solubility while controlling the polysaccharide breakdown to monomeric xylose without unwanted dehydration byproducts (Loow et al. 2018). Similarly, less severe alkaline pretreatments of biomass can also provide hemicellulose extraction without unwanted degradation to the polysaccharide. A two-stage alkaline pretreatment coupled with an oxidative delignification on bamboo has indicated promising results for hemicellulose recovery. The overall process was able to recover over 87% of the original sugars in the bamboo that led to a 95.7% utilization of sugars during ethanol fermentation using a pentose-hexose fermenting strain of *Saccharomyces cerevisiae* (Yuan et al. 2018). A similar process using corn stover and sugarcane bagasse determined that a mild NaOH treatment could remove 66% of acetyl groups from the hemicellulose structure which lowered the release of acetic acid during downstream liquid hot water pretreatment while increasing the final glucose and xylose content for fermentation (Wang et al. 2019).

The overall goal with the pretreatments discussed above is oriented around preserving hemicellulose fractions for downstream biochemical conversion. This strategy has the ability to improve the overall process economics for biorefinery processes. Techno-economic analysis for a biorefinery producing hemicellulosic sugars from lignocellulosic biomass determined that the minimum selling sugar price could be reduced from $446 per metric ton to $347 per metric ton when xylose is utilized for xylitol production (Ou et al. 2021). Although any biorefinery process that will incorporate additional hemicellulose sugar recovery and conversion will increase overall capital costs, the chemicals and co-products generated from these sugars should be of sufficient value to offset a portion of the cost increase (Ou et al. 2021; Kim et al. 2021). Following pretreatment, the xylose recovered can be biochemically converted to a wide array of products that provide high value within a biorefinery for improving revenue. Figure 4.1 outlines a generalized process for using xylose within a biorefinery. The products identified following pentose fermentation in Fig. 4.1 can either be utilized for biofuel applications or as high-value chemicals. Table 4.1 outlines each fermentation product and their potential applications.

This chapter focuses on current advances made on the utilization of hemicellulose from biomass feedstocks for further downstream biochemical conversion to value-added chemicals or bio-based fuels. Moreover, this chapter will predominantly focus on the utilization of xylan hemicellulose given that this polysaccharide is the most

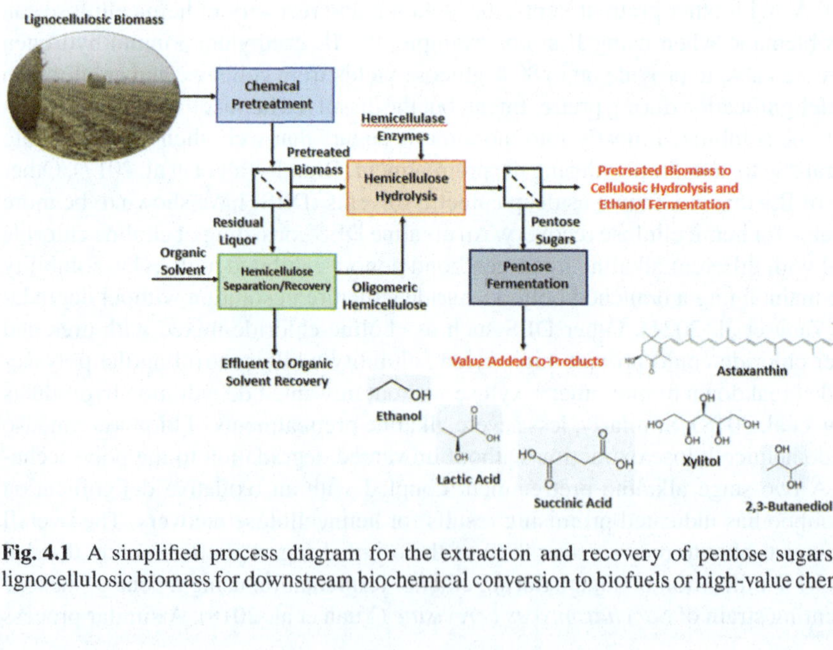

Fig. 4.1 A simplified process diagram for the extraction and recovery of pentose sugars from lignocellulosic biomass for downstream biochemical conversion to biofuels or high-value chemicals

Table 4.1 Microbial fermentation products from xylose and applications

Microbial fermentation product	Chemical type	Common organisms for fermentation	Applications	References
Ethanol	Alcohol	*Saccharomyces cerevisiae*, *Scheffersomyces stipitis*	Biofuels	Lopes et al. (2017), Lee et al. (2021), Song et al. (2019), Biazi et al. (2021)
Lactic acid	Alpha hydroxy acid	*Bacillus*, *Lactobacillus* family	Food additive, pharmaceuticals, precursor for bio-based polylactic acid (PLA) polymers	Alves de Oliveira et al. (2019), Parra-Ramírez et al. (2019)
Succinic acid	Dicarboxylic acid	*Actinobacillus succinogenes*, genetically engineered *Escherichia coli*	Food additive, pharmaceuticals, precursor for bio-based polybutylene succinate	Moncada et al. (2015), Harmsen et al. (2014), Ong et al. (2019)
Xylitol	Sugar alcohol	*Candida* family, genetically engineered *Saccharomyces cerevisiae*	Food additive	Yang et al. (2020), Narisetty et al. (2021)
Astaxanthin	Carotenoid	*Phaffia rhodozyma*	Aquaculture, nutritional supplements	Ytrestøyl and Bjerkeng (2007), Guerin et al. (2003)
2,3-Butanediol	Vic-Diol	*Klebsiella*, *Enterobacter* families, *Paenibacillus polymyxa*	Platform chemical for synthetic rubber, hydrocarbons for bio-jet fuel	Haveren et al. (2008), Ji et al. (2011)

abundant fraction in biomass feedstocks ranging from hardwoods, grasses, and agricultural byproducts such as corn stover or cereal straws (Deutschmann and Dekker 2012).

4.2 Hemicellulose Fermentation to Bioethanol

Ethanol fermentation for biofuel applications is still a primary focus for hemicellulose utilization. One of the primary challenges faced during ethanol fermentation of hemicellulose is the inability of wild-type *Saccharomyces cerevisiae* to ferment pentose sugars such as xylose. This barrier has been removed in part due to advances in genetic modification of native yeast strains to give them the ability to co-ferment glucose and xylose. The introduction of a fungal pathway that uses xylose reductase and xylitol dehydrogenase in *S. cerevisiae* generated haploid and diploid yeast strains that could metabolize xylose (Lopes et al. 2017). These modified strains were able to be used in fermentations with pretreated switchgrass hydrolysates that contained high levels of inhibitory compounds such as acetic acid, furfural, and hydroxymethylfurfural (HMF). Ethanol production in switchgrass hydrolysate maxed out at 38 g/L at a product yield of 0.48 g ethanol per g sugar (Lopes et al. 2017). Xylose fermentation by *S. cerevisiae* can also be achieved by using directed evolution techniques on respiration-deficient yeast strains (Lee et al. 2021). These strains of *S. cerevisiae* were grown in the presence of xylose and oxygen that induced genetic changes to the yeast to produce what is described as a Crabtree/Warburg effect for xylose. The evolved strains contained a pathway expressing xylose isomerase that allowed fermentation to ethanol in synthetic media and pretreated switchgrass hydrolysates both aerobically and anaerobically (Lee et al. 2021). Other routes to ethanol production from xylan hemicellulose include consolidated bioprocessing (CBP), which is a processing strategy using a microorganism that can provide enzyme production, polysaccharide hydrolysis, and fermentation in one step (Cripwell et al. 2019). Industrial strains of *S. cerevisiae* that exhibit traits such as tolerance to higher temperatures (40 °C) and resistance to inhibitors were adapted to include cell-surface production of hemicellulase enzymes (Cunha et al. 2020). The adapted strains were grown on corn cobs that underwent hydrothermal pretreatment and showed an ethanol production up to 11.1 g/L at a sugar yield of 0.328 g ethanol per g xylose and glucose (Cunha et al. 2020). Most of these strains have only been evaluated in a small-scale laboratory setting, but there are several commercialized yeast strains that co-ferment glucose and xylose at the industrial scale. The Danish company Terranol employs a recombinant, co-fermenting glucose/xylose *S. cerevisiae* strain cv-110, that was developed to express xylose isomerase during fermentation (Chandel et al. 2018). Fed-batch fermentation with *S. cerevisiae* cv-110 showed >90% ethanol yield on wheat straw hydrolysate with near-complete glucose and xylose utilization (Knudsen and Rønnow 2020). Terranol has also showed scalability for this strain from the 2 L bioreactor scale to the 270,000 L demonstration scale (https://www.terranol.com/). The Swedish company Taurus Energy AB has engineered a strain of *S. cerevisiae* that goes by the trade name XyloFerm T13 that rapidly ferments xylose by adding pathways for xylose reductase, xylose dehydrogenase, and xylulokinase (Drapcho et al. 2020). This strain generates ethanol titers above 100 g/L with ethanol yields ranging between 0.41 and 0.49 g per g sugar while also exhibiting inhibitor resistance to furfural, hydroxymethyfurfural, and acetic acid (https://taurusenergy.eu/en/about-taurus/product/). Mascoma, a division of the yeast company Lallemand based in Canada, has also developed a co-fermenting glucose/xylose strain of *S. cerevisiae* known as C5 Fuel that was engineered with improvements in the pentose phosphate pathway and increased xylose isomerase expression (Drapcho et al. 2020). This strain was cultivated in high xylose content hydrolysates from corn stover and wheat straw. Over 95% sugar consumption occurred for both glucose and xylose in each hydrolysate with ethanol yields above 0.46 g per g sugar (Drapcho et al. 2020).

While some companies have successfully implemented fermentation processes for co-fermenting glucose and xylose with engineered strains of *S. cerevisiae*, native microbial strains exist that already possess the ability to naturally ferment xylose to ethanol.

The most well-known microorganism that can naturally ferment xylose to ethanol is *Scheffersomyces stipitis* (formerly known as *Pichia stipitis*). *S. stipitis* is a haploid yeast strain that possesses the best xylose fermentation capacity of any native strain (Jeffries et al. 2007). Although this strain can easily ferment xylose, its performance suffers at the industrial scale due to a low tolerance for both sugar and ethanol (Song et al. 2019). Even with this limitation the performance of *S. stiptis* can be applied to various fermentation process strategies to boost ethanol production. A dual fermentation process with *S. cerevisiae* and *S. stipitis* was able to convert more sugars from pretreated hardwood species to increase overall ethanol output by 12% compared to a conventional fermentation process (Song et al. 2019). To overcome *S. stipitis* fermentation inhibition, glucose was first fermented to ethanol using *S. cerevisiae*, and the resulting ethanol was removed via pervaporation. This led to a secondary fermentation broth with lower ethanol content but still containing xylose. The secondary fermentation broth was inoculated with *S. stipitis* that resulted in 76.3% theoretical yield of ethanol from xylose (Song et al. 2019). A similar study has shown how further improvements in ethanol yield can be achieved by combining fermentation processes using *S. stipitis* as suspended cells and the bacteria *Zymomonas mobilis* as immobilized cells on polyvinyl alcohol (PVA) beads. This continuous process used hydrolysate from pretreated sugarcane bagasse to co-ferment glucose using *Z. mobilis* and *S. stipitis* to 81.18% theoretical yield and maximum ethanol productivity of 1.868 g/L * hr (Wirawan et al. 2020). A membrane bioreactor process set up to ferment xylose using suspended *S. stipitis* found that a high ethanol yield could be maintained, but cell activity by *S. stipitis* continually declined without additional nutrient supplementation (Wirawan et al. 2020). Other improvements to *S. stipitis* ethanol output can be realized using product separation strategies or microbial adaptions. Kans grass hydrolysate containing 60 g/L glucose and 40 g/L xylose was shown to produce an ethanol yield above 80% of the theoretical yield through a sequential fermentation process with in situ ethanol distillation (Gehlot et al. 2022). In this process developed by Gehlot et al., glucose was first fermented to ethanol using *Z. mobilis* for 12 h followed by immediate ethanol removal through distillation. This reduced the ethanol content in the fermentation broth to a suitable concentration in order for xylose fermentation using *S. stiptis* to occur. At 14 h, *S. stiptis* was inoculated and xylose was completely consumed after 32 h (Gehlot et al. 2022). This processing strategy led to an overall ethanol concentration of 45.73 g/L at a productivity rate of 1.45 g/L * h (Gehlot et al. 2022). The other route to improve a strain's fermentation performance can be accomplished through adaptation to improve tolerance to inhibitors. Recycled batches of *S. stipitis* cells were inoculated into increasing media levels of sugarcane bagasse hemicellulosic hydrolysate that contained xylose as a carbon source and acetic acid as the primary inhibitor (Biazi et al. 2021). The adapted strain of *S. stipitis* used in a fed-batch fermentation of hemicellulosic hydrolysate could tolerate up to 2.6 g/L of acetic acid compared to the non-adapted strain that only tolerated 0.7 g/L (Biazi et al. 2021). Although a substantial amount of glucose and xylose could be consumed with the adapted strain of *S. stipitis*, the overall ethanol output was only 16.87 g/L ethanol at a volumetric productivity of 0.16 g/L * h (Biazi et al. 2021). Although continued advances in fermenting xylose to ethanol using native or modified yeast strains can provide increased biofuel output, there is more value that can be gained by using different microorganisms for xylose fermentation to generate higher value chemicals.

4.3 Hemicellulose Fermentation to Organic Acids

Fermentations that generate organic acids from xylose can be incorporated into a biorefinery to improve revenue and profit margins. A number of organic acids are considered important platform chemicals for their ability to be converted into higher value products that can displace current petroleum-based products (Werpy, Petersen 2004). Organic acids are typically synthesized from petroleum, but they are also naturally generated by all microorganisms through the citric acid (TCA) cycle (Yin et al. 2015). With advances in genetic modifications, the TCA cycle in almost any microorganism can be altered to overexpress or downregulate certain genes to boost the production of any given organic acid. Most organic acids have value-added applications with respect to certain industries. For example, citric acid is not only widely used in the food and pharmaceutical industry but can also be applied as a biocompatible polymer in medical materials (Yin et al. 2015). Two other organic acids that are actively sought after as value-added co-products are lactic and succinic acids.

Lactic acid (LA) is commonly generated by bacterial strains and is associated with fermentation contamination with respect to ethanol production, but it is also a sought-after co-product for biorefineries. Unlike metabolism problems associated with yeast strains that produce ethanol, most bacterial strains are able to consume xylose during fermentation to organic acids. The bacterial strain *Bacillus coagulans* was effectively cultivated on hemicellulosic hydrolysate from sugarcane bagasse for the production of LA at high isomeric purity (Alves de Oliveira et al. 2019). Overall lactic acid production reached around 56 g/L at a yield of 0.87 g LA per g sugar with an isomeric purity of 99.4% (Alves de Oliveira et al. 2019). The hemicellulosic hydrolysate underwent roto-evaporation to concentrate xylose at 48 g/L and the cultivation of *B. coagulans* was not hindered by the high xylose content nor the presence of HMF and furfural. In this respect, *B. coagulans* can be considered a bio-detoxifying organism because it can actively consume HMF and furfural (Alves de Oliveira et al. 2019). In another study, the lactogenic bacteria *Escherichia coli* JU15 could be cultivated in simulated corn stover media containing over 30 g/L xylose to produce 40 g/L LA at a yield of 0.6 g LA per g total sugar (Parra-Ramírez et al. 2019). The potential also exists to improve a strain's uptake of xylose using adaptive evolution techniques. The strain *Lactobacillus pentosus* naturally ferments xylose, but through adaptive evolution, a new strain was produced when grown in xylose media with pH control (Cubas-Cano et al. 2019). This new strain showed vast improvement in xylose uptake, better tolerance to more acidic conditions, and a 1.4-fold increase in LA generation in wheat straw hydrolysate with co-consumption of xylose and other sugars (Cubas-Cano et al. 2019).

A second organic acid that is a valuable co-product for biorefinery production is succinic acid (SA). This organic acid is of considerable interest because of its ability to be upgraded to plastic materials such as polybutylene succinate and favorable sustainability performance that is on par with bioethanol production (Moncada et al. 2015; Harmsen et al. 2014). Pretreated sugarcane bagasse containing a high xylose concentration at 57.9 g/L can serve as a promising feedstock for fermentation with *Actinobacillus succinogenes* ATCC 55618 to generate 21.8 g/L succinic acid at a yield of 0.45 g SA per g sugar (Klein et al. 2017). The same strain could also ferment hemicellulosic sugars from pretreated sugar beet pulp to reach a slightly higher titer at 30 g/L SA and a yield of 0.8 g SA per g sugar (Alexandri et al. 2019). The co-fermentation of glucose and xylose can also be achieved with *Yarrowia lipolytica.* Pretreated sugarcane bagasse hydrolysate cultivated with *Y. lipotytica* could generate 33.2 g/L SA at a yield of 0.58 g SA per g sugar through the simultaneous fermentation of glucose and xylose (Ong et al. 2019). A different co-fermentation process can also be utilized that simultaneously cultivates *S. cerevisiae* for ethanol generation and *A. succinogenes* for SA production. This strategy takes advantage of glucose fermentation by *S. cerevisiae* and redirects the CO_2 off-gas back into the fermentation to synthesize SA production by *A. succinogenes* (Xu et al. 2021). In this process, the fermentable sugars of glucose and xylose could be fully utilized from pretreated sugarcane bagasse to generate 22.1 g/L SA and 22.0 g/L ethanol, which corresponded to 8.7 g SA and 8.6 g ethanol per 100 g of pretreated sugarcane bagasse (Xu et al. 2021). The wide applications of organic acids make their production via fermentation of xylose an attractive processing step.

4.4 Hemicellulose Fermentation to Other Value-Added Co-products

4.4.1 Astaxanthin

Fermentation processes that generate ethanol and organic acids from xylose are able to improve the process economics for a biorefinery, but different classes of value-added chemicals can also be generated using novel microorganisms that ferment xylose. One such chemical of interest is the carotenoid astaxanthin. This carotenoid is of particular interest due to its high-value applications in aquaculture feed and as nutritional supplement with reported benefits for heart and liver function (Ytrestøyl and Bjerkeng 2007; Guerin et al. 2003; Naguib 2000). The yeast strain *Phaffia rhodozyma* is a well-known producer of astaxanthin that possesses metabolic versatility with easy cultivation (Mussagy et al. 2021). *P. rhodozyma* also possesses the natural ability to consume xylose for astaxanthin fermentation. Higher specific cell content of astaxanthin was achieved on xylose compared to glucose (1.84 mg astaxanthin per g xylose against 1.24 mg astaxanthin per g glucose) using fermentable sugars from pretreated sugarcane bagasse and barley straw (Montanti et al. 2011). More recent research with *P. rhodozyma* showed that the yeast strain has more preference for xylose metabolism once glucose is depleted (Stoklosa et al. 2019a). Using detoxified hydrolysate from pretreated sugarcane bagasse *P. rhodozyma* could consume most of the glucose within 48 h which then resulted in faster xylose uptake and depletion within 72 h to produce an intracellular astaxanthin content of 2.07 mg astaxanthin per gram sugar consumed (Stoklosa et al. 2019a). To assess improved utilization of xylose in a biorefinery process, a recovered portion of xylose following ethanol fermentation and recovery could be redirected to a secondary fermentation process. A xylose-enriched post-ethanol fermentation stillage from pretreated sweet sorghum bagasse was detoxified and fermented with *P. rhodozyma* that was able to increase intracellular astaxanthin content to 3.9 mg astaxanthin per g sugar compared to only 1.0 mg astaxanthin per g sugar in non-detoxified stillage media (Stoklosa et al. 2019b).

4.4.2 Xylitol

Xylitol is a high-value sugar alcohol within the food industry where it can be utilized as a replacement for sucrose due to its low glycemic index (de Albuquerque et al. 2014). The strain *Candida mogii* efficiently utilizes xylose for biomass growth and cell maintenance, but a co-factor imbalance leads to excess xylitol production (de Albuquerque et al. 2014; Baptista et al. 2018). A separate fermentation process using the same xylose post-ethanol stillage media found that the yeast strain *C. mogii* also showed promising results for converting xylose to xylitol (Stoklosa et al. 2019b). It was determined using the xylose-enriched post-ethanol stillage media that *C. mogii* could be cultivated for high growth in non-detoxified stillage media (Stoklosa et al. 2019b). Following this fermentation, secondary cultivation using the recovered cells occurred in a synthetic xylose media to produce a yield of 0.55 g xylitol per g xylose consumed (Stoklosa et al. 2019b). This research indicated that xylose utilization could be improved by using the post-ethanol stillage media for microorganism growth, while a second-stage fermentation containing xylose from pretreated sweet sorghum bagasse using hemicellulases would be operated for product generation (Stoklosa et al. 2019b). While this dual-stage fermentation process with *C. mogii* can improve xylitol output, a single-stage fermentation process using more robust xylitol-producing yeast strains would ultimately be less cost intensive. Xylose from pretreated sugarcane bagasse hydrolysate was fermented with *Pichia fermentans* to produce 86.6 g/L xylitol and a conversion yield of 0.75 g xylitol per g sugar consumed (Narisetty et al. 2021). Although high xylitol titers were achieved, the fermentation with *P. fermentans* needed additional processing controls as glucose concentrations in the hydrolysate above 10 g/L were found to reduce xylitol generation even with an increase in biomass growth (Narisetty et al. 2021). Another route for xylitol production includes co-generating the sugar alcohol and ethanol with *S. cerevisiae*. An industrial strain of *S. cerevisiae* was engineered at different levels of xylose specificity by overexpressing the *XYL1* gene from *Candida tropicalis* that encodes for xylose reductase (Yang et al. 2020). In the fermentation of pretreated corn stover hydrolysate, the engineered strain of *S. cerevisiae* could produce 45.41 g/L xylitol and 50.19 g/L ethanol (Yang et al. 2020). This simultaneous production of ethanol and a co-product in the form of xylitol provides for efficient utilization of both glucose and xylose from lignocellulosic feedstocks.

4.4.3 2,3-Butanediol

Another promising platform chemical that is of more recent interest is 2,3-butanediol (2,3-BDO). This chemical can be upgraded to a multitude of products ranging from a form of synthetic rubber (Haveren et al. 2008), food-grade products such as acetoin (Xiao and Xu 2007), or hydrocarbons that can serve as bio-jet fuel (Ji et al. 2011). Most bacterial species that produce 2,3-BDO contain a metabolic pathway that can consume xylose and generate organic acids (Celińska and Grajek 2009). Detoxified hydrolysate from hydrothermally pretreated sugarcane bagasse containing glucose and xylose could be fermented with *Enterobacter ludwigii* to generate 63.5 g/L 2,3-BDO at a yield of 0.36 g 2,3-BDO per g sugar and productivity of 0.84 g/L * h (Amraoui et al. 2021). The 2,3-BDO output was reduced by about half when non-detoxified hydrolysate was used for fermentation, which was attributed to the presence of inhibitors and, primarily, an abundance of acetic acid (Amraoui et al. 2021). Another type of bacteria that can produce 2,3-BDO is *Klebsiella oxytoca*. Hemicellulose from corn cob was biologically pretreated with the fungus *Aspergillus niger* to generate a hydrolysate with high xylose concentration at 34.6 g/L (Sharma et al. 2018). An optimized fermentation process found that the newly isolated strain of *K. oxytoca* XF7 could generate 12.18 g/L 2,3-BDO with a xylose conversion efficiency of 96.65% (Sharma et al. 2018). Another recently isolated strain from an environmental consortium known as *Pantoea agglomerans* could convert >75% of xylose and arabinose from pretreated soybean hull hydrolysate and generate up to 14.02 g/L 2,3-BDO in 12 h of cultivation, which corresponded to a yield of 0.53 g 2,3-BDO per g sugar and productivity of 1.16 g/L * h (Ourique et al. 2020). A few challenges remain for microbial 2,3-BDO production at the industrial scale, but one of the more prominent limitations involves the choice of bacterial strain for fermentation. 2,3-BDO-producing strains that are in the *Enterobacter* and *Klebsiella* family are opportunistic pathogens in humans and any fermentation process that will require scale-up must consider the biohazardous implications. Other bacterial strains that are not pathogenic to humans are also available such as *Paenibacillus polymyxa*, which can simultaneously utilize the sugars glucose, xylose, and arabinose during fermentation. Pretreated wheat straw hydrolysate was fermented with *P. polymyxa* which could generate up to 23 g/L 2,3-BDO in non-detoxified hydrolysate and showed an added benefit of consuming the inhibitory compound HMF (Okonkwo et al. 2021).

The wide versatility for 2,3-BDO as a platform chemical coupled with most bacterial strains exhibiting native xylose consumption makes this compound attractive for incorporation into biorefinery processes.

4.5 Conclusion

The biochemical conversion of lignocellulosic biomass could improve bioeconomy applications through the simultaneous production of biofuels and value-added chemicals. Better process economics for biofuel production can be achieved by identifying biochemical conversion processes that utilize the entire contents of the plant cell wall. Hemicellulosic sugars, primarily the five-carbon sugar xylose, can be fermented to a wide array of products ranging from biofuels or high-value chemicals to displace similar petroleum products. Yeast strains that generate ethanol such as *S. cerevisiae* require some form of genetic modification or evolution to fully utilize xylose. Improved xylose utilization can be achieved for ethanol fermentation by using strains that naturally ferment xylose such as *S. stipitis*. Bacterial strains that generate organic acids such as lactic acid and succinic acid can also be included in biochemical processing options to produce intermediate chemicals for upgrading. More novel yeast strains could ferment xylose to astaxanthin for application in high-value aquaculture feed or the sugar alcohol xylitol as a sucrose replacement in certain food items. Similarly, bacterial strains that generate the platform chemical 2,3-butanediol can also be incorporated into a biochemical conversion strategy for their native xylose utilization. Overall, these fermentation products from xylose could improve the economic output of biorefinery processes and make them more competitive with their petroleum-based counterparts.

References

Alexandri M, Schneider R, Papapostolou H, Ladakis D, Koutinas A, Venus J (2019) Restructuring the conventional sugar beet industry into a novel biorefinery: fractionation and bioconversion of sugar beet pulp into succinic acid and value-added coproducts. ACS Sustain Chem Eng 7:6569–6579. https://doi.org/10.1021/acssuschemeng.8b04874

Alves de Oliveira R, Schneider R, Vaz Rossell CE, Maciel Filho R, Venus J (2019) Polymer grade l-lactic acid production from sugarcane bagasse hemicellulosic hydrolysate using Bacillus coagulans. Bioresour Technol Rep 6:26–31. https://doi.org/10.1016/j.biteb.2019.02.003

Amraoui Y, Narisetty V, Coulon F, Agrawal D, Chandel AK, Maina S, Koutinas A, Kumar V (2021) Integrated fermentative production and downstream processing of 2,3-butanediol from sugarcane bagasse-derived xylose by mutant strain of enterobacter ludwigii. ACS Sustain Chem Eng 9:10381–10391. https://doi.org/10.1021/acssuschemeng.1c03951

Baptista SL, Cunha JT, Romaní A, Domingues L (2018) Xylitol production from lignocellulosic whole slurry corn cob by engineered industrial Saccharomyces cerevisiae PE-2. Biores Technol 267:481–491. https://doi.org/10.1016/j.biortech.2018.07.068

Bär J, Phongpreecha T, Singh SK, Yilmaz MK, Foster CE, Crowe JD, Hodge DB (2018) Deconstruction of hybrid poplar to monomeric sugars and aromatics using ethanol organosolv fractionation. Biomass Convers Biorefinery 8:813–824

Biazi LE, Santos SC, Kaupert Neto AA, Sousa AS, Soares LB, Renzano E, Velasco J, Rabelo SC, Costa AC, Ienczak JL (2021) Adaptation strategy to increase the tolerance of scheffersomyces stipitis NRRL Y-7124 to inhibitors of sugarcane bagasse hemicellulosic hydrolysate through comparative studies of proteomics and fermentation. BioEnergy Res. https://doi.org/10.1007/s12155-021-10267-3

Brandt-Talbot A, Gschwend FJ, Fennell PS, Lammens TM, Tan B, Weale J, Hallett JP (2017) An economically viable ionic liquid for the fractionation of lignocellulosic biomass. Green Chem 19:3078–3102

Carpita NC, Whittern D (1986) A highly substituted glucuronoarabinoxylan from developing maize coleoptiles. Carbohyd Res 146:129–140

Celińska E, Grajek W (2009) Biotechnological production of 2,3-butanediol—current state and prospects. Biotechnol Adv 27:715–725. https://doi.org/10.1016/j.biotechadv.2009.05.002

Chandel AK, Antunes FAF, Terán-Hilares R, Cota J, Ellilä S, Silveira MHL, dos Santos JC, da Silva SS (2018) Chapter 5—Bioconversion of hemicellulose into ethanol and value-added products: commercialization, trends, and future opportunities. In: Chandel AK, Luciano Silveira MH (eds) Advances in sugarcane biorefinery. Elsevier, pp 97–134. https://doi.org/10.1016/B978-0-12-804534-3.00005-7

Cripwell RA, Rose SH, Favaro L, van Zyl WH (2019) Construction of industrial Saccharomyces cerevisiae strains for the efficient consolidated bioprocessing of raw starch. Biotechnol Biofuels 12:201. https://doi.org/10.1186/s13068-019-1541-5

Cubas-Cano E, González-Fernández C, Tomás-Pejó E (2019) Evolutionary engineering of Lactobacillus pentosus improves lactic acid productivity from xylose-rich media at low pH. Biores Technol 288:121540. https://doi.org/10.1016/j.biortech.2019.121540

Cunha JT, Romaní A, Inokuma K, Johansson B, Hasunuma T, Kondo A, Domingues L (2020) Consolidated bioprocessing of corn cob-derived hemicellulose: engineered industrial Saccharomyces cerevisiae as efficient whole cell biocatalysts. Biotechnol Biofuels 13:138. https://doi.org/10.1186/s13068-020-01780-2

Dale BE, Anderson JE, Brown RC, Csonka S, Dale VH, Herwick G, Jackson RD, Jordan N, Kaffka S, Kline KL, Lynd LR, Malmstrom C, Ong RG, Richard TL, Taylor C, Wang MQ (2014) Take a closer look: biofuels can support environmental, economic and social goals. Environ Sci Technol 48:7200–7203. https://doi.org/10.1021/es5025433

de Albuquerque TL, da Silva Jr IJ, de Macedo GR, Rocha MVP (2014) Biotechnological production of xylitol from lignocellulosic wastes: a review. Process Biochem 49:1779–1789

Deutschmann R, Dekker RFH (2012) From plant biomass to bio-based chemicals: latest developments in xylan research. Biotechnol Adv 30:1627–1640. https://doi.org/10.1016/j.biotechadv.2012.07.001

Drapcho CM, Nghiem N, Walker TH (2020) Biofuels engineering process technology, 2nd edn. McGraw Hill

Ebringerová A (2005) Structural diversity and application potential of hemicelluloses. In: Macromolecular symposia, vol 1. Wiley Online Library, pp 1–12. https://doi.org/10.1002/masy.200551401

Fang Z, Nikafshar S, Hegg EL, Nejad M (2020) Biobased divanillin as a precursor for formulating biobased epoxy resin. ACS Sustain Chem Eng 8:9095–9103. https://doi.org/10.1021/acssuschemeng.0c02351

Fulton LM, Lynd LR, Körner A, Greene N, Tonachel LR (2015) The need for biofuels as part of a low carbon energy future. Biofuels Bioprod Biorefin 9:476–483

Gehlot K, Sivakumar R, Ghosh S (2022) In situ distillation strategy to improve the sequential fermentation process using zymomonas mobilis and pichia stipitis for bioethanol production from Kans Grass biomass hydrolysate. BioEnergy Res. https://doi.org/10.1007/s12155-021-10383-0

Guerin M, Huntley ME, Olaizola M (2003) *Haematococcus* astaxanthin: applications for human health and nutrition. Trends Biotechnol 21:210–216. https://doi.org/10.1016/S0167-7799(03)00078-7

Harmsen PF, Hackmann MM, Bos HL (2014) Green building blocks for bio-based plastics. Biofuels Bioprod Biorefin 8:306–324

Haveren JV, Scott EL, Sanders J (2008) Bulk chemicals from biomass. Biofuels Bioprod Biorefin: Innov Sustain Econ 2:41–57

Helmerius J, von Walter JV, Rova U, Berglund KA, Hodge DB (2010) Impact of hemicellulose pre-extraction for bioconversion on birch Kraft pulp properties. Biores Technol 101:5996–6005. https://doi.org/10.1016/j.biortech.2010.03.029

https://www.terranol.com/. Accessed 22 February 2022

https://taurusenergy.eu/en/about-taurus/product/. Accessed 22 February 2022

Humbird D, Davis R, Tao L, Kinchin C, Hsu D, Aden A, Schoen P, Lukas J, Olthof B, Worley M (2011) Process design and economics for biochemical conversion of lignocellulosic biomass to ethanol: dilute-acid pretreatment and enzymatic hydrolysis of corn stover. National Renewable Energy Lab. (NREL), Golden, CO (United States)

Jeffries TW, Grigoriev IV, Grimwood J, Laplaza JM, Aerts A, Salamov A, Schmutz J, Lindquist E, Dehal P, Shapiro H (2007) Genome sequence of the lignocellulose-bioconverting and xylose-fermenting yeast Pichia stipitis. Nat Biotechnol 25:319

Ji X-J, Huang H, Ouyang P-K (2011) Microbial 2,3-butanediol production: a state-of-the-art review. Biotechnol Adv 29:351–364. https://doi.org/10.1016/j.biotechadv.2011.01.007

Khwanjaisakun N, Amornraksa S, Simasatitkul L, Charoensuppanimit P, Assabumrungrat S (2020) Techno-economic analysis of vanillin production from Kraft lignin: feasibility study of lignin valorization. Biores Technol 299:122559. https://doi.org/10.1016/j.biortech.2019.122559

Kim H, Lee S, Lee B, Park J, Lim H, Won W (2021) Improving revenue from lignocellulosic biofuels: an integrated strategy for coproducing liquid transportation fuels and high value-added chemicals. Fuel 287:119369. https://doi.org/10.1016/j.fuel.2020.119369

Kim S, Dale BE (2015) All biomass is local: the cost, volume produced, and global warming impact of cellulosic biofuels depend strongly on logistics and local conditions. Biofuels Bioprod Biorefin 9:422–434

Klein BC, Silva JFL, Junqueira TL, Rabelo SC, Arruda PV, Ienczak JL, Mantelatto PE, Pradella JGC, Junior SV, Bonomi A (2017) Process development and techno-economic analysis of bio-based succinic acid derived from pentoses integrated to a sugarcane biorefinery. Biofuels Bioprod Biorefin 11:1051–1064. https://doi.org/10.1002/bbb.1813

Knudsen JD, Rønnow B (2020) Extended fed-batch fermentation of a C5/C6 optimised yeast strain on wheat straw hydrolysate using an online refractive index sensor to measure the relative fermentation rate. Sci Rep 10:6705. https://doi.org/10.1038/s41598-020-63626-z

Lee S-B, Tremaine M, Place M, Liu L, Pier A, Krause DJ, Xie D, Zhang Y, Landick R, Gasch AP, Hittinger CT, Sato TK (2021) Crabtree/Warburg-like aerobic xylose fermentation by engineered Saccharomyces cerevisiae. Metab Eng 68:119–130. https://doi.org/10.1016/j.ymben.2021.09.008

Liu Z-H, Le RK, Kosa M, Yang B, Yuan J, Ragauskas AJ (2019) Identifying and creating pathways to improve biological lignin valorization. Renew Sustain Energy Rev 105:349–362. https://doi.org/10.1016/j.rser.2019.02.009

Loow Y-L, Wu TY, Yang GH, Ang LY, New EK, Siow LF, Md. Jahim J, Mohammad AW, Teoh WH (2018) Deep eutectic solvent and inorganic salt pretreatment of lignocellulosic biomass for improving xylose recovery. Bioresour Technol 249:818–825. https://doi.org/10.1016/j.biortech.2017.07.165

Lopes DD, Rosa CA, Hector RE, Dien BS, Mertens JA, Ayub MAZ (2017) Influence of genetic background of engineered xylose-fermenting industrial Saccharomyces cerevisiae strains for ethanol production from lignocellulosic hydrolysates. J Ind Microbiol Biotechnol 44:1575–1588. https://doi.org/10.1007/s10295-017-1979-z

Moncada J, Posada JA, Ramírez A (2015) Early sustainability assessment for potential configurations of integrated biorefineries. Screening of bio-based derivatives from platform chemicals. Biofuels Bioprod Biorefin 9:722–748

Montanti J, Nghiem NP, Johnston DB (2011) Production of astaxanthin from cellulosic biomass sugars by mutants of the yeast *Phaffia rhodozyma*. Appl Biochem Biotechnol 164:655–665. https://doi.org/10.1007/s12010-011-9165-7

Mussagy CU, Pereira JF, Dufossé L, Raghavan V, Santos-Ebinuma VC, Pessoa Jr A (2021) Advances and trends in biotechnological production of natural astaxanthin by Phaffia rhodozyma yeast. Crit Rev Food Sci Nutrit 1–15

Naguib YM (2000) Antioxidant activities of astaxanthin and related carotenoids. J Agric Food Chem 48:1150–1154. https://doi.org/10.1021/jf991106k

Nan Y, Jia L, Yang M, Xin D, Qin Y, Zhang J (2018) Simplified sodium chlorite pretreatment for carbohydrates retention and efficient enzymatic saccharification of silvergrass. Biores Technol 261:223–231. https://doi.org/10.1016/j.biortech.2018.03.106

Naran R, Black S, Decker SR, Azadi P (2009) Extraction and characterization of native heteroxylans from delignified corn stover and aspen. Cellulose 16:661–675

Narisetty V, Castro E, Durgapal S, Coulon F, Jacob S, Kumar D, Kumar Awasthi M, Kishore Pant K, Parameswaran B, Kumar V (2021) High level xylitol production by Pichia fermentans using non-detoxified xylose-rich sugarcane bagasse and olive pits hydrolysates. Biores Technol 342:126005. https://doi.org/10.1016/j.biortech.2021.126005

Okonkwo CC, Ujor V, Ezeji TC (2021) Production of 2,3-Butanediol from non-detoxified wheat straw hydrolysate: impact of microbial inhibitors on Paenibacillus polymyxa DSM 365. Ind Crops Prod 159:113047. https://doi.org/10.1016/j.indcrop.2020.113047

Ong KL, Li C, Li X, Zhang Y, Xu J, Lin CSK (2019) Co-fermentation of glucose and xylose from sugarcane bagasse into succinic acid by Yarrowia lipolytica. Biochem Eng J 148:108–115. https://doi.org/10.1016/j.bej.2019.05.004

Ou L, Dou C, Yu J-H, Kim H, Park Y-C, Park S, Kelley S, Lee EY (2021) Techno-economic analysis of sugar production from lignocellulosic biomass with utilization of hemicellulose and lignin for high-value co-products. Biofuels Bioprod Biorefin 15:404–415. https://doi.org/10.1002/bbb.2170

Ourique LJ, Rocha CC, Gomes RCD, Rossi DM, Ayub MAZ (2020) Bioreactor production of 2,3-butanediol by Pantoea agglomerans using soybean hull acid hydrolysate as substrate. Bioprocess Biosyst Eng 43:1689–1701. https://doi.org/10.1007/s00449-020-02362-0

Parra-Ramírez D, Martinez A, Cardona CA (2019) Lactic acid production from glucose and xylose using the lactogenic Escherichia coli strain JU15: experiments and techno-economic results. Biores Technol 273:86–92. https://doi.org/10.1016/j.biortech.2018.10.061

Raj T, Gaur R, Lamba BY, Singh N, Gupta RP, Kumar R, Puri SK, Ramakumar SSV (2018) Characterization of ionic liquid pretreated plant cell wall for improved enzymatic digestibility. Biores Technol 249:139–145. https://doi.org/10.1016/j.biortech.2017.09.202

Rogers JN, Stokes B, Dunn J, Cai H, Wu M, Haq Z, Baumes H (2017) An assessment of the potential products and economic and environmental impacts resulting from a billion ton bioeconomy. Biofuels Bioprod Biorefin 11:110–128. https://doi.org/10.1002/bbb.1728

Saraeian A, Aui A, Gao Y, Wright MM, Foston M, Shanks BH (2020) Evaluating lignin valorization via pyrolysis and vapor-phase hydrodeoxygenation for production of aromatics and alkenes. Green Chem 22:2513–2525

Sharma A, Nain V, Tiwari R, Singh S, Nain L (2018) Optimization of fermentation condition for co-production of ethanol and 2,3-butanediol (2,3-BD) from hemicellolosic hydrolysates by Klebsiella oxytoca XF7. Chem Eng Commun 205:402–410. https://doi.org/10.1080/00986445.2017.1398743

Shen X-J, Wen J-L, Mei Q-Q, Chen X, Sun D, Yuan T-Q, Sun R-C (2019) Facile fractionation of lignocelluloses by biomass-derived deep eutectic solvent (DES) pretreatment for cellulose enzymatic hydrolysis and lignin valorization. Green Chem 21:275–283

Song Y, Cho EJ, Park CS, Oh CH, Park B-J, Bae H-J (2019) A strategy for sequential fermentation by Saccharomyces cerevisiae and Pichia stipitis in bioethanol production from hardwoods. Renew Energy 139:1281–1289. https://doi.org/10.1016/j.renene.2019.03.032

Stoklosa RJ, Hodge DB (2014) Integration of (hemi)-cellulosic biofuels technologies with chemical pulp production. In: Biorefineries. Elsevier, pp 73–100. https://doi.org/10.1016/B978-0-444-59498-3.00004-X

Stoklosa RJ, Hodge DB (2015) Fractionation and improved enzymatic deconstruction of hardwoods with alkaline delignification. Bioenergy Res 8:1224–1234

Stoklosa RJ, Johnston DB, Nghiem NP (2019a) Phaffia rhodozyma cultivation on structural and non-structural sugars from sweet sorghum for astaxanthin generation. Process Biochem 83:9–17. https://doi.org/10.1016/j.procbio.2019.04.005

Stoklosa RJ, Nghiem NP, Latona RJ (2019b) Xylose-enriched ethanol fermentation stillage from sweet sorghum for xylitol and astaxanthin production. Fermentation 5:84

Usmani Z, Sharma M, Gupta P, Karpichev Y, Gathergood N, Bhat R, Gupta VK (2020) Ionic liquid based pretreatment of lignocellulosic biomass for enhanced bioconversion. Biores Technol 304:123003. https://doi.org/10.1016/j.biortech.2020.123003

Wang Q, Tian D, Hu J, Shen F, Yang G, Zhang Y, Deng S, Zhang J, Zeng Y, Hu Y (2018) Fates of hemicellulose, lignin and cellulose in concentrated phosphoric acid with hydrogen peroxide (PHP) pretreatment. RSC Adv 8:12714–12723

Wang Z, Dien BS, Rausch KD, Tumbleson ME, Singh V (2019) Improving ethanol yields with deacetylated and two-stage pretreated corn stover and sugarcane bagasse by blending commercial xylose-fermenting and wild type Saccharomyces yeast. Biores Technol 282:103–109. https://doi.org/10.1016/j.biortech.2019.02.123

Werpy T, Petersen G (2004) Top value added chemicals from biomass: volume I—results of screening for potential candidates from sugars and synthesis gas. National Renewable Energy Lab., Golden, CO (US)

Wirawan F, Cheng C-L, Lo Y-C, Chen C-Y, Chang J-S, Leu S-Y, Lee D-J (2020) Continuous cellulosic bioethanol co-fermentation by immobilized Zymomonas mobilis and suspended Pichia stipitis in a two-stage process. Appl Energy 266:114871. https://doi.org/10.1016/j.apenergy.2020.114871

Xiao Z, Xu P (2007) Acetoin metabolism in bacteria. Crit Rev Microbiol 33:127–140

Xu C, Alam MA, Wang Z, Peng Y, Xie C, Gong W, Yang Q, Huang S, Zhuang W, Xu J (2021) Co-fermentation of succinic acid and ethanol from sugarcane bagasse based on full hexose and pentose utilization and carbon dioxide reduction. Biores Technol 339:125578. https://doi.org/10.1016/j.biortech.2021.125578

Yang B-X, Xie C-Y, Xia Z-Y, Wu Y-J, Li B, Tang Y-Q (2020) The effect of xylose reductase genes on xylitol production by industrial Saccharomyces cerevisiae in fermentation of glucose and xylose. Process Biochem 95:122–130. https://doi.org/10.1016/j.procbio.2020.05.023

Yang J, Wang Y, Zhang W, Li M, Peng F, Bian J (2021) Alkaline deep eutectic solvents as novel and effective pretreatment media for hemicellulose dissociation and enzymatic hydrolysis enhancement. Int J Biol Macromol. https://doi.org/10.1016/j.ijbiomac.2021.10.223

Yin X, Li J, Shin H-d, Du G, Liu L, Chen J (2015) Metabolic engineering in the biotechnological production of organic acids in the tricarboxylic acid cycle of microorganisms: advances and prospects. Biotechnol Adv 33:830–841. https://doi.org/10.1016/j.biotechadv.2015.04.006

Ytrestøyl T, Bjerkeng B (2007) Dose response in uptake and deposition of intraperitoneally administered astaxanthin in Atlantic salmon (Salmo salar L.) and Atlantic cod (Gadus morhua L.). Aquaculture 263:179–191. https://doi.org/10.1016/j.aquaculture.2006/10.021

Yuan Z, Wen Y, Kapu NS (2018) Ethanol production from bamboo using mild alkaline pre-extraction followed by alkaline hydrogen peroxide pretreatment. Biores Technol 247:242–249. https://doi.org/10.1016/j.biortech.2017.09.080

Chapter 5
Biochemical Conversion of Lignin

Nhuan Phu Nghiem

Abstract Lignin is one of the three major components of lignocellulosic biomass. It is highly resistant to either chemical or biochemical degradation due to the presence of the ether and C–C linkages in its heterogeneous structure. In the early attempts to develop processes for the bioconversion of biomass to fuels and chemicals, lignin was considered as a waste by-product and burned to supply heat for internal uses. Recently, it has been realized that in order to make a biorefinery economically feasible, lignin must also be used as a feedstock for the production of high-value products, in addition to cellulose and hemicellulose. Processes for the bioconversion of lignin subsequently were developed. A typical lignin bioconversion process consists of three steps, which include depolymerization, funneling, and product formation. In the first step, depolymerizing enzymes are used to break down lignin to its monomer and oligomer units. These monomers and oligomers then are converted to metabolic intermediates by a process normally referred to as funneling. Finally, the intermediates, which can enter the central metabolism, are converted to the desired products by various microbial species in fermentation processes. This chapter discusses the recent developments in the bioconversion of lignin and the potential commercial products that can be made. The future research directions that are needed to develop a complete biorefinery, which includes a component for lignin bioconversion to high-value products, are also discussed.

5.1 Overview

Lignin is a complex aromatic heterogeneous biopolymer, which is composed of phenylpropanoid aryl-C_3 units. These units in lignin are linked together via various types of ether and C–C bonds. The ether and C–C linkages in lignin are not susceptible to hydrolytic attack, either chemically or biologically, which makes lignin highly

N. P. Nghiem (✉)
Department of Environmental Engineering and Earth Sciences, Clemson University, Clemson, SC, USA
e-mail: nnghiem@g.clemson.edu

N. P. Nghiem et al. (eds.), *Biomass Utilization: Conversion Strategies*,
https://doi.org/10.1007/978-3-031-05835-6_5

resistant to degradation (Bugg et al. 2011). In nature, however, several microorganisms, which include fungi and bacteria, have emerged with the capability of degrading lignin. The potential applications of these organisms for biotransformation of lignin derived from biomass in a biorefinery recently have attracted considerable attention.

Lignin consists of three major subunits, namely *p*-hydroxyphenyl (H unit), guaiacyl (G unit), and syringyl (S unit). The corresponding precursors that form these subunits during lignin synthesis are *p*-coumaryl alcohol, coniferyl alcohol, and sinapyl alcohol, respectively. The chemical structures of the lignin subunits and their precursors are shown in Fig. 5.1. The inter-connections of the subunits in a typical lignin structure are shown in Fig. 5.2. The relative abundances of the subunits in lignin from various sources are shown in Table 5.1. The data in Table 5.1 were obtained by a 2D-NMR technique (Mansfield et al. 2012), which were in general agreement with those obtained by other methods such as thioacidolysis (TA) (Rolando et al. 1992), nitrobenzene oxidation (NBO) (Yamamura et al. 2010), and derivatization followed by reductive cleavage (DFRC) (Lu and Ralph 1997).

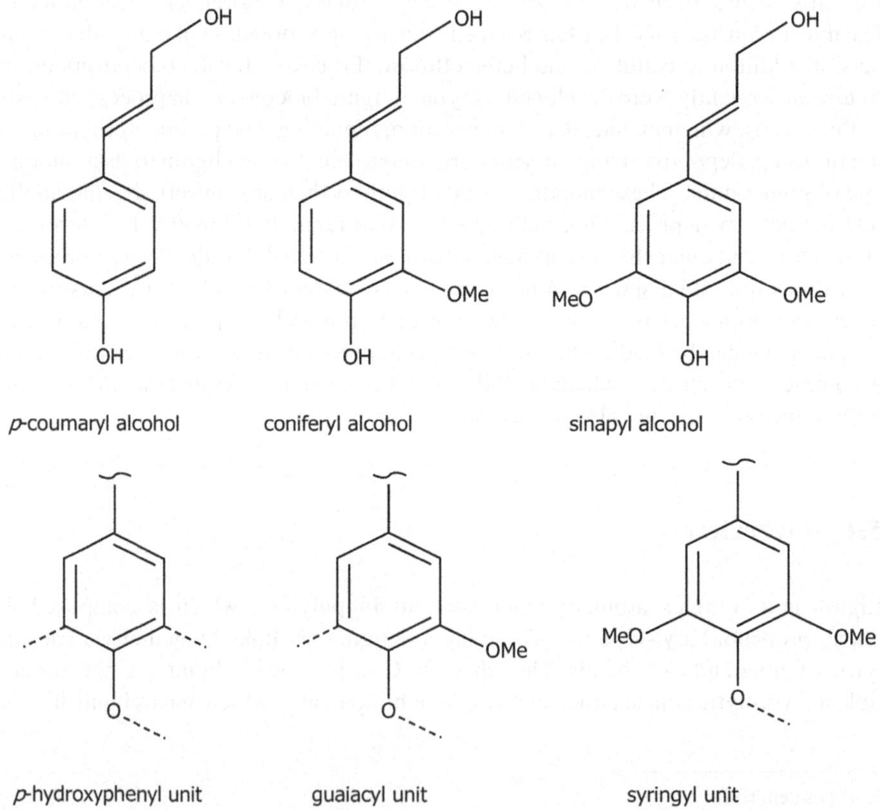

Fig. 5.1 Chemical structures of lignin subunits and their precursors (adapted from Li and Zheng 2020)

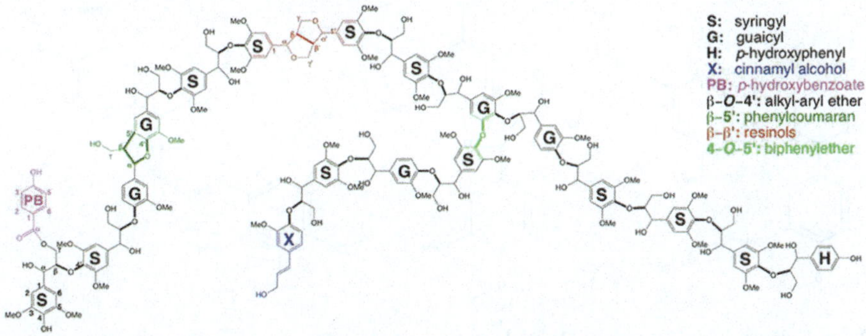

Fig. 5.2 The inter-connections of the subunits in the structure of poplar lignin (Li et al. 2016)

Table 5.1 The relative abundances of subunits in lignin from various sources (Mansfield et al. 2012)

Species	Lignin units		
	S (%)	G (%)	H (%)
Poplar (hardwood)	61.9	37.8	0.3
Pine (softwood)	0	98.3	1.7
Corn stalks	58.9	38.3	2.8
Arabidopsis	20.1	77.1	2.8

The major linkages that connect the subunits in lignin are shown in Fig. 5.3 and their relative occurrences are shown in Table 5.2.

The key steps in the biochemical conversion of lignin are illustrated in Fig. 5.4. The first step is depolymerization where lignin is broken up into the three subunits H, S, and G by enzymes, which include laccase, lignin peroxidase (LiP), manganese peroxidase (MnP), versatile peroxidase (VP), and dye-decolorizing peroxidase (DyP) (Li and Zheng 2020; Kamimura et al. 2019). The next step, which is referred to as funneling, involves a series of reactions that convert the three lignin subunits to vanillic acid, protocatechuic acid, and gallic acid. These products subsequently go through various ring-cleavage pathways to form the intermediates, which can enter the central metabolism.

5.2 Lignin Biotransformation

5.2.1 Depolymerization

Lignin depolymerization is caused by enzymes produced and secreted by fungi, which include white-rot and brown-rot fungi, and bacteria. White-rot fungi, so named because when they degrade lignin, they leave the delignified wood in a fibrous state

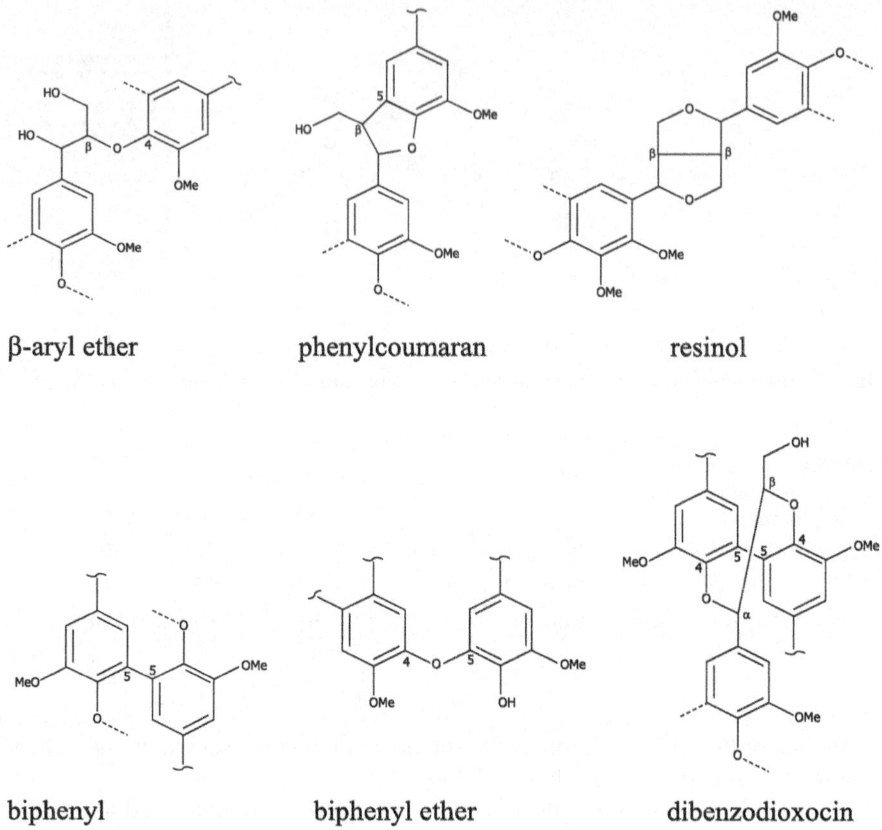

Fig. 5.3 Major linkages in lignin (adapted from Li and Zheng 2020)

Table 5.2 Relative occurrences of major linkages in lignins from various sources (Rinaldi et al. 2016)

	Type of linkages and their occurrence (%)				
	B-aryl ether	Phenylcoumaran	Resinol	Dibenzodioxocin	Biphenylether
Softwood	45–50	9–12	2–6	5–7	2
Hardwood	60–62	3–11	3–12	<1	2
Grasses	74–84	5–11	1–7	Not determined	Not determined

and whitish in color, are the most efficient lignin degraders. The white-rot fungi that have been studied include *Phanerochaete chrysosporium*, *Pleurotus ostreatus*, *Coriolus versicolor*, *Cyathus stercoreus*, *Ceriporiopsis subvermispora*, *Phellinus pini*, *Phlebia* spp., *Pleurotus* spp., *Trametes versicolor*, *Heterobasidion annosum*, and *Irpex lacteus*. Brown-rot fungi degrade wood polysaccharides and partially modify

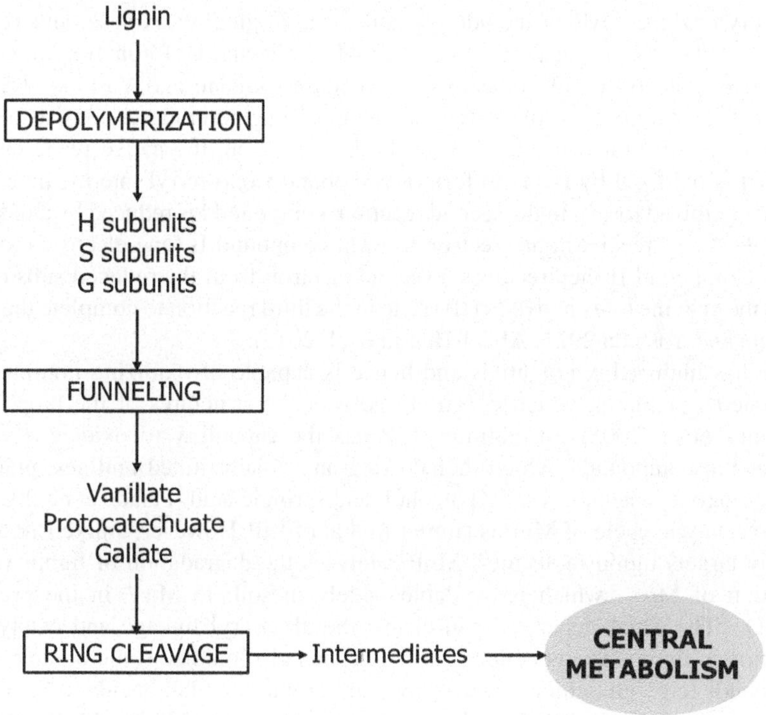

Fig. 5.4 The key steps in the biochemical conversion of lignin (adapted from Li and Zheng 2020; Kamimura et al. 2019)

lignin, leaving the decayed wood in a distinct brown color due to the oxidized lignin (Abdel-Hamid et al. 2013).

5.2.1.1 Fungal Enzymes

The fungal lignin-degrading enzymes include the phenol oxidase laccase and the four heme-containing peroxidases LiP, MnP, VP, and DyP. These fungal lignin-degrading enzymes have been the subjects of several reviews (Falade et al. 2016; Sugano and Yoshida 2021; Hofrichter 2002; Arregui et al. 2019; Ravichandran and Sridhar 2016). Peroxidases catalyze the oxidation of various substrates in the presence of hydrogen peroxide according to the general reaction below

$$2S + H_2O_2 + 2e^- \rightarrow 2S_{ox} + 2H_2O \tag{5.1}$$

where S is an electron-donating substrate and S_{ox} is the oxidized substrate (Falade et al. 2017). The H_2O_2 required by LiP, MnP, VP, and DyP is generated by other accessory enzymes, which include glyoxal oxidase, glucose oxidase, and veratryl alcohol oxidase. The four peroxidases LiP, MnP, VP, and DyP function in a cyclic manner, similar to that of horseradish peroxidase (Sugano and Yoshida 2021). In general, the catalytic cycle of the four heme-containing peroxidases consists of three reactions, as illustrated in Fig. 5.5. In the first reaction, the native ferric enzyme [Fe(III)] is oxidized by H_2O_2 to form a compound I oxo-ferryl intermediate (two-electron oxidized form). In the second reaction, compound I is reduced by a reducing substrate A and receives one electron to form compound II (one-electron oxidized form). Compound II then receives a second electron from the reduced substrate to return the enzyme to its native Fe(III) state in the third reaction to complete the cycle (Sugano and Yoshida 2021; Abdel-Hamid et al. 2013).

LiP has high redox potentials and hence is capable of oxidizing non-phenolic components of lignin, which constitute 90% of the structures of the biopolymer (Martinez et al. 2005). In addition, LiP has the capability to oxidize a variety of phenolic compounds, which include ring-and N-substituted anilines, guaiacol, acetosyringone, catechol, vanillyl alcohol, and syringic acid (Falade et al. 2016).

The catalytic cycle of MnP is similar to that of LiP. However, unlike LiP, which directly targets lignin molecules, MnP catalyzes the degradation of lignin via the oxidation of Mn^{2+}, which is available widely in soil, to Mn^{3+} in the presence of H_2O_2. The Mn^{3+} generated can cleave the alkyl-aryl linkage and catalyze α-oxidation in lignin. The generated Mn^{3+} ions can also oxidize some organic sulfur compounds (e.g., glutathione, L-cysteine) and unsaturated fatty acids (e.g., linoleic acid) to sustain the lignin degradation process (Hofrichter 2002). MnP, however, cannot oxidize the more recalcitrant non-phenolic compounds like LiP (Li and Zheng 2020). As the name suggests, VP has the catalytic capability of both LiP and MnP. It can oxidize both phenolic and non-phenolic compounds. Although the catalytic mechanism of VP is similar to that of LiP and MnP, VP can oxidize compounds over a wider range of redox potentials. The protein structure of VP provides binding sites

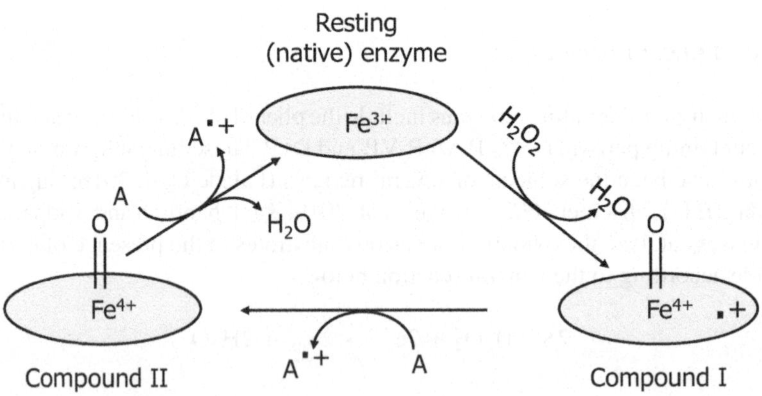

Fig. 5.5 The general steps in the catalytic cycle of LiP, MnP, VP, and DyP (adapted from Sugano and Yoshida 2021)

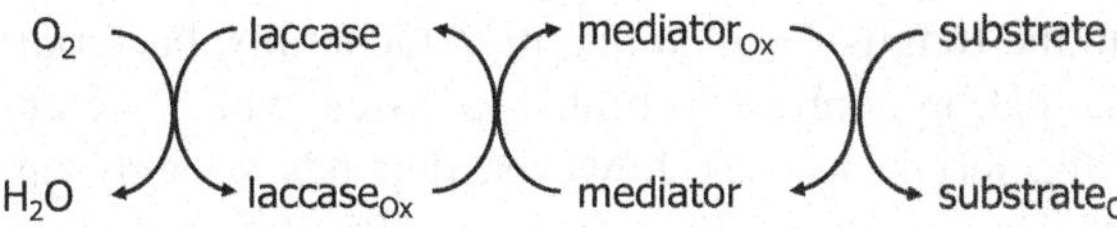

Fig. 5.6 Proposed mechanism of substrate oxidation by laccase in the presence of a mediator (adapted from Christopher et al. 2014)

for different substrates, and hence it can perform its catalytic function effectively without the need for mediators (Li and Zheng 2020). DyP is produced by both fungi and bacteria. The role of this type of enzyme in the biodegradation of lignin by fungi in nature, however, is still not clear.

Laccase is a copper-containing phenol oxidase which can oxidize both phenolic and non-phenolic substrates. This enzyme does not require hydrogen peroxide and catalyzes the oxidation reaction with the concomitant reduction of molecular oxygen to water by a radical-mediated mechanism (Bourbonnais et al. 1997). Laccase can effectively degrade phenolic lignin compounds, but the presence of redox mediators is required for the degradation of the non-phenolic compounds. A mediator is a small molecule that can be continuously oxidized by the enzyme and subsequently reduced by the substrate. The role of the mediator is to serve as an electron carrier, which shuttles electrons between the substrate and the enzyme active sites (Christopher et al. 2014). The proposed mechanism of substrate oxidation by laccase in the presence of a mediator is illustrated in Fig. 5.6.

Both synthetic and naturally occurring mediators are effective in assisting laccase catalytic function. Naturally occurring mediators, however, are preferred in practice since they are less expensive and non-toxic. Some of these naturally occurring mediators include 3-hydroxyanthranilic acid, 4-hydroxybenzoic acid, phenolsulfonphthalein, acetosyringone, syringaldehyde, vanillin, and methyl syringate (Christopher et al. 2014).

5.2.1.2 Bacterial Enzymes

Bacteria, which include strains of *Streptomyces*, *Rhodococcus*, *Pseudomonas*, and *Bacillus*, have been found to have the capability of lignin degradation (Lee et al. 2019). Bacterial LiP, MnP, DyP, and laccase have been identified (Chauhan 2020; Atiwesh et al. 2021). Among these, DyP and laccase are more commonly found and believed to play a more important role in lignin degradation by bacteria (Lee et al. 2019). Compared to fungal DyP, bacterial DyP has lower oxidizing power, and its substrates are limited to the less recalcitrant phenolic moieties in lignin (de Gonzalo et al. 2016). Bacterial laccases have been known to be stable under various conditions of pH, temperature, organic solvents, and salt concentrations (Guan et al. 2018). Laccase from *Bacillus subtilis* exhibited a half-life of 250 min at 70 °C, and laccase from *Streptomyces viridochromeogenes* showed a half-life of 30 min at 80 °C compared to the half-life of 10 min at the same temperature observed with laccase from the fungus *Cerrena unicolor.* Generally, bacterial laccases prefer neutral or alkaline pH, in contrast to fungal laccases, which prefer acidic pH. The optimum pH for bacterial laccases, however, depends strongly on the substrates. Laccases from *Marinomonas mediterranea* and *Bacillus halodurans* can tolerate 1 M or higher NaCl concentrations. Laccase from *Bacillus pumilus* W3 showed high tolerance to mixtures of water and organic solvents including ethanol, methanol, dimethylformamide, acetonitrile, acetone, and dimethylsulfoxide (Arregui et al. 2019).

5.2.2 Monomer Funneling

It is generally accepted that lignin is depolymerized by fungi to monomers, which subsequently are funneled by bacteria into metabolism intermediates for further use in the biosynthesis of final products (Xu et al. 2019). The G-lignin is depolymerized to ferulic acid-type monomers. Ferulic acid, therefore, is considered as a standard model compound for G-lignin. The degradation of ferulic acid in bacteria follows four pathways, which include the nonoxidative decarboxylation pathway, the coenzyme A (CoA)-dependent non-β-oxidation pathway, the CoA-dependent β-oxidation pathway, and the side chain reduction pathway. These four pathways all converge at vanillic acid, which subsequently is converted to protocatechuic acid and then to catechol. The H-lignin is depolymerized to *p*-coumaric acid-type monomers. Similar to the case of ferulic acid and G-lignin, *p*-coumaric acid is considered as a model compound to represent H-lignin. The degradation of *p*-coumaric acid in bacteria follows three pathways, which include the CoA-dependent β-oxidation pathway, the CoA-dependent non-β-oxidation pathway, and the CoA-independent pathway. These three pathways all converge at *p*-hydroxybenzoic acid, which subsequently is converted to protocatechuic acid and then to catechol (Xu et al. 2019). The pathways for the degradation of ferulic acid and *p*-coumaric acid are summarized in Fig. 5.7. Upon formation, catechol and protocatechuate are subjected to the aromatic ring cleavage process, which is catalyzed by the O_2-dependent dioxygenase. The ring cleavage is the first step of a series of reactions required to bring the carbon from lignin into the central metabolism. The degradation pathways of catechol are shown in Fig. 5.8. In the *meta*-cleavage, catechol is ultimately degraded to acetaldehyde and pyruvate. In the *ortho*-cleavage, catechol is degraded via the β-ketoadipate pathway (Hardwood and Parales 1996). The product of this pathway, i.e., β-ketoadipate, is further converted to succinate and acetyl-CoA. The conversion of β-ketoadipate to succinate and acetyl-CoA is shown in Fig. 5.9. The degradation of protocatechuate follows three pathways, which are the *ortho*-cleavage pathway, the 2,3-*meta*-cleavage pathway, and the 4,5-*meta*-cleavage pathway. The three pathways of protocatechuate degradation are shown in Fig. 5.10. The product of the *ortho*-cleavage of protocatechuate is β-ketoadipate, which is subsequently converted to succinate and acetyl-CoA (Fig. 5.9). In the 2,3-*meta*-cleavage of protocatechuate, the final products are pyruvate and acetyl-CoA. Pyruvate is the sole product of the 4,5-*meta*-cleavage of

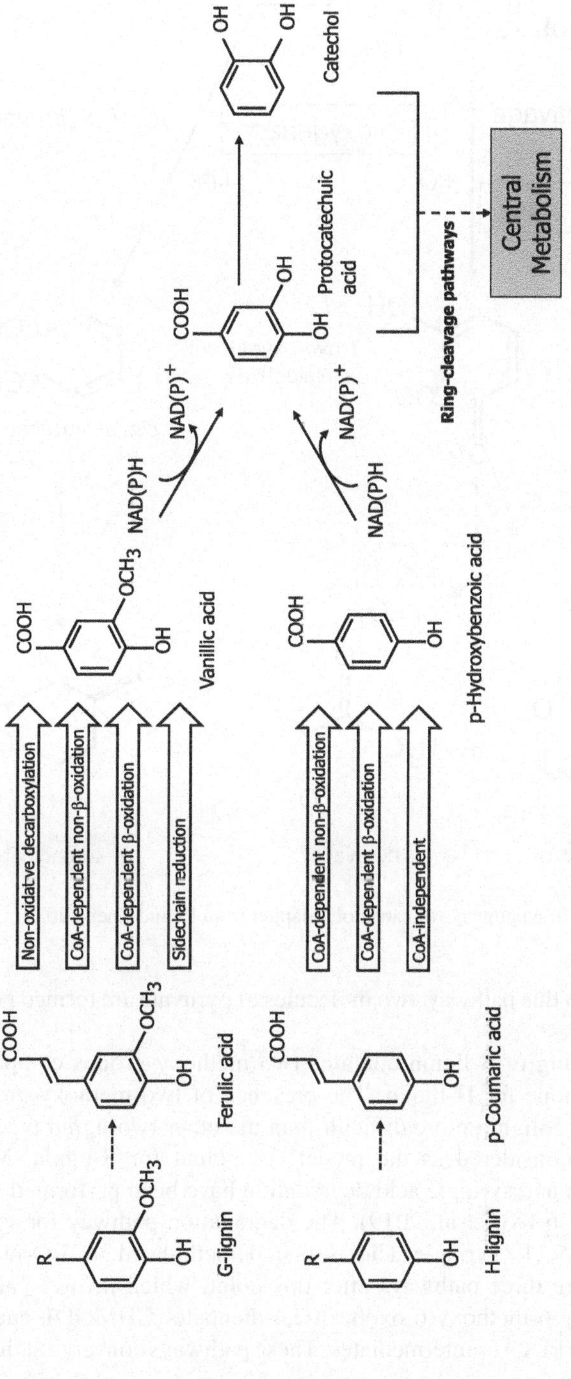

Fig. 5.7 The pathways for the degradation of G-lignin-derived ferulic acid and H-lignin-derived *p*-coumaric acid (adapted from Xu et al. 2019)

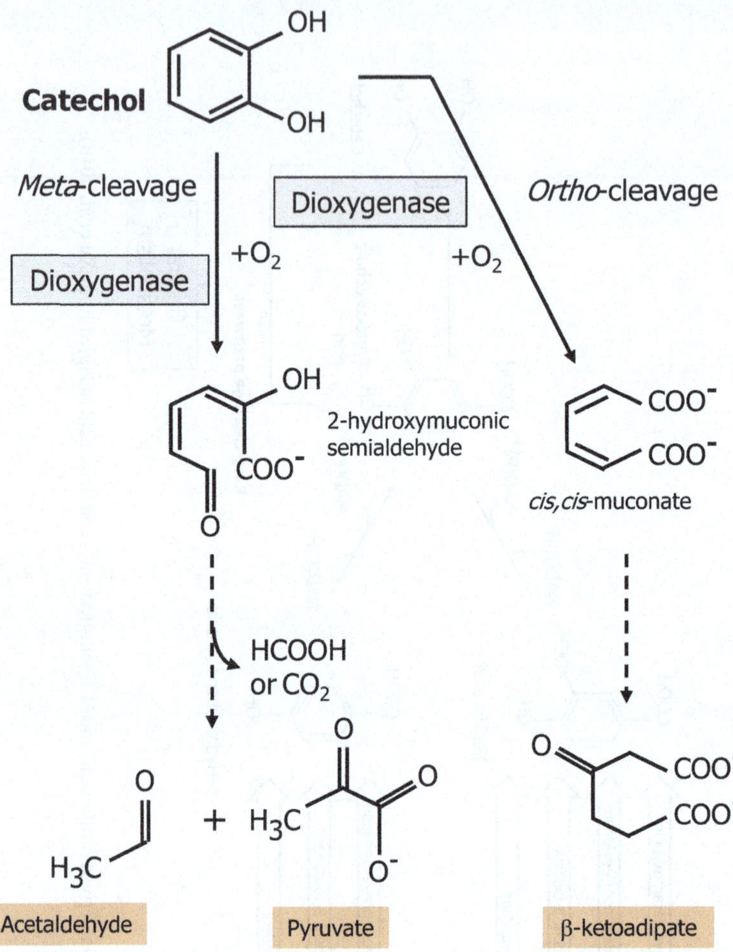

Fig. 5.8 The degradation pathways of catechol (adapted from Li and Zheng 2020)

protocatechuate. In this pathway, two molecules of pyruvate are formed per molecule of protocatechuate.

The aromatic ring of S-lignin contains two methoxy groups compared to one for G-lignin and none for H-lignin. The presence of two methoxy groups makes the degradation of S-lignin more difficult than the other two lignin types. Syringic acid is normally considered as the model compound for S-lignin. Most of the studies on S-lignin and syringic acid degradation have been performed with *Sphingomonas* sp. SYK-6 (Xu et al. 2019). The degradation pathway for syringic acid is shown in Fig. 5.11. Syringic acid is first demethylated to 3-O-methylgallate (3MGA). There are three pathways after this point, which involve gallic acid, 4-carboxy-2-hydroxy-6-methoxy-6-oxohexa-2,4-dienoate (CHMOD), and 2-pyrone-4,6-dicarboxylate (PDC) as intermediates. These pathways converge at the formation

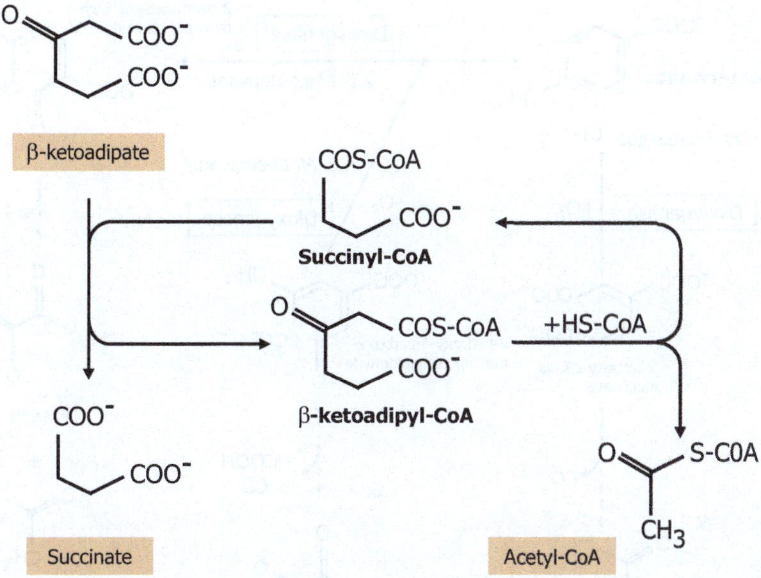

Fig. 5.9 The conversion of β-ketoadipate to succinate and acetyl-CoA (adapted from Li and Zheng 2020)

of 4-oxalomesaconate (OMA), which is also an intermediate of the 4,5-meta-cleavage pathway of protocatechuate. The degradation pathway of syringic acid and the 4,5-cleavage pathway of protocatechuate, therefore, converge at 4-oxalomesaconate.

All of the compounds formed at the end of the pathways for degradation of lignin monomers can enter the central metabolism and be utilized for the synthesis of cellular matters and target products. The yields of these products, therefore, are dependent on the recovery of carbon in the degradation pathways. From this perspective, it can be stated that in the degradation of catechol, the *ortho*-cleavage pathway is more efficient than the *meta*-cleavage pathway since in the *ortho*-cleavage pathway, all six carbons in the starting compound are recovered in the products, i.e., succinate and acetyl-CoA, compared to only five carbons recovered in the products of the *meta*-cleavage pathway, i.e., acetaldehyde and pyruvate. Similarly, in the degradation of protocatechuate, the *ortho*-cleavage and 4,5-*meta*-cleavage pathways are more efficient than the 2,3-*meta*-cleavage pathway since in the first two pathways six carbons are recovered from seven carbons in the starting compound compared to five carbons recovered in the other pathway. In the degradation of syringic acid, seven carbons are recovered from nine carbons in the starting compound.

The lignin degradation products can also be fed directly into subsequent pathways for the production of value-added products. For example, pyruvate can be used as a substrate for the production of ethanol, lactic acid, and succinic acid. It has been shown that when the catechol *ortho* degradation pathway, which is endogenous to *Pseudomonas putida* KT2440, was replaced with an exogenous *meta*-cleavage

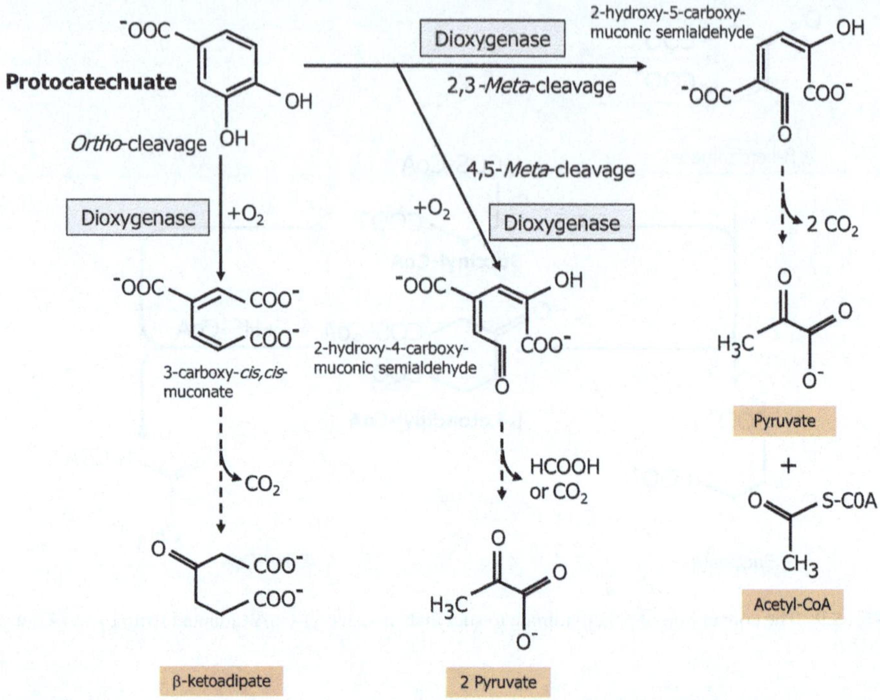

Fig. 5.10 The degradation pathways of protocatechuate (adapted from Li and Zheng 2020; Xu et al. 2019)

pathway from *Sphingobium* sp. SYK-6, the yield of pyruvate from benzoic acid, which is the precursor of catechol, was increased by almost five-fold (Johnson and Beckham 2015). This study demonstrated that the degradation pathways of lignin monomers could be modified to improve the yield of the desired end product for use as a substrate for the subsequent production of value-added products.

5.2.3 *Potential Products*

Various products can be produced from the lignin monomers. Some of these products are discussed in this section.

Dicarboxylic acids. Cis,cis-muconic acid (MA) is a dicarboxylic acid with important applications. It can be chemically converted to adipic acid and terephthalic acid, which are monomers for plastics and also are used in many cosmetic, pharmaceutical, and food applications. Attempts have been made to develop microbial strains that can synthesize MA from lignin monomers. The general strategy is to enhance the

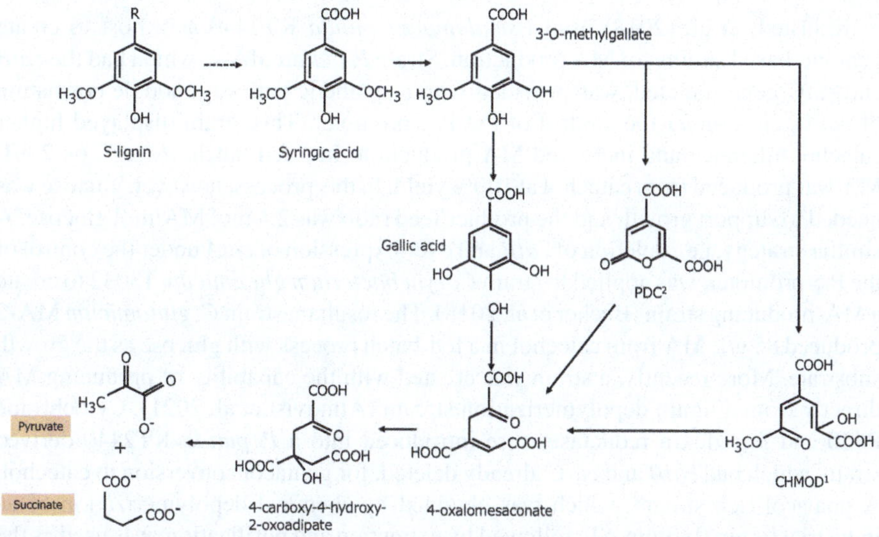

Fig. 5.11 The degradation pathway of syringic acid (adapted from Xu et al. 2019). Notes (1) CHMOD: 4-carboxy-2-hydroxy-6-methoxy-6-oxohexa-2,4-dienoate; (2) PDC: 2-pyrone-4,6-dicarboxylic acid

formation of MA from catechol and prevent its conversion to other metabolites (Fig. 5.12).

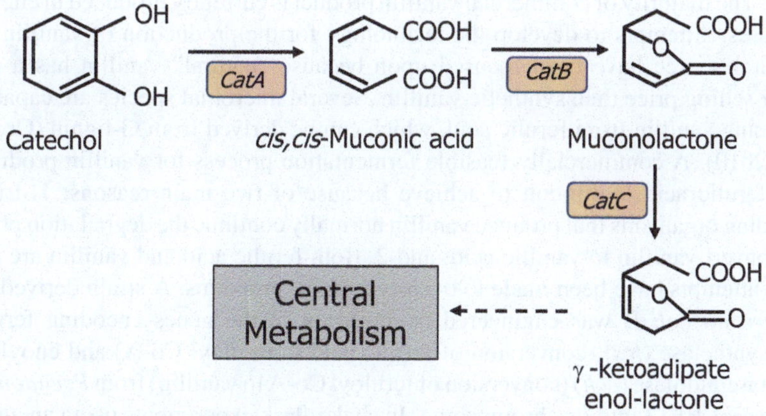

Fig. 5.12 Formation and conversion of cis,cis-muconic acid

Kohlstedt et al. (2018) used *Pseudomonas putida* KT2440 as a host to create genome-based strains for MA production. Strain *P. putida* MA-6, which had the *catB* and *catC* genes deleted, was provided with a synthetic pathway module consisting of *catA* genes under the control of the P_{cat} promoter. This strain displayed higher catechol tolerance and increased MA production. In a fed-batch process, 64.2 g/L MA was produced from catechol at 100% yield. In this process, however, glucose was needed to support growth and the product/feed ratio was 2.4 mol MA/mol glucose. A similar strategy, i.e., deletion of *catB* and over-expression of *catA* under the control of the P_{tuf} promoter, was applied to strain *Corynebacterium glutamicum* 13032 to create a MA-producing strain (Becker et al. 2018). The resultant strain *C. glutamicum* MA-2 produced 85 g/L MA from catechol in a fed-batch process with glucose as the growth substrate. More recently, a strain was created with the capability of producing MA directly from a lignin depolymerization stream (Almqvist et al. 2021). Cytochrome P450 and ferredoxin reductase were introduced into a *P. putida* KT2440-derived strain, which had *catB* and *catC* already deleted, for guaiacol conversion to catechol. A guaiacol-rich stream, which was obtained by chemical depolymerization of an industrial lignin (Indulin AT) followed by extraction and purification, was used as the feedstock for MA production. In shake-flask experiments using media supplemented with glucose to support growth, 14 g/L MA was produced.

PDC is a potential precursor of polyesters. A strain of *Novosphingobium aromaticivorans* DSM12444 was genetically engineered for PDC production (Perez et al. 2019). In the engineered strain, the genes responsible for the conversion of PDC to OMA and the diversion of carbon in 3MGA from PDC synthesis were deleted (see the pathway that is illustrated in Fig. 5.11). In batch experiments using defined media, the engineered strain converted several lignin degradation products obtained by chemical depolymerization of poplar lignin, which included some non-natural compounds, to PDC with yields from 22 to 100%.

Vanillin. Vanillin is an intermediate of the degradation pathway of G-lignin and is a high-value chemical that has many food and polymer applications (Fache et al. 2015). The majority of commercial vanillin product is currently produced in chemical processes. Attempts to develop the technology for the production of vanillin from a natural source have been focused upon because "natural" vanillin has a much higher selling price than synthetic vanillin. Several microbial species are capable of producing vanillin from ferulic acid, which can be derived from G-lignin (Converti et al. 2010). A commercially feasible fermentation process for vanillin production from ferulic acid is difficult to achieve because of two main reasons: 1. Lignin-degrading organisms that produce vanillin normally continue the degradation process and convert vanillin to vanillic acid; and 2. Both ferulic acid and vanillin are toxic. Many attempts have been made to overcome these problems. A strain derived from *Escherichia coli* B was engineered by insertion of the genes encoding feruloyl-CoA synthetase (*fcs*) (conversion of ferulic acid to feruloyl Co-A) and enoyl-CoA hydratase/aldolase (*ech*) (conversion of feruloyl Co-A to vanillin) from *Pseudomonas fluorescens* BF13 into its chromosome. In shake-flask experiments using an agarose gel matrix entrapping ferulic acid to control its release, the engineered *E. coli* strain was able to produce vanillin at 4.3 g/L (Luziatelli et al. 2019). *Streptomyces* sp. strain V-1 was used for vanillin production from ferulic acid in shake-flask experiments where an adsorbent was added at 8% (w/v) to remove some of the produced vanillin and thus reducing its toxicity. In these experiments, 45 g/L ferulic acid could be added continually and 19.2 g/L vanillin was obtained in 55 h (Hua et al. 2007). When the *vdh* gene encoding vanillin dehydrogenase in *Amycolatopsis* sp. ATCC 39116, which was selected for its high tolerance of vanillin, was deleted, the conversion of vanillin to vanillic acid was reduced by 90%. Upon further improvement was made via constitutive expression of *fcs* and *ech*, 19.3 g/L vanillin with a molar yield of 94.9% was obtained. In a fed-batch mode, 22.3 g/L vanillin was obtained but at a lower yield (Fleige et al. 2016). Although the final concentrations of vanillin were relatively low, they were actually above the minimum concentration that allows effective crystallization for recovery, which was reported to be only 10 g/L (Converti et al. 2010).

Microbial lipids. Several *Rhodococcus* strains, for example, *R. opacus* and *R. jostii* can utilize lignin-derived compounds to produce lipids (He et al. 2017), which in turn can be used as a feedstock for the production of biodiesel and in other applications. The liquid stream from the two-stage NaOH and NaOH/H_2O_2 pretreatment of corn stover, which contained extracted lignin, was used as the substrate for lipid production by *R. opacus* PD 630, *R. opacus* DSM 1069, and *R. jostii* DSM 44719^{T}. The highest yield was obtained with *R. opacus* PD 630, which converted 6.2% of organic content to produce lipid at 1.3 g/L and 42.1% of the cell dry mass after 48 h (Le et al. 2017). When laccase was added to the detoxified liquid stream obtained from the two-stage 1% H_2SO_4/1% NaOH pretreatment of corn stover, 1.83 g/L lipid was produced by *R. opacus* PD 630 in a fed-batch fermentation (Liu et al. 2018).

Polyhydroxyalkanoates (PHAs). PHAs are a class of biodegradable and biocompatible polyesters that can be produced by several microbial species (Kumar et al. 2019). In an investigation using *P. putida* KT2440 for PHA production in a defined medium containing p-coumarate, ferulate, and benzoate, it was found that PHA production is favored by high C/N ratio (60) and highly aerobic conditions (oxygen transfer rate > 20 mmol/L–h) (Ramirez-Morales et al. 2021). PHA production by *P. putida* KT2440 using a lignin-containing stream obtained from a sequential dilute H_2SO_4 pretreatment/cellulase and xylanase hydrolysis/NaOH treatment of corn stover and supplemented with laccase, which also contained low levels of glucose, was investigated. The highest yield of PHA at 1.5 g/L was obtained with a lignin stream containing 6 g/L glucose (Liu et al. 2019). In the follow-up study employing a similar procedure, the lignin stream was analyzed before and after the fermentation. The results indicated that *p*-coumaric acid was the major lignin component that was utilized (Arreola-Vargas et al. 2021). In addition to *P. putida*, other microorganisms are capable of producing PHA from lignin-derived compounds. *Cupriavidus necator* DSM 545 was investigated for polyhydroxybutyrate (PHB) production using a lignin-rich stream obtained from the NaOH pretreatment of corn stover. In a fed-batch fermentation performed in a 1.7-L fermentor, 4.5 g/L was obtained (Li 2019).

5.3 Future Research and Potential Commercial Implementation

The process for the production of high-value products from lignin must be designed to be a component that can be integrated into the biorefinery. The majority of the research efforts in the area of lignin bioconversion, however, usually focused on lignin only and did not examine how this component could be integrated with cellulose and hemicellulose bioprocessing. These efforts also had their scopes usually narrowed to the individual steps, i.e., depolymerization, funneling, and product formation. More often than not, the investigations on monomer funneling and product formation used standard compounds that are considered as representatives of the three types of lignin as starting materials and rarely extended the findings to the real feedstocks, i.e., the fractions obtained by lignin depolymerization. In a few cases, the entire process, which included biomass pretreatment, enzymatic hydrolysis for production of fermentable sugars, and lignin recovery, depolymerization, and bioconversion, was investigated (Liu et al. 2019, 2021; Arreola-Vargas et al. 2021). To successfully develop a commercially feasible technology for lignin bioconversion that can be integrated with other components of the biorefinery, much more research is needed in the areas discussed below.

5.3.1 *Feedstock Development*

Feedstock development and characterization probably is the most important area for future research, not only because it is the starting point but also because the characteristics of the lignin feedstock will strongly influence the decisions on the target products and the processes for their production. For example, since softwood has very high contents of G-lignin, which can be depolymerized and funneled to catechol, muconic acid will be a good candidate to consider in the development of lignin-based value-added products for this type of feedstock. In a biorefinery, the characteristics of the lignin streams generated from the biomass feedstock are linked very closely to the processes employed for the production of fermentable sugars and their subsequent fermentation. In Renmatix's Plantrose® process, sub- and super-critical water were used to extract C5 and C6 sugars from biomass. A highly pure lignin product, which was sulfur-free and contained <1% sugars and <0.1% ash, was extracted from the solid residue obtained after the sugar extraction of a hard-wood mixture (Capanema and Balakshin 2015). The lignin feedstocks obtained in other biomass processing schemes may not have that high level of purity. The lignin obtained after enzymatic hydrolysis and subsequent fermentation of the pretreated biomass may contain sugars (from the carbohydrate fractions) and proteins (from enzymes and cell mass). Even the lignin obtained prior to enzymatic hydrolysis, for example, in an H_2SO_4-catalyzed organosolv extraction process, may contain carbo-hydrate impurities (Huijgen et al. 2014). These impurities, however, may actually be helpful in the downstream. Sugars can serve as readily fermentable substrates, and proteins can supply nitrogen for the growth of the microorganisms used for conversion of the lignin-derived compounds to the target products. The potential negative effects of sugar and protein impurities on the enzymatic depolymerization of the lignin feedstock, however, need to be investigated. The structural and chemical modifications of the lignin feedstocks that occur in the extraction process should also be considered. When Scots pine (*Pinus sylvestris*) and silver/white birch (*Betula pendula/pubescens*) wood chips were subjected to various alkaline pretreatments, it was found that increasing the alkaline charge substantially increased the average molecular weights of the solubilized lignin (Lehto et al. 2015). The effects of molecular weight on the depolymerization of lignin and subsequent funneling of the monomers need studies. The intensity of process parameters used in biomass pretreatment also has significant effects on the characteristics of the extracted lignin. Low pH in acid pretreatment and excessive base charge in alkaline pretreatment caused lignin condensation, which may negatively affect depolymerization and funneling. The chemical structures of the extracted lignin also are significantly changed in both acid and alkaline pretreatments (Narron et al. 2016). In addition to purity, the lignin products obtained in the aforementioned Plantrose® process were thoroughly analyzed by various techniques (Capanema and Balakshin 2015, 2016). The reported characteristics of the lignin products can serve as a good template for the characterization of potential lignin feedstocks in future research.

5.3.2 Production of Lignin-Depolymerizing Enzymes

Several lignin-depolymerizing enzymes are commercially available. The purchasing costs, however, may be economically prohibitive. This has been one of the key issues encountered by the cellulosic ethanol industry. The current commercial cellulosic ethanol plants have been designed to include a unit dedicated to on-site enzyme production using some of the sugars obtained by hydrolysis of the biomass feedstock. This option should be considered for lignin bioconversion as a component of the biorefinery. Some recent studies on the production of lignin-depolymerizing enzymes have been reviewed (Iram et al. 2021). Except for one study, all of the reported investigations used expensive nitrogen sources such as peptone and yeast extract. A less expensive nitrogen source such as urea, ammonium sulfate, corn steep liquor, or hydrolysates obtained by acid or thermal hydrolysis of the spent microbial biomass in the biorefinery should be examined. A suitable recovery and purification process for the enzymes produced should also be investigated. It may not be necessary to purify these enzymes to very high purity. A partial purification actually may allow some of the residual nutrients to be used in the subsequent fermentation steps, i.e., funneling and conversion.

5.3.3 Lignin Depolymerization Kinetics

As discussed previously, there are several lignin-depolymerizing enzymes. The suitability of each class of enzymes for depolymerization of specific types of lignin should be investigated. The effects of pH, temperature, salt concentration, and sugar and protein impurities on enzyme activity and stability should be studied. The potential feedback inhibition by the resultant oligomers and monomers on these enzymes should also be investigated. If feedback inhibition is significant, a simultaneous depolymerization and funneling/conversion process probably is preferred over a process that employs two separate steps. This situation is similar to the choice between the simultaneous saccharification and fermentation (SSF) process and the separate hydrolysis and fermentation (SHF) process in cellulosic ethanol production.

5.3.4 Fermentation Development

The development of fermentation processes for the production of lignin-derived products should be performed using the real substrates instead of the lignin model compounds. Model compounds when used as substrates in a fermentation process almost always yield better results compared to the real substrates, especially at the laboratory scale. The potential toxicity toward the selected microorganisms of the compounds formed during the depolymerization and funneling stages, although may be at low concentrations, should be examined.

5.3.5 Process Integration

Integration of lignin bioconversion with the other processes in the biorefinery is also an important area that deserves more research. For example, research should be performed to determine if and how much of the sugar streams obtained from cellulose and hemicellulose hydrolysis should be used to support the growth of the microorganisms used for the production of lignin depolymerizing enzymes and the production of the target value-added products. The case of softwood utilization as a biomass feedstock is another example of process integration. After G-lignin is used to generate catechol, which can be used for muconic acid production, research should be performed to determine whether the remaining lignin fractions should be used for the production of a second value-added product or sent to the anaerobic digestion unit for the production of methane to be used for internal energy requirement.

5.4 Conclusion

Lignin valorization by biochemical processes has attracted significant attention recently. These efforts have generated many important and useful discoveries on the bioconversion of lignin to various value-added products. To successfully bring these achievements to implementation in a commercial biorefinery, much more research is needed, both in basic science and process engineering technology.

References

Abdel-Hamid et al (2013) Insights into lignin degradation and its potential industrial applications. Adv Appl Microbiol 82:1–27

Almqvist H et al (2021) Muconic acid production using engineered *Pseudomonas putida* KT2440 and a guaiacol-rich fraction derived from Kraft lignin. ACS Sustain Chem Eng 9:8097–8106

Arregui L et al (2019) Laccases: structure, function, and potential application in water bioremediation. Microb Cell Fact 18:200

Arreola-Vargas J et al (2021) Enhanced medium chain length polyhydroxyalkanoate production by co-fermentation of lignin and holocellulose hydrolysates. Green Chem 23:8226–8237

Atiwesh G et al (2021) Lignin degradation by microorganisms: a review. Biotechnol Prog 1:e3226. https://doi.org/10.1002/btpr.3226

Becker J et al (2018) Metabolic engineering of *Corynebacterium glutamicum* for the production of cis, cis-muconic acid from lignin. Microb Cell Fact 17:115

Bourbonnais R, Paice MG, Freiermuth B, Bodie E, Borneman S (1997) Reactivities of various mediators and laccases with kraft pulp and lignin model compounds. Appl Environ Microbiol 63:4627–4632

Bugg TDH, Ahmad M, Hardiman EM, Rahman Rahmanpour R (2011) Pathways for degradation of lignin in bacteria and fungi. Nat Prod Rep 28:1883–1896

Capanema EA, Balakshin M (2015) Plantrose® lignins: a new type of technical lignins. Proceedings of the 18th international symposium on wood, fiber and pulping chemistry, Vienna, Austria, September 09–11th, 2015. Available on-line. https://www.researchgate.net/publication/305222916_Plantrose_lignins_a_new_type_of_technical_lignins. Accessed 16 January 2022

Capanema EA, Balakshin M (2016) High purity lignin, lignin compositions, and higher structured lignin. European Patent EP2970595A1; published 01-20-2016

Chauhan PS (2020) Role of various bacterial enzymes in complete depolymerization of lignin: a review. Biocatal Agric Biotechnol 23:101498

Christopher LP, Yao B, Ji Y (2014) Lignin biodegradation with laccase-mediator systems. Front Energy Res 2, Article 12. https://doi.org/10.3389/fenrg.2014.00012

Converti A et al (2010) Microbial production of biovanillin. Braz J Microbiol 41:519–530

De Gonzalo G et al (2016) Bacterial enzymes involved in lignin degradation. J Biotechnol 236:110–119

Fache M, Boutevin B, Caillon S (2015) Vanillin, a key-intermediate of biobased polymers. Eur Polymer J 68:488–502

Falade AO et al (2017) Lignin peroxidase functionalities and prospective applications. MicrobiologyOpen. 6:e00394

Fleige C, Meyer F, Steinbüchel A (2016) Metabolic engineering of the actinomycete *Amycolatopsis* sp. strain ATCC 39116 towards enhanced production of natural vanillin. Appl Environ Microbiol 82:3410–3419

Guan ZB, Liao Q, Wang HR, Chen Y, Liao XR (2018) Bacterial laccases: promising biological green tools for industrial applications. Cell Mol Life Sci 75(3569–3592):106

Hardwood CS, Parales RE (1996) The beta-ketoadipate pathway and the biology of self-identity. Annu Rev Microbiol 50:553–590

He Y et al (2017) Lipid production from dilute alkali corn stover lignin by *Rhodococcus* strains. ACS Sustain Chem Eng 5:2302–2311

Hofrichter M (2002) Review: lignin conversion by manganese peroxidase (MnP). Enzyme Microb Technol 30:454–466

Hua D et al (2007) Enhanced vanillin production from ferulic acid using adsorbent resin. Appl Microbiol Biotechnol 74:783–790

Huijgen WJJ et al (2014) Characteristics of wheat straw lignins from ethanol-based organosolv treatment. Ind Crops Prod 59:85–95

Iram A, Berenjian A, Demirci A (2021) A review on the utilization of lignin as a fermentation substrate to produce lignin-modifying enzymes and other value-added products. Molecules 26:2960

Johnson CW, Beckham GT (2015) Aromatic catabolic pathway selection for optimal production of pyruvate and lactate from lignin. Metab Eng 28:240–247

Kohlstedt M et al (2018) From lignin to nylon: Cascaded chemical and biochemical conversion using metabolically engineered *Pseudomonas putida*. Metab Eng 47:279–293

Kumar P et al (2019) Bioconversion of lignin and its derivatives into polyhydroxyalkanoates: Challenges and opportunities. Biotechnol Appl Biochem 66:153–162

Kamimura N et al (2019) Advances in microbial lignin degradation and its applications. Curr Opin Biotechnol 56:179–186

Le R et al (2017) Conversion of corn stover alkaline pre-treatment waste streams into biodiesel via *Rhodococci*. RSC Adv 7:4108

Lee S et al (2019) Bacterial valorization of lignin: strains, enzymes, conversion pathways, biosensors, and perspectives. Front Bioeng Biotechnol 7: Article 209. https://doi.org/10.3389/fbioe.2019.00209

Lehto J, Pakkanen H, Alén R (2015) Characterization of lignin dissolved during alkaline pretreatment of softwood and hardwood. J Wood Chem Technol 35:337–347

Li M, Pu Y, Ragaukas AJ (2016) Current understanding of the correlation of lignin structure with biomass recalcitrance. Front Chem 4:45

Li M (2019) Adding value to lignocellulosic biorefinery: Efficient process development of lignocellulosic biomass conversion into polyhydroxybutyrate. PhD dissertation, University of Nebraska-Lincoln

Li X, Zheng Y (2020) Biotransformation of lignin: mechanisms, applications and future work. Biotechnol Prog 36:e2922

Liu Z-H et al (2018) Combinatorial pretreatment and fermentation optimization enabled a record yield on lignin bioconversion. Biotechnol Biofuels 11:21

Liu Z-H et al (2019) Cooperative valorization of lignin and residual sugar to polyhydroxyalkanoate (PHA) for enhanced yield and carbon utilization in biorefineries. Sustain Energy Fuels 3:2024

Liu Z-H et al (2021) Transforming biorefinery designs with 'Plug-In Processes of Lignin' to enable economic waste valorization. Nat Commun 12:3912

Lu F, Ralph J (1997) The DFRC method for lignin analysis. Part 1. A new method for β-aryl ether cleavage: lignin model studies. J Agr Food Chem 45:4655–4660

Luziatelli F et al (2019) Maximizing the efficiency of vanillin production by biocatalyst enhancement and process optimization. Front Bioeng Biotechnol 7: Article 279

Mansfield SD et al (2012) Whole plant cell wall characterization using soluble-state 2D NMR. Nat Protoc 7(9):1579–1589

Martinez AT, Speranza M, Ruiz-Dueñas FJ, Ferreira P, Camarero S, Guillen F, Martinez MJ, Gutierrez A et al (2005) Biodegradation of lignocellulosics: microbial, chemical and enzymatic aspects of the fungal attack of lignin. Int Microbiology 8:195–204

Narron RH et al (2016) Biomass pretreatments capable of enabling lignin valorization in a biorefinery process. Curr Opin Biotechnol 38:39–46

Perez JM et al (2019) Funneling aromatic products of chemically depolymerized lignin into 2-pyrone-4-6- dicarboxylic acid with *Novosphingobium aromaticivorans*. Green Chem 21:1340
Ramirez-Morales JE et al (2021) Lignin romatics to PHA polymers: nitrogen and oxygen are the key factors for *Pseudomonas*. ACS Sustain Chem Eng 9:10579–10590
Ravichandran A, Sridhar M (2016) Versatile peroxidases: super peroxidases with potential biotechnological applications-a mini review. J Dairy Vet Anim Res 4(2):00116
Rinaldi R et al (2016) Paving the Way for Lignin Valorisation: recent advances in bioengineering, Biorefining and catalysis. Angew Chem Int Ed 55:8164–8215
Rolando C, Monties B, Lapierre C (1992) Thioacidolysis in methods in lignin chemistry (eds. Dence CW, Lin SY). Springer, pp 334–349
Sugano Y, Yoshida T (2021) DyP-type peroxidases: recent advances and perspectives. Int J Mol Sci 22:5556
Xu Z, Lei P, Zhai R, Wen Z, Jin M (2019) Recent advances in lignin valorization with bacterial cultures: microorganisms, metabolic pathways, and bio-products. Biotechnol Biofuels 12:32
Yamamura M, Hattori T, Suzuki S, Shibata D, Umezawa T (2010) Microscale alkaline nitrobenzene oxidation method for high-throughput determination of lignin aromatic components. Plant Biotechnol 27:305–310

Chapter 6
Thermochemical Conversion of Cellulose and Hemicellulose

Anh Quynh Nguyen and Ly Thi Phi Trinh

Abstract Thermochemical conversion process is an important and potential route to transform biomass feedstocks into powers, fuels, and a variety of chemical platforms. The chapter describes general characteristics of thermochemical processes of biomass including combustion, pyrolysis, liquefaction, and gasification, with the focus on the thermal decomposition of individual biomass components such as cellulose and hemicellulose. Thermochemical processes occur at a wide range of temperatures and pressures with or without catalysts, in which cellulose and hemicellulose undergo serial primary and secondary reactions to form a variety of product types and yields. Primary reactions of cellulose and hemicellulose are associated with the dehydration and depolymerization process to smaller fragments, monosaccharides units, and volatiles which further decompose to low molecular weight compounds at severe conditions of temperature, times, pressures, and catalysts. Thermochemical decomposition of cellulose and hemicellulose typically produces various fuel sources including bio-char, bio-oil, bio-crude, and syngas, along with diverse substances such as anhydrosugars (levoglucosan, mannosan, galactosan), furans (furfural, 5-hydroxymethylfurfural), organic acids (acetic acid, formic acid, levulinic acid), ketones, and aldehydes. Cellulose and hemicellulose are the most abundant constituents in lignocellulosic biomass. Understanding the mechanism of thermochemical conversion of cellulose and hemicellulose leads to the choice of suitable biomass feedstocks and the transformation process for targeted production.

A. Q. Nguyen (✉) · L. T. P. Trinh
Khai Minh Technologies Group – KMTG, Ho Chi Minh City, Vietnam
e-mail: kmtganhnq@gmail.com; quynhanhnqhc@gmail.com

L. T. P. Trinh
e-mail: phily@hcmuaf.edu.vn

L. T. P. Trinh
Research Institute for Biotechnology and Environment, Nong Lam University Ho Chi Minh City, Ho Chi Minh City, Vietnam

N. P. Nghiem et al. (eds.), *Biomass Utilization: Conversion Strategies*,
https://doi.org/10.1007/978-3-031-05835-6_6

6.1 Introduction

Thermochemical conversion processing is the use of heat to decompose and transform biomass feedstocks into power, fuels, and chemical products. Thermochemical processing occurs at high temperatures from several hundred to 1000 °C or even higher. Therefore, thermochemical processes take place in a short time, in seconds or minutes, compared to hours or days for biochemical processing. Thermochemical processes can be classified into combustion, pyrolysis, gasification, and liquefaction. Biomass combustion has long been used to supply heat and power in the industry. Combustion is the complete oxidation of all organic matters in biomass using sufficient oxygen, while gasification of biomass is performed by partial oxidation of solid biomass feedstocks to produce a mixture of gases at high temperatures using a controlled amount of oxygen. Pyrolysis is the thermal decomposition of biomass into bio-char, bio-oil, and syngas in the absence of oxygen. Liquefaction produces liquid fuels and various chemical platforms from biomass that occurs at mild temperatures in the presence of pressurized water or solvent (Buendia-Kandia et al. 2020; Robert and Kaige 2017). Each process has its own characteristics and uses different reaction conditions, such as temperature, heating rate, residence time, pressure, purge gas flow rate, and catalyst. The yield and composition of the product depend not only on the operating parameters but also on the physicochemical properties of biomass feedstock. Lignocellulosic biomass is mainly composed of cellulose, hemicelluloses, and lignin. Cellulose is rigid and dense because of its highly ordered structure. Hemicelluloses are heterogeneous polymers of hexose and pentose sugars and are less stable than cellulose. Lignin is much more difficult to completely decompose as compared to cellulose and hemicellulose due to its complex aromatic structure. These components that vary from one biomass to another are able to make interactions with each other during thermochemical processes, thus resulting in diverse products with different yields and qualities (Patel et al. 2020). For a better understanding of the mechanism of thermochemical biomass conversion, the processing of the individual components such as cellulose and hemicellulose have been widely performed and investigated. Understanding the decomposition mechanism of each biomass component can provide important information for the effective transformation of biomass into target products, such as energy, power, or chemical products. In this chapter, we discuss the characteristics and behaviors of thermochemical conversion processes of cellulose and hemicellulose including combustion, pyrolysis, liquefaction, and gasification, and their main products (Fig. 6.1).

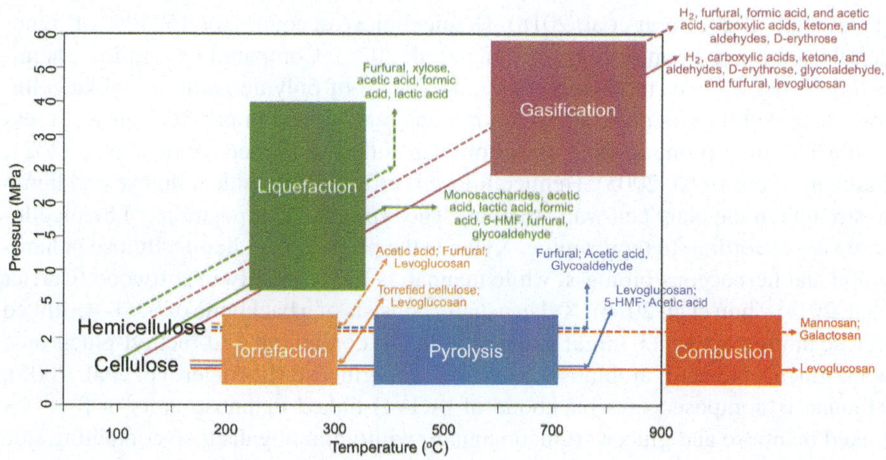

Fig. 6.1 General behaviors of thermochemical conversion technologies and their main products. Dotted line: conversion of hemicellulose; solid line: conversion of cellulose

6.2 Cellulose

Cellulose is the most abundant organic polymer on earth, which is an important structural component of the plant cell wall (Börjesson and Westman 2015). Cellulose is a linear homo-polysaccharide composed of D-glucose monomers linked together by β-1,4 glycosidic bonds. The degree of polymerization of cellulose depends on its source and ranges from several hundred to over ten thousand. Structurally, cellulose is made up of highly ordered crystalline regions and amorphous parts (Amenaghawon et al. 2021; Collard and Blin 2014; Hendriks and Zeeman 2009). These crystalline regions give mechanical stability, hydrophobicity, and chemical recalcitrance to cellulose microfibrils. These microfibrils are arranged and bound to biomass matrices such as hemicellulose and lignin to form bundles or macrofibrils. In addition, multiple hydroxyl groups in the cellulose can form intermolecular hydrogen bonds among different cellulose chains or intramolecular hydrogen bonds within the polymer itself. The high crystallinity and high degree of hydrogen bonds in the cellulose microfibrils give cellulose fibers strength and stiffness (Börjesson and Westman 2015; Pinkert et al. 2009).

6.3 Hemicellulose

Hemicellulose is a heterogeneous and complex polymer of different sub-constituents such as pentoses (xylose and arabinose), hexoses (glucose, mannose, galactose), hexuronic acids (4-O-methyl-D-glucuronic acid, D-glucuronic acid, and D-galacturonic acid), acetyl groups, and small amounts of L-rhamnose and L-fucose

(Patel et al. 2020; Zhou et al. 2018). Hemicellulose accounts for 15–30% of lignocellulosic biomass compositions (Wang et al. 2021). Compared to cellulose, hemicellulose has a lower molecular mass and degree of polymerization. Unlike cellulose, hemicellulose is made up of amorphous regions and branched, making it less stable and more prone to decomposition than cellulose (Amenaghawon et al. 2021; Laureano-Perez et al. 2005). Hemicellulose is cross-linked with cellulose and lignin to strengthen the plant cell wall structure. The types and compositions of hemicellulose vary according to their source. Xylan is the predominant hemicellulose in hardwood and herbaceous biomass, while mannan is found mainly in softwood (Carrier et al. 2011; Zhou et al. 2016). Xylan usually consists of a backbone of β -(1–4)-linked xylose monomers which might attach side chains containing 4-O-methyl-glucuronic acid, glucuronic acid, arabinose, galactose, and glucose (Ebringerovμ et al. 2005). Mannan is composed of a backbone of β-(1–4)-linked mannose units or β-(1–4)-linked mannose and glucose (glucomannan) with/without galactose-containing side chains (Amenaghawon et al. 2021; Zhou et al. 2016) (Table 6.1).

6.4 Combustion

Combustion is the most feasible and conventional way to utilize biomass as a renewable energy source by cleaving the chemical bonds of fuel and generating a series of reactions in the presence of air or oxygen under the heat. To evaluate combustion performance, it is important to recognize the chemical composition and physical characteristics of biomass. Thermogravimetric analysis (TGA) is typically used to investigate thermal decomposition characteristics of solid materials based on weight change at the determined heating rate as a function of time or temperature. The structure and chemical composition of lignocellulosic fibers (cellulose, hemicellulose, and lignin) greatly influence their thermal degradation, and the differential thermal analysis is an effective method to investigate the thermal behavior of the biomass (Chen et al. 2017; Jiang et al. 2017; Sefain et al. 1985).

6.4.1 Combustion of Cellulose

TGA analysis showed that the combustion of pure cellulose showed only one peak, which could be explained by the occurrence of only the combustion of the volatile fraction of cellulose (Boukaous et al. 2018). Combustion kinetics can be gained by using a multi-Gaussian-distributed activation energy model and density-functional theory (Wang et al. 2021; Yu et al. 2021). For combustion of cellulose, reactive force field molecular dynamics (ReaxFF-MD) simulations were applied to study the thermal decomposition of amorphous and partially crystalline cellulose (i.e., microfibrils). By following the complete transformation of cellulose into low molecular weight products, scientists found the decomposition begins with glycosidic bond

Table 6.1 Major mechanisms and recent applications of thermochemical conversion technologies

Technologies	Major mechanisms	Applications	
		Cellulose	Hemicellulose
Combustion	1. Heat (high, > 900 °C) 2. Low pressure 3. With oxygen	– Electrode materials, metal oxides, and ceramics (Kumar 2019) – DOPO–Cinnamoyl cellulose–DCC (Liu et al. 2021) – High-performance lithium-ion battery (Zhu et al. 2016)	– Mannosan and galactosan (Kuo et al. 2008; Li et al. 2019; Ruan et al. 2020; Segato et al. 2021)
Pyrolysis	1. Heat (>300 °C) 2. Pressure (<5 MPa) 3. No oxygen	– Anhydrosugars (levoglucosan etc.) (Junior et al. 2020) – Low molecular weight (LMW) compounds (glycoaldehyde, pyruvaldehyde, hydroxyacetone, and glyceraldehyde) (Patwardhan et al. 2011; Wang et al. 2012, 2020b) – Furans (5-hydroxymethylfurfural, furfural, 2-furan methanol, 3-furan methanol, and 5-methyl furfural) (Patwardhan et al. 2011; Yang et al. 2020b)	– Char and bio-oils (Chen et al. 2019b; Zhou et al. 2016, 2018) – Light oxygenated compounds: glycoaldehyde, acetaldehyde, 1-hydroxy-2-propanone, 4-hydroxy-5,6-di-hydro-(2 H)-pyran-2-one, and 1-hydroxy-2-butanone (Usino et al. 2020; Werner et al. 2014) – Furfurals (Chen et al. 2019b; Wang et al. 2015)
Liquefaction	1. Temperature (280–370 °C) 2. Pressure (10–25 MPa) 3. No oxygen 4. Aqueous liquefaction (using water) 5. Nonaqueous liquefaction (methanol, ethanol, isopropanol, phenol, etc.) 6. Required catalyst (Brønsted or Lewis acids, MnO, CaO, CeZrOx, Raney Ni-NaOH, Na_2CO_3, Fe-zeolite, Na_2CO_3-Fe)	– Monosaccharides and further to acids, aldehydes, ketones (Gagic et al. 2018; Wang et al. 2020a; Jasiukaitytė-Grojzdek et al. 2021; Liu et al. 2020; Sun et al. 2020a) – Lactic and formic acid (Xu and Li 2021) – Photothermal plastic (Sun et al. 2021; Zimmermann et al. 2021) – Biopolyols and polyurethane (Ge et al. 2018; Kosmela et al. 2018)	– Oligosaccharides, monosaccharides, furfural, glyceraldehyde, acetic acid, lactic acid, etc. (Ghimire et al. 2021; Song et al. 2020; Yang et al. 2020a) – Xylose and 4-O-methyl glucuronic (Ghimire et al. 2021)

(continued)

Table 6.1 (continued)

Technologies	Major mechanisms	Applications	
		Cellulose	Hemicellulose
Gasification	1. Temperature (> 700 °C) 2. Pressure (> 25 MPa) 3. Oxygen 4. Steam 5. Required catalyst (MnO, CaO, Raney Ni-NaOH, SiO_2, Fe-zeolite, Na_2CO_3-Fe)	– Carboxylic acids, ketone, and aldehydes (Ong et al. 2019; Soomro et al. 2018; Yang et al. 2020a; Yu et al. 2018) – Syngas (Craven et al. 2020; Zou et al. 2018; Minami et al. 2018)	– Xylose, 4-O-methylglucuronic furfural, formic acid, and acetic acid (Berthet et al. 2016; Hassan et al. 2020) – Hydrogen (García-Jarana et al. 2020; Okolie et al. 2020)

cleavage. Particularly, the crystallinity has no appreciable effects on the mechanism or kinetics of chain scission (which is generated by glycosidic bond cleavage occurs during cellulose decomposition), the evolution of the molecular weight distribution, or the low molecular weight products (Paajanen et al. 2021).

Stochastic reactive molecular dynamics (RMD) simulations were used to identify and analyze the primary thermal decomposition reactions of an isolated cellulose molecule at a range of temperatures from 1400 to 2200 K (1127–1927 °C), and the results suggest that the decomposition occurs primarily through random cleavage of the (1–4)-glycosidic bonds by an activation energy of (171 ± 2) kJ.mol-1(Paajanen and Vaari 2017). When cellulose is combusted at temperatures above 300 °C, levoglucosan (1,6-anhydro-β-D-glucopyranose) is the most abundant monosaccharide anhydride that was released. Levoglucosan is considered as a molecular marker for the combustion of cellulose (Kuo et al. 2008; Li et al. 2019; Ruan et al. 2020; Segato et al. 2021).

Cellulose combustion typically releases carbon monoxide and carbon dioxide. The mechanism of carbon monoxide release in cellulose combustion was investigated by using molecular dynamics (MD) simulations together with ReaxxFF to analyze reactions of cellobiose oxidation at different temperatures. Basically, this approach uses molecular dynamics simulations equipped with a reactive force field to study the formation of cellobiose from the cellulose oxidation process. The production of carbon monoxide is highly dependent on the abundance of formyl and carboxyl groups, which are formed through cellobiose decomposition. Elevated temperatures cause more carbon monoxide to be released. Subsequently, the formed carbon monoxide is oxidized into carbon dioxide, where reaction steps for the formation and decomposition of the carboxyl group are involved. The simulation results help to identify critical reaction steps and lead to the development of a method to reduce the concentration of free radicals, which then allows the formation of carbon monoxide to be reduced (Barzegar et al. 2020; Hao et al. 2020; Luo et al. 2018).

6.4.2 Recent Applications of Cellulose Combustion

A new synthetic method known as combustion synthesis (CS) has emerged and gained considerable research attention due to it being fast and economic and involving simple synthesis steps. Using a thin cellulose paper, the technique called "Cellulose Assisted Combustion Synthesis" (CACS) or "Impregnated Layer Combustion Synthesis" (ILCS) has been applied for the synthesis of electrode materials, metal oxides, ceramics, catalysts, plus other products (Kumar 2019). Cellulose-based materials with their excellent film-forming properties, mechanical properties, and flame-retardancy, have found application in the battery industry. In this process, cellulose was prepared in a membrane, with excellent infiltration of the electrolyte and flame-retardant performance, namely DOPO–Cinnamoyl Cellulose (DCC), then assembled into a lithium-ion battery. The corresponding battery characterization was then tested for its cycle and rate stability. The test results showed excellent characteristics and enhanced battery performance making the cellulose-based composite materials a very promising separator for high power applications, which broadens their uses in the field of energy storage devices (Liu et al. 2021).

Cellulose combustion can also be applied in making Ni–MgO catalysts with cellulose paper being impregnated with $Mg(NO_3)_2$, $Ni(NO_3)_2$, glycine solutions, and their different combinations. It was established that the combustion mechanism changes as a function of the impregnated media composition, and after combusting, the resulting materials had a highly porous, sponge-like microstructure (Danghyan et al. 2020). Using a single-step nitrate-cellulose combustion synthesis, a novel method was proposed to produce MnO/carbon composites, in which the MnO nanoparticles were embedded into a porous carbon matrix, that resulted in a MnO/carbon composite with enhanced cycling performance and capacity retention, as it had potential to be an anode alternative for high-performance lithium-ion battery (Zhu et al. 2016).

To have a deep understanding of the mechanism, scientists developed a novel model for the combustion of reactive solutions impregnated into a cellulose carrier, which were shown to be effective in the synthesis of metallic oxides with a nanoscale microstructure, that made the cellulose-assisted combustion materials suitable for catalyst applications. Basically, the model involved three reactions: (1) combustion of the carrier matrix, (2) an endothermic reaction related to the decomposition or gasification of the synthesis reaction precursors, and (3) the exothermic oxide synthesis reaction. The results indicated that manipulation of the cellulose burning reaction was the most favorable to increase the reaction yield of the composite materials (Lennon et al. 2011).

6.4.3 Combustion of Hemicellulose

Hemicellulose combustion occurs via a two-step process: (1) Reduced degree of polymerization in the first step; and (2) Decomposed into volatiles and char in the second step. The ignition temperature is the temperature at which the combustion reaction begins, while the final temperature indicates the end of the combustion reaction, which is essential to ensure the perfect design of the combustor and avoid unburned solid fuel at the outlet of the reactor. In the first step, hemicellulose requires the lowest ignition temperature and also gains the lowest final temperature. However, in the second step, the ignition and final temperature of hemicellulose are higher than those required for the other components and just only lower than that of lignin. TGA characteristic curves showed the presence of three peaks, two of them were completely overlapped and could be explained by the heterogeneity of hemicellulose, which is mainly constituted of xylose along with small parts of glucuronic acid and other sugars (Boukaous et al. 2018). Through the process, mannosan and galactosan are considered as molecular markers for the combustion of hemicellulose (Kuo et al. 2008; Li et al. 2019; Ruan et al. 2020; Segato et al. 2021).

6.5 Pyrolysis

Pyrolysis is the thermochemical conversion process of biomass feedstock in the absence of oxygen to break down the large complex polymers present in the biomass into smaller fragments and molecules. The pyrolysis process is initiated by the evaporation of water, followed by primary decomposition and secondary reactions. Primary decomposition consists of three pathways including char formation, depolymerization, and fragmentation (Collard and Blin 2014; Lu et al. 2011; Van de Velden et al. 2010; Wooten et al. 2003). The charring process involves the formation of aromatic polycyclic structures of chars in which benzene rings are formed, combined, and rearranged (Cho et al. 2010; McGrath et al. 2003). Depolymerization breaks down the linkages in the biomass polymers, resulting in a decrease in the degree of polymerization of the chains and releasing volatile compounds (Azeez et al. 2011; Madhua et al. 2020). Fragmentation occurs in the linkage of the polymers even within the monomer units, leading to the formation of small condensable organic compounds and incondensable gases (Collard and Blin 2014; Kostetskyy and Broadbelt 2020). Secondary reactions include cracking or recombination (Chen et al. 2019a). Cracking reactions involve the breaking of chemical bonds within the unstable and volatile compounds, which results in the formation of lower molecular weight molecules (Collard and Blin 2014; Neves et al. 2011). In contrast, released volatile compounds can be recombined together to yield higher molecular weight molecules (Orcid et al. 2017; Wei et al. 2006). The products of pyrolysis are divided into three groups including carbon-rich solids (char), liquid products of condensable vapors (tars and oils), and non-condensable species (Abhijeet et al. 2020; Patel et al. 2020; Robert and Kaige 2017). The yield of each fraction depends on the heating rate and residence time. The slow heating rates and long residence times favor the production of solid char, while high heating rates and short residence times facilitate the production of vapor products. It is considered as slow pyrolysis when the heating rate is below 10 °C/s, fast pyrolysis when it is higher than 100 °C/s, and flash pyrolysis when the process is performed at higher than 500 °C/s. Biomass pyrolysis has been used for centuries and continually improved since it provides many benefits and serves as a sustainable means of producing biofuels, biochemicals, and other commodities (Amenaghawon et al. 2021; Robert and Kaige 2017).

6.5.1 *Slow Pyrolysis*

Slow pyrolysis is the oldest form of biomass pyrolysis that usually occurs over a long period of time using slow heating rates to produce charcoal or bio-char as the main product. Lower heating rates and long residence times facilitate the re-polymerization reactions of the biomass constituents, which may generate a polycyclic carbon structure and maximize the yield of solid bio-char. Most of the slow pyrolysis literature focused primarily on the production of carbon-rich solids, thus the process is called carbonization (Gabhane et al. 2020). Carbonization is carried out over a wide temperature range from 300 to 900 °C.

Temperature plays the most important role in the production of bio-char by slow pyrolysis. The processing temperature determines the structural and physicochemical properties of bio-char such as surface area, pore structure, surface functional groups and elemental composition (Tag et al. 2016). It was found that high pyrolysis temperature resulted in increased specific surface area, pore-volume, and carbon content of bio-char but reduced bio-char yield, nitrogen, oxygen, and hydrogen content (Chatterjee et al. 2020; Dhar et al. 2020; Ferraro et al. 2021; Moradi-Choghamarani et al. 2019; Salam et al. 2020). Bio-char with a high specific surface area, large pore size, and elevated carbon content have potential applications in pollution remediation, soil fertility improvement, and carbon sequestration (Agnieszka et al. 2020; Tag et al. 2016).

Torrefaction is another type of slow pyrolysis which is performed at mild temperatures between 200 and 300 °C in the absence of oxygen to produce torrefied biomass (Boateng, 2020). During the torrefaction process, water and volatile compounds in the biomass feedstock are released resulting in the formation of a dark solid material (torrefied biomass) that exhibits higher energy density and greater homogeneity compared to the original feedstock (Amenaghawon et al. 2021). Currently, torrefaction has been investigated intensively and is gaining a high capability to be implemented at an industrial production scale.

6.5.1.1 Torrefaction of Cellulose

Previous studies mainly investigated the effects of torrefaction pretreatment on the chemical structure and pyrolysis behaviors of whole biomass, while a few studies focused on the structural transformation of cellulose. The reaction temperature significantly affects the distribution of the products from cellulose pyrolysis. The torrefaction of cellulose slowly occurred at 200 °C to produce solid char as the main product, along with a small amount of bio-oil, but no gaseous products were generated at below 250 °C. As the temperature was increased, the yield of bio-char decreased while bio-oil production was elevated (Zhang et al. 2021). The formation of an intermediate product called "active cellulose" or anhydrocellulose was reported during the processing below 300 °C (Paajanen and Vaari 2017). Active cellulose is obtained from partial depolymerization of the cellulose, whereas anhydrocellulose is formed after the dehydration reactions. For application, cellulose can be directly used for the production of hydrocarbon-rich bio-oil using integrated microwave torrefaction and ZSM-5 catalyst, but the results are not equal in comparison to those from biomass (Bu et al. 2021; Shen and Gu 2009; Zhou et al. 2020). Significant influences of metal salts, such as chlorides ($CaCl_2$, $ZnCl_2$, $MgCl_2$, and Al_2O3), hydroxides ($Ca(OH)_2$ and $Mg(OH)_2$), and acetates ($Ca(CH_3COO)_2$) on the torrefaction of cellulose, indicated that the conversion mechanism of torrefaction process needed to be intently considered before scaling up to the industrial application (Atienza-Martínez et al. 2017; Tancredi et al. 2017; Zhao et al. 2020; Zhou et al. 2020).

6.5.1.2 Torrefaction of Hemicellulose

According to the TGA profile, the thermal decomposition of hemicellulose occurs at a temperature range of 200–350 °C (Collard and Blin 2014; Zhou et al. 2016). The lower thermal stability of hemicellulose compared to cellulose is attributed to its amorphous structure and its lower degree of polymerization. Therefore, hemicellulose requires lower temperature and activation energy for thermal decomposition (Chen et al. 2019b; Negi et al. 2020; Zhou et al. 2016). The structural changes in hemicellulose after torrefaction, even disappeared at higher torrefaction temperatures (250–300 °C), probably is caused by the removal of hydroxyls in hemicellulose resulting in the generation of carboxyl and conjugated ketone, based on two-dimensional perturbation correlation analysis using Fourier transform infrared spectroscopy (Chen et al. 2015; Wang et al. 2016a). After torrefaction, oxygen content decreases significantly, leading to the increase of a high heating value. The results obtained from analyzing two-dimensional perturbation correlation based on diffuse reflectance infrared Fourier transform spectroscopy showed that the dehydration of hydroxyls and the dissociation of branches were the main reactions at low torrefaction temperature, but when the temperature was increased, the depolymerization of hemicellulose and the fragmentation of monosaccharide residues occurred. Via the activation energy model, the results showed that hemicellulose torrefaction process enhanced the activation energy but decreased the yields of torrefied products (Cahyanti et al. 2021; Wang et al. 2016b). The torrefaction process strongly affects the degradation of hemicellulose, that leads to the increase of carbon content, decreased H/C and O/C ratios, increased mass energy density, higher heating value, better grindability, higher hydrophobicity, and resistance to biodegradation (Niu et al. 2019; Zheng et al. 2021).

6.5.2 *Fast Pyrolysis*

In contrast to slow pyrolysis, fast pyrolysis involves the use of a high heating rate, short residence time (<2 s), and rapid vapor cooling to obtain bio-oil, mixed gases, and solid char (Pawar et al. 2020). During the fast pyrolysis process, rapid decomposition of the biomass occurs with the formation of vapors and aerosols, which are condensed after quenching to recover a darkish brown liquid known as bio-oil. Slow pyrolysis gives high solid yields with low liquid yields, while fast pyrolysis gives high liquid yields with low solid yields (Amenaghawon et al. 2021). Fast pyrolysis is the most popular technique for the production of liquid fuels and various commodity chemicals.

Recently, bio-oil produced through the fast pyrolysis process has attracted considerable attention since it provides potential uses as renewable biofuels source, biofuel additives, and as a precursor for the production of specialty chemicals (Patel et al. 2020). The quality and yield of bio-oil produced by fast pyrolysis are affected by not only processing conditions but also the chemical composition of the biomass feedstock. Many studies have shown that the optimum temperatures for obtaining high liquid yields were between 450 and 550 °C, and bio-oil yields varied according to the type of biomass used. It was reported that biomass feedstocks with moisture content less than 10% and particle size of 2–3 mm were the primary requirements for achieving a high heating rate and heat transfer during fast pyrolysis (Pandey et al. 2015).

Moreover, the pyrolysis reactor must be designed and controlled to heat up the biomass rapidly and cool the vapor phase to make it condense to form the bio-oil products in seconds (Robert and Kaige 2017). Several configurations of reactors are used for pyrolysis including bubbling fluidized bed, circulating fluidized bed, conical spouted bed, ablative reactor, rotating cone, vacuum pyrolysis reactor, entrained flow reactor, wire mesh reactor, and auger (twin screw) reactor. The fluidized bed is the most widely used pyrolysis reactor because it is simple to design, construct, and operate. Moreover, it is proven to exhibit a high heat transfer rate, good temperature control, temperature uniformity, and large heat storage capacity (Boateng 2020; Pandey et al. 2015). A circulating fluidized bed reactor is similar to a bubbling fluidized bed except for the velocity of the gas used to fluidize the bed. In the circulating fluidized beds, the gas velocity is set high enough to transport char and heat carrier particles (e.g., sands) to the second combustor. The sand stream is reheated through the char combustion process and then recirculated to the fluidized bed to heat up the biomass feedstock. Circulating fluidized bed reactors have been used for obtaining a high yield of bio-oil from sugarcane bagasse (78%) (Treedet and Suntivarakorn 2018), sawdust, giant Miscanthus, and empty fruit bunch (60%) at pilot scale (Park et al. 2019). Another technology is the rotating cone reactor in which biomass feedstock and hot sand are fed near the bottom of a cone at the same time. The centrifugal forces generated by the rotation of the cone enable the particles to move upwards without the need for large volumes of carrier gas. Hot sand is then recycled back to the rotating cone reactor. The Biomass Technology Group (BTG) has commercialized this reactor that converts biomass feedstock into bio-oil as the main product in just 2 s. It has been reported that a conical spouted bed reactor is suitable for the fast pyrolysis to obtain high bio-oil yields from rice husk (70%) (Alvarez et al. 2014) and eucalyptus waste (75.4%) (Maider Amutio et al. 2015). The outstanding features of the spouted bed reactor include high heating and mass transfer rates, short residence time, and continuous char removal, which accelerate de-volatilization reactions and minimize the cracking of these components (Alvarez et al. 2014; Amutio et al. 2012).

6.5.3 Flash Pyrolysis

Efforts have been carried out to increase the heating rate and minimize the residence time of the vapor phase to improve the production yield of bio-oil from biomass pyrolysis. High bio-oil yields of 75–80% can be achieved by flash pyrolysis occurring at high temperatures (>800 °C) using very high heating rates (>1000 °C/s) and very short residence times (<0.5 s) (Pawar et al. 2020). Feedstock particle size less than 0.2 mm is a primary requirement to achieve such a high heating rate and heat transfer rate (Amenaghawon et al. 2021; Balat et al. 2009; Dada et al. 2021). Rotating cone and conical spouted bed reactors are considered good configurations for the flash pyrolysis process due to their high heat transfer rates and short residence time (Amutio et al. 2012; Papari and Hawboldt 2015).

6.5.4 *Pyrolysis of Cellulose*

TGA is the most common analytical method used to study the mechanism of polymer degradation. TGA of cellulose showed that the cellulose is thermally degraded at a temperature range of 300–400 °C with the highest decomposition rate between 330 and 370 °C (Collard and Blin 2014; Zhou et al. 2016). The reaction temperature significantly affects the distribution of the products from cellulose pyrolysis. Bio-oil yield increased with increasing temperatures and was maximized at 450–500 °C. Further increase in the temperature resulted in a gradual decrease in the yield of tar and bio-oil. In contrast, the conversion of cellulose into gaseous products was dominant at above 600 °C and the yield of syngas including CO, CO_2, CH_4, and H_2 significantly increased with increasing temperatures (Zhang et al. 2021).

Pyrolysis of cellulose produces several products classified into three groups: anhydrosugars, low molecular weight (LMW) compounds, and furans. Anhydrosugars are formed by transglycosylation reactions (Junior et al. 2020). Levoglucosan (LG) is the most abundant anhydrosugar (up to 80% in relative peak area) from cellulose pyrolysis (Collard and Blin 2014; Junior et al. 2020; Yang et al. 2020a). The yield of levoglucosan is affected by processing temperatures, residence times, and the degree of polymerization of the cellulose chains (Wang et al. 2020a; Yang et al. 2020a). The subsequent decomposition of LG produces LMW compounds, such as 1-pentene-3,4-dione, acetaldehyde, 2,3 dihydroxypropanal, and propanedialdehyde from C –O bond breaking reactions (Zhang et al. 2012). The levoglucosan vaporized above 500 °C contributes mainly to gaseous and liquid streams rather than solid char formation (Banyasz et al. 2001; Basu 2013). Other anhydrosugars were found in much smaller proportions than LG, such as levoglucosenone, 1,4:3,6-dianhydro-α-D-glucopyranose, 2,3-anhydro-D-mannosan, and 1,6-anhydro-β-D-glucofuranose. LMW compounds from cellulose pyrolysis include glycoaldehyde, pyruvaldehyde, hydroxyacetone, and glyceraldehyde (Patwardhan et al. 2011; Yang et al. 2020a). Organic acids such as formic acid, acetic acid, and propionic acid also were detected among the LMW pyrolytic products (Patwardhan et al. 2011; Wamg et al. 2012, 2020b). Furans are formed by open-ring reactions and dehydration from the glucopyranose structure (Junior et al. 2020). The yields of furans (5-hydroxymethyl furfural, furfural, 2-furan methanol, 3-furan methanol, and 5-methyl furfural) varied with operating conditions. It is noted that the furan ring is very stable, but its side groups (methyl or oxygenated groups) are less stable and are readily cleaved or rearranged with increasing temperatures. Decomposition of furans led to an increase in the formation of low molecular weight compounds and gaseous species. 5-Hydroxymethyl furfural (5-HMF) and furfural, the attractive platform chemicals, were found to be the most abundant furans (Patwardhan et al. 2011; Yang et al. 2020a). In a previous study, the furans content reached the highest value, about 35% (in relative peak area), and changed little with increasing temperature. It is proven that the thermal degradation of pentose sugars (xylose and arabinose) tends to yield furfural, whereas the processing of hexose sugars (glucose, mannose, and galactose) tends to produce 5-HMF (Zhou et al. 2016). Further decomposition of 5-HMF via fast pyrolysis at 600 °C was performed and resulted in the release of 72.8% furfural. This evidence demonstrated the presence of furfural as a predominant secondary product of cellulose pyrolysis reported in the literature (Collard and Blin 2014; Wang et al. 2012).

6.5.5 Pyrolysis of Hemicellulose

According to the TGA profiles, the entire thermal degradation of hemicellulose could take place in three steps. The first step includes the dehydration and cleavage of the side chains of hemicellulose with slight mass loss. The dissociation of side chains in xylan occurs readily due to its relatively low energy of activation (Dai et al. 2021). Acetic acid was reported as the major product in xylan pyrolysis which was formed by the cleavage of acetyl groups attached to the backbone of xylan (Zhou et al. 2016). The main pyrolysis, which occurs in the second step, is responsible for the most mass loss through a sequence of reactions such as dehydration, decarboxylation, and decarbonylation. Finally, in the third step, further decomposition of hemicellulose occurs to release volatiles from the residue resulting in a slight mass loss (Collard and Blin 2014; Peng and Wu 2010).

Xylan is the most abundant hemicellulose in nature. Xylan is typically used as a model compound for understanding the mechanism of hemicellulose pyrolysis. Pyrolysis of xylan typically yields 20–30% char, 10–20% non-condensable gas species, and 40–60% bio-oil (Zhou et al. 2018). The non-condensable gaseous products include H_2, CO, CO_2, and light hydrocarbons such as CH_4, C_2H_6, and C_3H_8. Bio-oil mainly includes acids (acetic acid and formic acid), furans (furfural), anhydrosugars (anhydroxylose and dianhydroxylose), aldehydes, ketones, and minor aromatic compounds (Zhou et al. 2016, 2018).

Hemicellulose is composed of heterogeneous and diverse monomers. Pyrolysis of hemicellulose generates more complicated and various product distributions compared to cellulose pyrolysis. Hemicellulose pyrolytic products are mainly categorized into two groups including light oxygenated compounds and furans. Light oxygenated compounds mainly include glycoaldehyde, acetaldehyde, 1-hydroxy-2-propanone, 4-hydroxy-5,6-di-hydro-(2 H)-pyran-2-one, and 1-hydroxy-2-butanone, among which glycoaldehyde is detected as the most abundant compound (Werner et al. 2014). Furfural is widely reported as a major furan in hemicellulose pyrolysis (Zhou et al. 2018). Furfural is derived from thermal degradation of pentose units or by secondary decomposition of 5-hydroxymethyl furfural (5-HMF) (Usino et al. 2020). Hemicellulose rich in acetyl groups attached to its backbone yields more acetic acid as a major product (Wang et al. 2015). Furans and acetic acid contents decrease with increasing temperatures (Wang et al. 2015).

6.6 Liquefaction

Liquefaction (hydrothermal liquefaction) is a process responsible for the conversion of lignocellulosic biomass into bio-liquid and/or crude oil-like products at temperatures of 280–370 °C and pressures of 10–25 MPa in the absence of oxygen (Khatiwada et al. 2021; Yang et al. 2020a). This technology has much attraction to produce renewable fuels due to its low operating temperatures and fast reaction rates, and the use of wet feedstocks without the need for an energy-intensive drying process (Amarasekara and Reyes 2020; Khatiwada et al. 2021; Song et al. 2020). There are two types of liquefaction, which include aqueous liquefaction (using water) and non-aqueous liquefaction (using organic solvents such as methanol, ethanol, isopropanol, phenol, and others). The aqueous liquefaction is typically carried out in subcritical water and requires catalysts such as Brønsted or Lewis acids and metal oxides such as MnO, CaO, CeZrOx, Raney Ni-NaOH, Na_2CO_3, Fe-zeolite, Na_2CO_3-Fe. Aqueous liquefaction generally gives higher quality and quantity of bio-oil compared to non-aqueous liquefaction (Amarasekara and Reyes 2020; Feng et al. 2018; Song et al. 2020; Yang et al. 2020a).

6.6.1 *Liquefaction of Cellulose*

Under thermochemical liquefaction, cellulose is mainly hydrolyzed into monosaccharides and further to acids, aldehydes, ketones, and other products. Cellulose is decomposed into soluble sugars (cellobiose, lactose, and glucose) as primary products in subcritical water at 200–300 °C. Soluble sugars subsequently undergo secondary reactions of isomerization, dehydration, and retro-aldol condensation to form D-erythrose, glycolaldehyde, and furfural, which are further degraded to smaller species. A significant decrease in soluble products was observed with increasing reaction times because the sugars were readily degraded to carboxylic acids, ketone, and aldehydes (Gagic et al. 2018; Wang et al. 2020a). These products then can be upgraded to liquid fuels, platform biochemicals, and commodity chemicals such as ethanol, liquid alkanes, 5-hydroxymethylfurfural (5-HMF), furfural, and acetic acid. The results obtained in several studies showed that cellulose could be liquefied with high efficiency (Li et al. 2020; Peng et al. 2018; Wang et al. 2020b; Xu et al. 2019).

Catalyst is one of the key factors that affect the quantity and quality of liquefaction products. In addition to acidic, alkali, and metal catalysts, glycol organosolv and supercritical ethanol (with 2,2,6,6-Tetramethylpiperidinooxy-TEMPO) processes have recently received much attention, due to their environmental benefits, low viscosity, and high solubility (Jasiukaitytė-Grojzdek et al. 2021; Liu et al. 2020; Sun et al. 2020a). Hydrothermal liquefaction with metallic-Fe catalyst is still the most effective process for producing water-soluble fraction, thereby enhancing the hydrocarbon yield (Hirano et al. 2020).

Cellulose liquefaction shows great promise for using cellulose as a supporting substrate or template material in photothermal plastic due to its renewability, degradability, abundant availability, and low cost. In fact, cellulose composite materials have been successfully developed and applied in various fields such as antibacterial compounds, UV shielding, catalysis, flame retardant, fluorescence, metal ion adsorption, and supercapacitor (Sun et al. 2021b; Zimmermann et al. 2021). In liquefaction technology, solvent is one of the key factors. Water, low-carbon alcohols, supercritical fluids, and hydrocarbon-based solvents such as n-alkanes are often employed in cellulose liquefaction (Li et al. 2021).

The increasing trends of using biomass as an alternative to fossil fuels lead to the formation of the biorefinery concept, which now utilizes a range of biomass and known conversion technologies to produce green chemicals and polymers without impacting food and feed security. Among them, biopolyols and polyurethane obtained by direct conversion of separated cellulose or lignocellulosic biomass through liquefaction technology were listed among the "Top 10" platform chemicals in a report prepared by the Pacific Northwest National Laboratory (PNNL) and the National Renewable Energy Laboratory (NREL) (Ge et al. 2018). Interestingly, polyurethane foams were produced via cellulose liquefaction in the presence of crude glycerol, which indicated that cellulose could be a good alternative material to produce polyurethane (Ge et al. 2018; Kosmela et al. 2018).

6.6.2 Liquefaction of Hemicellulose

Due to its heterogeneous and complex structure, the product profile of hydrothermal liquefaction of hemicellulose is complicated and includes a variety of chemicals such as oligosaccharides, monosaccharides, furfural, glyceraldehyde, acetic acid, lactic acid, plus others (Ghimire et al. 2021; Song et al. 2020; Yang et al. 2020a). Hemicellulose decomposition in subcritical water at below 220 °C yielded monosaccharide sugars, mainly xylose and 4-O-methylglucuronic acid, which were further degraded into small molecule products such as furfural, formic acid, and acetic acid at temperatures higher than 220 °C. Xylan usually is used as a model compound in many studies. Xylan liquefaction resulted in different product distributions through hydrolysis and oxidation reactions depending on operating conditions. Hydrolysis occurs via the depolymerization of xylan to soluble sugars, while oxidation promotes the formation of organic acids. The released organic acids subsequently catalyze the degradation of the hydrolyzed products into small molecules. Catalysts are typically used in liquefaction. The presence of a high concentration of oxidizing agent (H_2O_2) and/or pressurization facilitates strong oxidative reaction and/or acidic hydrolysis, yielding xylo-oligosaccharides, xylose, arabinose, glucose, acetic acid, and their decomposition compounds, which include furans and organic acids (Phaiboonsilpa et al. 2020; Zhou et al. 2016). In the presence of ethylene glycol, xylan was decomposed and transferred to the liquid phase, with the average molecular weight of xylan significantly decreasing after liquefaction. The products from the liquefaction of xylan in ethylene glycol are ethylene glycol derivatives, alcohols, aldehydes, ketones, some acids, and their esters (Wang et al. 2016a).

6.7 Gasification

Gasification is a thermochemical conversion process that involves complex reactions, pressure changes, and heat and mass transfer. In general, gasification is used to convert solid fuels (coal or biomass) into value-added products or to release heat for heating and power generation at high temperatures without combustion. Gasification requires oxygen, air, and steam to convert carbonaceous materials into gaseous fuels. Fixed bed and fluidized bed gasifiers are common technologies. Basically, gasification consists of four steps: (1) Drying or dehydration, in which evaporation of moisture occurs under 150 °C; (2) Pyrolysis, in which devolatilization occurs at a temperature range of 150–700 °C; (3) Combustion, in which fuel constituents oxidize and exothermic reactions occur in the temperature range of 700–1500 °C; and (4) Reduction, in which endothermic reactions occur in the temperature range of 800–1100 °C (Ong et al. 2019; Soomro et al. 2018; Yang et al. 2020a; Yu et al. 2018).

6.7.1 Gasification of Cellulose

Cellulose consists of several hundred to many thousands of β-glycosidic bonds linking D-glucose units, which are not stable and tend to cleave at high temperatures. It was observed that cellulose began to decompose at 180 °C and the reaction rate was much faster with increasing temperature and residence time, leading to sugar degradation to carboxylic acids, ketone, and aldehydes. Glucose was completely decomposed in 60 s at 300 °C, but only in 0.5 s at 460 °C (Yang et al. 2020a). Degradation of levoglucosan, the major volatile intermediate being produced from cellulose gasification, was completed at a temperature range of 550–700 °C within only 0.11–0.45 s (Fukutome et al. 2017).

Normally, gasification requires a catalyst to increase efficiency while reducing temperature and time. Among these, alkali and alkaline earth metals and Ni-based catalysts are commonly used in the gasification process (Ong et al. 2019). Cellulose gasification produces the largest amount of H_2 in the absence of steam and catalyst, but adding the Ni-based catalyst significantly increases the gas yield, particularly for H_2 production (Hassan et al. 2020). Moreover, nickel-based catalysts were used and gained the highest H_2 yield from gasification, especially in the presence of SiO_2 (Sun et al. 2020b; Taylor et al. 2020). To enhance H_2 production, scientists have proposed using pure cellulose instead of lignocellulosic biomass and also suggested that a combination of gasification of cellulose and dark fermentation may give a higher hydrogen yield (Hassan et al. 2020). Ca-Fe oxygen carrier is a potential material for efficient lignocellulose conversion and hydrogen-enriched syngas production, acting as catalysts to promote cellulose decomposition (Tang et al. 2021). CaO sorption enhanced the gasification efficiency of biomass for hydrogen-rich syngas production, demonstrating its potential as an inexpensive catalyst for the gasification process (Mbeugang et al. 2021). In addition, the highest carbon monoxide (CO) concentration was found in cellulose gasification compared to those originated from hemicellulose and lignin, which may be due to the abundance of C–O compounds in cellulose (Hassan et al. 2020).

Reliance on catalysts is one of the bottlenecks of gasification, so scientists have proposed other options, in which physical catalysts (Ce/Fe, Ni, etc.) were not used but replaced by a plasma-catalyst system to produce a cleaner syngas. The results provided an alternative and cleaner way for gasification, although it still needs to be evaluated further (Craven et al. 2020; Zou et al. 2018). Recently, scientists proposed a very low temperature gasification system for cellulose (around 50 °C) by using glow-discharge plasma. Due to the low temperature, pyrolysis did not occur, but a very long retention time (90 h) was required. Interestingly, no tar formation was observed, indicating that all of the cellulose was decomposed into gaseous products. This achievement demonstrated that clean and complete gasification of cellulose and/or hemicellulose could be achieved with plasma technology (Minami et al. 2018).

6.7.2 *Gasification of Hemicellulose*

Xylose and 4-O-methylglucuronic acid were mainly detected in the degradation of hemicellulose at a temperature below 220 °C, but low molecule weight products such as furfural, formic acid, and acetic acid were observed at higher temperatures (Berthet et al. 2016; Hassan et al. 2020). Xylose was completely decomposed in less than 5 s at a temperature of 300–450 °C and 25 MPa. The decomposition behavior was similar to that of glucose as they have a homoeologous molecule structure, except that xylose has one less CH-OH group than glucose. The formation of furfural via dehydration of xylose was considered to be analogous to that of dehydration of glucose (Fukutome et al. 2015; Hassan et al. 2020; Tang et al. 2021).

Hydrothermal gasification in either subcritical or supercritical water is an attractive approach to produce hydrogen from cellulose and hemicellulose, in which hydrogen yields from hemicellulose normally were higher than those from cellulose (García-Jarana et al. 2020; Okolie et al. 2020). Moreover, the use of hemicellulose isolated from biomass not only helps to understand and analyze the catalytic gasification characteristics of natural biomass material with high hemicellulose content, but also gives a considerable method to produce H_2 from only hemicellulose, with the highest amount of H_2 produced but a lower amount of gasification solid residues than the whole biomass used (Gökkaya et al. 2020).

6.8 Conclusion

For efficient use of biomass, technologies that employ thermochemical conversions, including combustion, pyrolysis, liquefaction, and gasification, have demonstrated high efficiencies and improved environmental performance, leading to their recognition worldwide. Process mechanisms and characteristics of each technology applying to cellulose and hemicellulose, which are two main components of biomass, have been investigated and the reported results have been helpful in identifying potential applications and useful products. Before they can be implemented in industrial applications, the related technologies need more research and development efforts followed by a demonstration at semi-commercial scales. Among the thermochemical conversion processes, combustion offers potential direct applications, especially in the innovation of new materials, whereas liquefaction may become a new route for the production of liquid fuels. Pyrolysis, particularly torrefaction, has been applied widely for heat and electricity generation, but the technology required sophisticated designs for specific equipment in industrial scales, while gasification often consumes extensive energy to reach the required temperatures and pressures. Advances in biomass processing such as pretreatments and fractionation as well as in system and engineering process design may enhance the effectiveness of biomass thermochemical conversion into fuels and value-added products with better performance and cost competitiveness.

References

Abhijeet P, Swagathnath G, Rangabhashiyam S, Asok Rajkumar M, Balasubramanian P (2020) Prediction of pyrolytic product composition and yield for various grass biomass feedstocks. Biomass Convers Biorefinery 10(3):663–674

Agnieszka T, Zofia S, Patrycja B (2020) Biochar physicochemical properties: pyrolysis temperature and feedstock kind effects. Rev Environ Sci Biotechnol 19:24

Alvarez J, Lopez G, Amutio M, Bilbao J, Olazar M (2014) Bio-oil production from rice husk fast pyrolysis in a conical spouted bed reactor. Fuel 128:8

Amarasekara AS, Reyes CDG (2020) Pd/C catalyzed room-temperature, atmospheric pressure hydrogenation of furanic bio-oils from acidic ionic liquid catalyzed liquefaction of biomass in acetone. Fuel Process Technol 200:106320

Amenaghawon AN, Anyalewechi CL, Okieimen CO, Kusuma HS (2021) Biomass pyrolysis technologies for value-added products: a state-of-the-art review. Environ Dev Sustain

Amutio M, Lopez G, Alvarez J, Olazar M, Bilbao J (2015) Fast pyrolysis of eucalyptus waste in a conical spouted bed reactor. Biores Technol 194:8

Amutio M, Lopez G, Artetxe M, Elordi G, Olazar M, Bilbao J (2012) Influence of temperature on biomass pyrolysis in a conical spouted bed reactor. Resour Conserv Recycl 59:9

Atienza-Martínez M, Rubio I, Fonts I, Ceamanos J, Gea G (2017) Effect of torrefaction on the catalytic post-treatment of sewage sludge pyrolysis vapors using γ-Al2O3. Chem Eng J 308:264–274

Azeez A, Meier D, Odermatt J (2011) Temperature dependence of fast pyrolysis volatile products from European and African biomasses. J Anal Appl Pyrolysis 90:12

Balat M, Balat M, Kirtay E, Balat H (2009) Main routes for the thermo-conversion of biomass into fuels and chemicals. Part 1: pyrolysis systems. Energy Convers Manag 50:11

Banyasz J, Li S, Lyons-Hart J, Shafer K (2001) Cellulose pyrolysis: the kinetics of hydroxyacetaldehyde evolution. J Anal Appl Pyrol 57:223–248

Barzegar R, Yozgatligil A, Olgun H, Atimtay AT (2020) TGA and kinetic study of different torrefaction conditions of wood biomass under air and oxy-fuel combustion atmospheres. J Energy Inst 93(3):889–898

Basu P (2013) Biomass gasification, pyrolysis and torrefaction: practical design and theory, 2nd edn. Elsevier

Berthet M-A, Commandré J-M, Rouau X, Gontard N, Angellier-Coussy HJM (2016) Torrefaction treatment of lignocellulosic fibres for improving fibre/matrix adhesion in a biocomposite. Mater Design 92:223–232

Boateng AA (2020) Pyrolysis of biomass for fuels and chemicals. Elsevier

Börjesson M, Westman G (2015) Crystalline nanocellulose—preparation, modification, and properties. In: Cellulose: fundamental aspects and current trends. InTechOpen, pp 35

Boukaous N, Abdelouahed L, Chikhi M, Meniai A-H, Mohabeer C, Bechara TJE (2018) Combustion of flax shives, beech wood, pure woody pseudo-components and their chars: a thermal and kinetic study 11(8):2146

Bu Q, Cao M, Wang M, Zhang X, Mao H (2021) The effect of torrefaction and ZSM-5 catalyst for hydrocarbon rich bio-oil production from co-pyrolysis of cellulose and low density polyethylene via microwave-assisted heating. Sci Total Environ 754:142174

Buendia-Kandia F, Brosse N, Petitjean D, Mauviel G, Rondags E, Guedon E, Dufour A (2020) Hydrothermal conversion of wood, organosolv, and chlorite pulps. Biomass Convers Biorefinery 10(1):1–13

Cahyanti MN, Doddapaneni TRKC, Madissoo M, Pärn L, Virro I, Kikas T (2021) Torrefaction of agricultural and wood waste: comparative analysis of selected fuel characteristics. Energies 14(10):2774

Carrier M, Loppinet-Serani A, Denux D, Lasnier JM, Ham-Pichavant F, Cansell F, Aymonier C (2011) Thermogravimetric analysis as a new method to determine the lignocellulosic composition of biomass. Biomass Bioenerg 35(1):10

Chatterjee R, Sajjadi B, CheN W-Y, Mattern LD, Hammer N, Raman V, Dorris A (2020) Effect of pyrolysis temperature on physicochemical properties and acoustic-based amination of biochar for efficient CO2 adsorption. Front Energy Res 8(85):18

Chen C, Zhao L, Wang J, Lin S (2017) Reactive molecular dynamics simulations of biomass pyrolysis and combustion under various oxidative and humidity environments. Ind Eng Chem Res 56(43):12276–12288

Chen D, Zheng Z, Fu K, Zeng Z, Wang J, Lu M (2015) Torrefaction of biomass stalk and its effect on the yield and quality of pyrolysis products. Fuel 159:27–32

Chen W-H, Wang C-W, Ong HC, Show PL, Hsieh T-H (2019a) Torrefaction, pyrolysis and two-stage thermodegradation of hemicellulose, cellulose and lignin. Fuel 258:116168

Chen Y, Fang Y, Yang H, Xin S, Zhang X, Wang X, Chen H (2019b) Effect of volatiles interaction during pyrolysis of cellulose, hemicellulose, and lignin at different temperatures. Fuel 248:7

Cho J, Davis JM, Huber GW (2010) The Intrinsic Kinetics and Heats of Reactions for Cellulose Pyrolysis and Char Formation. Chemsuschem 3:4

Collard F-X, Blin J (2014) A review on pyrolysis of biomass constituents: Mechanisms and composition of the products obtained from the conversion of cellulose, hemicelluloses and lignin. Renew Sustain Energy Rev 38:15

Craven M, Wang Y, Yang H, Wu C, Tu X (2020) Integrated gasification and non-thermal plasma-catalysis system for cleaner syngas production from cellulose. IOP SciNotes 1(2):024001
Dada TK, Sheehan M, Murugavelh S, Antunes E (2021) A review on catalytic pyrolysis for high-quality bio-oil production from biomass. Biomass Convers Biorefinery
Dai G, Wang G, Wang K, Zhou Z, Wang S (2021) Mechanism study of hemicellulose pyrolysis by combining in-situ DRIFT, TGA-PIMS and theoretical calculation. Proc Combust Inst 38(3):9
Danghyan V, Orlova T, Roslyakov S, Wolf E, Mukasyan A (2020) Cellulose assisted combustion synthesis of high surface area Ni-MgO catalysts: mechanistic studies. Combust Flame 221:462–475
Dhar SA, Sakib TU, Hilary LN (2020) Effects of pyrolysis temperature on production and physicochemical characterization of biochar derived from coconut fiber biomass through slow pyrolysis process. Biomass Convers Biorefinery
Ebringerovµ A, Hromµdkovµ Z, Heinze T, Valantinavičius M (2005) Polysaccharides I, vol 186. Springer, Berlin, pp 67
Feng S, Wei R, Leitch M, Xu CC (2018) Comparative study on lignocellulose liquefaction in water, ethanol, and water/ethanol mixture: roles of ethanol and water. Energy 155:234–241
Ferraro G, Pecori G, Rosi L, Bettucci L, Fratini E, Casini D, Rizzo AM, Chiaramonti D (2021). Biochar from lab-scale pyrolysis: influence of feedstock and operational temperature. Biomass Convers Biorefinery
Fukutome A, Kawamoto H, Saka S (2015) Processes forming gas, tar, and coke in cellulose gasification from gas-phase reactions of levoglucosan as intermediate. ChemSusChem 8(13):2240–2249
Fukutome A, Kawamoto H, Saka S (2017) Kinetics and molecular mechanisms for the gas-phase degradation of levoglucosan as a cellulose gasification intermediate.J Anal Appl Pyrolysis 124:666–676
Gabhane JW, Bhange VP, Patil PD, Bankar ST, Kumar S (2020) Recent trends in biochar production methods and its application as a soil health conditioner: a review. SN Appl Sci 2(7):1307
Gagic T, Perva-Uzunalic A, Knez Z, Skerget M (2018) Hydrothermal degradation of cellulose at temperature from 200 to 300 °C. Ind Eng Chem Res 57:9
García-Jarana MB, Portela JR, Sánchez-Oneto J, Martinez de la Ossa EJ, Al-Duri B (2020) Analysis of the supercritical water gasification of cellulose in a continuous system using short residence times. Appl Sci 10(15):5185
Ge X, Chang C, Zhang L, Cui S, Luo X, Hu S, Qin Y, Li Y (2018) Conversion of lignocellulosic biomass into platform chemicals for biobased polyurethane application. In: Advances in bioenergy, vol 3. Elsevier, pp 161–213
Ghimire N, Bakke R, Bergland WH (2021) Liquefaction of lignocellulosic biomass for methane production: a review. Bioresour Technol 125068
Gökkaya DS, Sert M, Sağlam M, Yüksel M, Ballice L (2020) Hydrothermal gasification of the isolated hemicellulose and sawdust of the white poplar (Populus alba L.). J Supercrit Fluids 162:104846
Hao H, Chow CL, Lau D (2020) Carbon monoxide release mechanism in cellulose combustion using reactive forcefield. Fuel 269:117422
Hassan N, Jalil A, Vo D, Nabgan W (2020) An overview on the efficiency of biohydrogen production from cellulose. Boimass Convers Biorefinery 1–23
Hendriks ATWM, Zeeman G (2009) Pretreatments to enhance the digestibility of lignocellu- losic biomass. Bioresour Technol 100(1):9
Hirano Y, Miyata Y, Taniguchi M, Funakoshi N, Yamazaki Y, Ogino C, Kita Y (2020) Fe-assisted hydrothermal liquefaction of cellulose: effects of hydrogenation catalyst addition on properties of water-soluble fraction. J Anal Appl Pyrolysis 145:104719
Jasiukaitytė-Grojzdek E, Vicente FA, Grilc M, Likozar B (2021) Ambient-pressured acid-catalysed ethylene glycol organosolv process: liquefaction structure-activity relationships from model cellulose-lignin mixtures to lignocellulosic wood biomass. Polymers 13(12):1988

Jiang X, Chen D, Ma Z, Yan J (2017) Models for the combustion of single solid fuel particles in fluidized beds: a review. Renew Sustain Engery Rev 68:410–431
Junior II, Do Nascimento MA, de Souza ROMA, Dufourd A, Wojcieszaka R (2020) Levoglucosan: a promising platform molecule? Green Chem 22:22
Khatiwada JR, Shrestha S, Sharma HK, Qin W (2021) Bioconversion of hemicelluloses into hydrogen. In: Sustainable bioconversion of waste to value added products. Springer, pp 267–280
Kosmela P, Hejna A, Formela K, Haponiuk J, Piszczyk Ł (2018) The study on application of biopolyols obtained by cellulose biomass liquefaction performed with crude glycerol for the synthesis of rigid polyurethane foams. J Polym Environ 26(6):2546–2554
Kostetskyy P, Broadbelt LJ (2020) Progress in modeling of biomass fast pyrolysis: a review. Energy Fuels 34(12):22
Kumar A (2019) Current trends in cellulose assisted combustion synthesis of catalytically active nanoparticles. Ind Eng Chem Res 58(19):7681–7689
Kuo L-J, Herbert BE, Louchouarn P (2008) Can levoglucosan be used to characterize and quantify char/charcoal black carbon in environmental media? Organic Geochem 39(10):1466–1478
Laureano-Perez L, Teymouri F, Alizadeh H, Dale BE (2005) Understanding factors that limit enzymatic hydrolysis of biomass. Appl Biochem Biotechnol 124(1–3):19
Lennon E, Tanzy M, Volpert V, Mukasyan A, Bayliss AJ (2011) Combustion of reactive solutions impregnated into a cellulose carrier: modeling of two combustion fronts. Chem Eng 174(1):333–340
Li J, Qin G, Tang X, Xiang C (2021) Fe/HZSM-5 catalyzed liquefaction of cellulose assisted by glycerol. Catal Commun 151:106268
Li Q, Hu X, Liu D, Song L, Yan Z, Li M, Liu Q (2020) Comprehensive evaluation of hydro-liquefaction characteristics of lignocellulosic subcomponents. J Energy Inst 93(4):1705–1712
Li Q, Wang N, Barbante C, Kang S, Callegaro A, Battistel D, Argiriadis E, Wan X, Yao P, Pu T (2019) Biomass burning source identification through molecular markers in cryoconites over the Tibetan Plateau. Environ Pollut 244:209–217
Liu J, Qin D, Chen K, Wang B, Bian H, Shao Z (2021) The preparation of intrinsic dopo-cinnamic flame-retardant cellulose and its application for lithium-ion battery separator. Mater Res Express 8(7):076404
Liu Y, Wu S, Zhang H, Xiao R (2020) Preparation of carbonyl precursors for long-chain oxygenated fuels from cellulose ethanolysis catalyzed by metal oxides. Fuel Process Technol 206:106468
Lu Q, Yang X, Dong C (2011) Influence of pyrolysis temperature and time on the cellulose fast pyrolysis products: analytical Py-GC/MS study. J Anal Appl Pyrolysis 92:9
Luo J, Li Q, Meng A, Long Y, Zhang Y (2018) Combustion characteristics of typical model components in solid waste on a macro-TGA. J Therm Anal Calorim132(1):553–562
Madhua P, Dhanalakshmia CS, Mathew M (2020) Multi-criteria decision-making in the selection of a suitable biomass material for maximum bio-oil yield during pyrolysis. Fuel 277:12
Mbeugang CFM, Li B, Lin D, Xie X, Wang S, Wang S, Zhang S, Huang Y, Liu D, Wang Q (2021) Hydrogen rich syngas production from sorption enhanced gasification of cellulose in the presence of calcium oxide. Energy 228:120659
McGrath T, Chan W, Hajaligol M (2003) Low temperature mechanism for the formation of polycyclic aromatic hydrocarbons from the pyrolysis of cellulose. J Anal Appl Pyrolysis 66:20
Minami E, Fujimoto S, Saka SJ (2018) Complete gasification of cellulose in glow-discharge plasma. J Wood Sci 64(6):854–860
Moradi-Choghamarani F, Moosavi AA, Baghernejad M (2019) Determining organo-chemical composition of sugarcane bagasse-derived biochar as a function of pyrolysis temperature using proximate and Fourier transform infrared analyses. J Therm Anal Calorim 138
Negi S, Jaswal G, Dass K, Mazumder K, Elumalai S, Roy JK (2020) Torrefaction: a sustainable method for transforming of agri-wastes to high energy density solids (biocoal). Rev Environ Sci Biotechnol 19(2)
Neves D, Thunman H, Matos A (2011) Characterization and prediction of biomass pyrolysis products. Progr Energy Combust Sci 37:20

Niu Y, Lv Y, Lei Y, Liu S, Liang Y, Wang D (2019) Biomass torrefaction: properties, applications, challenges, and economy. Renew Sustain Energy Rev 115:109395

Okolie JA, Nanda S, Dalai AK, Kozinski JA. J. I. J. o. H. E. (2020). Optimization and modeling of process parameters during hydrothermal gasification of biomass model compounds to generate hydrogen-rich gas products. Int J Hydrog Energy 45(36):18275–18288

Ong HC, Chen W-H, Farooq A, Gan YY, Lee KT, Ashokkumar V (2019) Catalytic thermochemical conversion of biomass for biofuel production: a comprehensive review. Renew Sustain Energy Rev 113:109266

Orcid ER, Debiagi PEA, Frassoldati A (2017) Mathematical modeling of fast biomass pyrolysis and bio-oil formation. Note II: secondary gas-phase reactions and bio-oil formation. ACS Sustain Chem Eng 5(4):15

Paajanen A, Rinta-Paavola A, Vaari J (2021) High-temperature decomposition of amorphous and crystalline cellulose: reactive molecular simulations. Cellulose 1–19

Paajanen A, Vaari J (2017) High-temperature decomposition of the cellulose molecule: a stochastic molecular dynamics study. Cellulse 24(7):2713–2725

Pandey A, Stöcker M, Bhaskar T, Sukumaran RK (2015) Recent advances in thermochemical conversion of biomass. Elsevier

Papari S, Hawboldt K (2015) A review on the pyrolysis of woody biomass to bio-oil: Focus on kinetic models. Renew Sustain Energy Rev 52:16

Park JY, Kima J-K, Oh C-H, Park J-W, Kwon EE (2019) Production of bio-oil from fast pyrolysis of biomass using a pilot-scale circulating fluidized bed reactor and its characterization. J Environ Manag 234:7

Patel A, Agrawal B, Rawal BR (2020) Pyrolysis of biomass for efficient extraction of biofuel. Energy Sources Part A Recover Util Environ Effects 42:13

Patwardhan PR, Dalluge DL, Shanks BH, Brown RC (2011) Distinguishing primary and secondary reactions of cellulose pyrolysis. Bioresour Technol 102(8):5

Pawar A, Panwar NL, Salvi BL (2020) Comprehensive review on pyrolytic oil production, upgrading and its utilization. J Mater Cycles Waste Manag 22:11

Peng J, Liu X, Bao Z (2018) Experimental study on the liquefaction of cellulose in supercritical ethanol. Paper presented at the IOP conference series: materials science and engineering

Peng Y, Wu S (2010) The structural and thermal characteristics of wheat straw hemicellulose. J Anal Appl Pyrolysis 88(2):6

Phaiboonsilpa N, Champreda V, Laosiripojana N (2020) Comparative study on liquefaction behaviors of xylan hemicellulose as treated by different hydrothermal methods. Energy Rep 6:714–718

Pinkert A, Marsh KN, Pang S, Staiger MP (2009) Ionic liquids and their interaction with cellulose. Chem Rev 109:17

Robert CB, Kaige W (2017) Fast pyrolysis of biomass advances in science and technology

Ruan Y, Mohtadi M, Dupont LM, Hebbeln D, van der Kaars S, Hopmans E, Schouten S, Hyer EJ, Schefuß E (2020) Interaction of fire, vegetation, and climate in tropical ecosystems: a multiproxy study over the past 22,000 years. Global Biogeochem Cycles 34(11):e2020GB006677

Salam A, Bashir S, Khan I, Hu H (2020) Biochar production and characterization as a measure for effective rapeseed residue and rice straw management: an integrated spectroscopic examination. Biomass Convers Biorefinery

Sefain MZ, El-Kalyoubi SF, Shukry N (1985) Thermal behavior of holo-and hemicellulose obtained from rice straw and bagasse 23(5):1569–1577

Segato D, Villoslada Hidalgo MDC, Edwards R, Barbaro E, Vallelonga P, Kjær HA, Simonsen M, Vinther B, Maffezzoli N, Zangrando R, Turetta C (2021) Five thousand years of fire history in the high North Atlantic region: natural variability and ancient human forcing. Clim Past 17(4):1533–1545

Shen D, Gu S (2009) The mechanism for thermal decomposition of cellulose and its main products. Biores Technol 100:6496–6504

Song C, Zhang C, Zhang S, Lin H, Kim Y, Ramakrishnan M, Du Y, Zhang Y, Zheng H, Barcelo D (2020). Thermochemical liquefaction of agricultural and forestry wastes into biofuels and chemicals from circular economy perspectives. Sci Total Environ 141972
Soomro A, Chen S, Ma S, Xu C, Sun Z, Xiang W (2018) Elucidation of syngas composition from catalytic steam gasification of lignin, cellulose, actual and simulated biomasses. Biomass Bioenergy 115:210–222
Sun J, Xie X-A, Fan D, Wang X, Liao W (2020a) Effect of TEMPO and characterization of bio-oil from cellulose liquefaction in supercritical ethanol. Renew Energy 145:1949–1956
Sun J, Xu L, Dong G-H, Nanda S, Li H, Fang Z, Kozinski JA, Dalai AK (2020b) Subcritical water gasification of lignocellulosic wastes for hydrogen production with Co modified Ni/Al2O3 catalysts. J Supercrit Fluids 162:104863
Sun L, Li L, An X, Qian X (2021) Mechanically strong, liquid-resistant photothermal bioplastic constructed from cellulose and metal-organic framework for light-driven mechanical motion. Molecules 26(15):4449
Tag A-T, Duman G, Ucar S, Yanik J (2016a) Effects of feedstock type and pyrolysis temperature on potential applications of biochar. J Anal Appl Pyrolysis 120:7
Tancredi N, Gabús M, Yoshida MI, Suárez AC (2017) Thermal studies of wood impregnated with ZnCl 2. Eur J Wood Wood Prod 75(4):633–638
Tang G, Gu J, Huang Z, Yuan H, Chen Y (2021) Cellulose gasification with Ca-Fe oxygen carrier in chemical-looping process. Energy
Taylor MJ, Michopoulos AK, Zabaniotou AA, Skoulou V (2020) Probing synergies between lignin-rich and cellulose compounds for gasification. Energies 13(10):2590
Treedet W, Suntivarakorn R (2018) Design and operation of a low cost bio-oil fast pyrolysis from sugarcane bagasse on circulating fluidized bed reactor in a pilot plant. Fuel Process Technol 179:25
Usino D, Supriyanto Ylitervo P, Pettersson A, Richards T (2020) Influence of temperature and time on initial pyrolysis of cellulose and xylan. J Anal Appl Pyrol 147:104782
Van de Velden M, Baeyens J, Brems A (2010) Fundamentals, kinetics and endothermicity of the biomass pyrolysis reaction. Renew Energy 35:11
Wang C, Zhang H, Guo H, Shi S, Xiong L, Chen X (2016a) Part A: Recovery, Utilization,, & Effects, E. (2016a). Real-time monitoring of hemicellulose liquefaction in ethylene glycol. Energy Sources, Part A Recover Util Environ Effects 38(7):2517–2523
Wang S, Dai G, Ru B, Zhao Y, Wang X, Zhou J, Luo Z, Cen K (2016b). Effects of torrefaction on hemicellulose structural characteristics and pyrolysis behaviors. Bioresour Technol 218:1106–1114
Wang C, Yang S, Xu DJ, Li Y, Li J, Zhang Y (2020a). Hydrothermal liquefaction and gasification of biomass and model compounds: a review. Green Chem 22, 24, 22, 24.
Wang Q, Song H, Pan S, Dong N, Wang X, Sun S (2020b) Initial pyrolysis mechanism and product formation of cellulose: An Experimental and Density functional theory(DFT) study. Sci Rep 10:18
Wang S, Ru B, Lin H, Sun W (2015) Pyrolysis behaviors of four O-acetyl-preserved hemicelluloses isolated from hardwoods and softwoods. Fuel 150:9
Wang S, Guo X, Liang T, Zhou Y, Luo Z (2012) Mechanism research on cellulose pyrolysis by Py-GC/MS and subsequent density functional theory studies. Bioresour Technol 104(2012):722–728, 104, 7
Wang S, Zou C, Yang H, Lou C, Cheng S, Peng C, Wang C, Zou H (2021) Effects of cellulose, hemicellulose, and lignin on the combustion behaviours of biomass under various oxygen concentrations. Bioresour Technol 320:124375
Wei L, Xu S, Zhang L (2006) Characteristics of fast pyrolysis of biomass in a free fall reactor. Fuel Process Technol 9
Werner K, Pommer L, Broström M (2014) Thermal decomposition of hemicelluloses. J Anal Appl Pyrolysis 110:8

Wooten JB, Seeman JI, Hajaligol MR (2003) Observation and Characterization of cellulose pyrolysis intermediates by 13C CPMAS NMr. A new mechanistic model. Energy Fuels 2003(18):15

Xu Y-H, Li M-F (2021) Hydrothermal liquefaction of lignocellulose for value-added products: Mechanism, parameter and production application. Bioresour Technol 342:126035

Xu Z-X, Cheng J-H, He Z-X, Wang Q, Shao Y-W, Hu X (2019) Hydrothermal liquefaction of cellulose in ammonia/water. Boiresour Technol 278:311–317

Yang C, Wang S, Yang J, Xu D, Li Y, Li J, Zhang Y (2020a) Hydrothermal liquefaction and gasification of biomass and model compounds: a review. Green Chem 22:24

Yang X, Fu Z, Han D, Zhao Y, Li R, Wu Y (2020b) Unveiling the pyrolysis mechanisms of cellulose: Experimental and theoretical studies. Renew Energy 147:11

Yu C, Ren S, Wang G, Xu J, Teng H, Li T, Huang C, Wang C (2021) Kinetic analysis and modeling of maize straw hydrochar combustion using multi-Gaussian-distributed activation energy model. Int J Miner Metall Mater

Yu H, Wu Z, Chen G (2018) Catalytic gasification characteristics of cellulose, hemicellulose and lignin. Renew Energy 121:559–567

Zhang C, Li C, Zhang Z, Zhang L, Li Q, Fan H, Zhang S, Liu Q, Qiao Y, Tian Y, Wang Y, Hu X (2021) Pyrolysis of cellulose: Evolution of functionalities and structure of bio-char versus temperature. Renew Sustain Energy Rev 135(2021):110416, 135, 20

Zhang X, Yang W, Blasiak W (2012) Thermal decomposition mechanism of levoglucosan during cellulose pyrolysis. J Anal Appl Pyrolysis 96:10

Zhao L, Zhou H, Xie Z, Li J, Yin Y (2020) Effects of potassium on solid products of peanut shell torrefaction. Energy Sources Part A Recovery Util Environ Effects 42(10):1235–1246

Zheng A, Li L, Zhao Z, Tian Y, Li H (2021). Effect of torrefaction pretreatment on chemical structure and pyrolysis behaviors of Cellulose. Paper presented at the IOP conference series: earth and environmental science

Zhou H, Zhao L, Fu X, Zhu Y, Pan Y (2020) Effects of typical alkaline earth metal salts on cellulose torrefaction: solid products and thermal gravimetric analysis. Bioresources 15(1):1678–1691

Zhou X, Li W, Mabon R, Broadbelt LJ (2016) A critical review on hemicellulose pyrolysis. Energy Technol 4:29

Zhou X, Li W, Mabonb R, Broadbelt LJ (2018) A mechanistic model of fast pyrolysis of hemicellulose. Energy Environ. Sci 11:1240, 11, 21

Zhu C, Han C-G, Saito G, Akiyama T (2016) Facile synthesis of MnO/carbon composites by a single-step nitrate-cellulose combustion synthesis for Li ion battery anode. J Alloys Compd 689:931–937

Zimmermann A, Visscher C, Kaltschmitt M (2021) Plant-based fructans for increased animal welfare: provision processes and remaining challenges. Biomass Convers Biorefinery 1–19

Zou J, Oladipo J, Fu S, Al-Rahbi A, Yang H, Wu C, Cai N, Williams P, Chen H (2018). Hydrogen production from cellulose catalytic gasification on CeO2/Fe2O3 catalyst. Energy Convers Manag 171:241–248

Chapter 7
Thermochemical and Catalytic Conversion of Lignin

Charles A. Mullen

Abstract The development of efficient processes to produce high-value chemicals from lignin remains one of the largest technical hurdles in the development of competitive, sustainable biorefineries. This chapter details progress in thermo-catalytic lignin conversion with a focus on processes that produce liquid products largely comprising high-value aromatic monomers. The processes examined include both pyrolysis and solvent liquefaction. The pyrolysis section covers both non-catalytic fast pyrolysis and catalytic pyrolysis, while the solvent liquefaction section covers the liquefaction process catalyzed by acids, bases, and active metals. Both reductive and oxidative processes will be covered. Also, the recently developed lignin-first biorefinery concept, where *in situ* lignin depolymerization occurs concurrently with its fractionation from the biomass, leaving a carbohydrate-rich by-product, is covered. Finally, a summary of the remaining technical challenges in lignin depolymerization is presented.

7.1 Introduction

Lignocellulosic biomass is the most abundant source of renewable carbon on Earth, and its efficient conversion to fuels, chemicals, and materials is the foundation of a bioeconomy that can displace our society's dependence on fossil fuels. In the United States alone, more than one billion tons of lignocellulosic biomass can be sourced annually without interfering with food production (Langholtz et al. 2016). This biomass is composed primarily of cellulose, hemicellulose, and lignin along with minor amounts of other constituents such as protein, lipids, resins, and minerals (ash). Existing biorefineries, including pulp mills and biochemical conversion-based cellulosic ethanol plants, can efficiently convert the carbohydrate fractions (particularly cellulose) of the biomass to products, but both generate an underutilized lignin-rich by-product. Pulp mills, which have operated for a much longer time than more

C. A. Mullen (✉)
USDA, Agricultural Research Service, Eastern Regional Research Center, 600 E. Mermaid Lane, Wyndmoor, Pennsylvania 19038, United States
e-mail: charles.mullen@usda.gov

N. P. Nghiem et al. (eds.), *Biomass Utilization: Conversion Strategies*,
https://doi.org/10.1007/978-3-031-05835-6_7

modern biorefineries, utilize only 2% of their lignin for higher value products, and the remainder is used only for heat generation (Gosselink et al. 2004; Ha et al. 2019).

Lignin is the most abundant natural source of aromatic compounds in the biosphere, which makes it an attractive renewable feedstock to produce biofuels, commodity chemicals, and other value-added products. However, given its function in living plants, to provide rigid structure, strength, and defense, lignin is understandably difficult to convert. Lignin is an amorphous, three-dimensional biopolymer composed of phenolic units cross-linked by C-C and ether bonds. Unfortunately, unlike cellulose, few effective biochemical and/or enzymatic methods have been developed for the depolymerization of lignin. Similarly, even the thermochemical conversion process utilizing whole biomass achieves conversion to fuel and chemical products from the carbohydrate fractions, but the lignin carbon is often untimely found in biochar or coke. These deficiencies in lignin conversion leave at least 20% of biomass carbon extremely underutilized, and this represents an enormous opportunity that must be realized for biorefineries to improve their economic outlook and for many bio-products to become cost-competitive with their fossil-based counterparts.

This chapter will cover the thermochemical and catalytic conversion of lignin. The focus will be on the utilization of isolated (technical) lignin or lignin-rich fractions, in contrast to the processing of whole biomass resulting in a simultaneous reaction of lignin and carbohydrate fractions because the former comes with its own unique opportunities and challenges. However, it will also cover "lignin-first" routes to depolymerization, where native lignin within the biomass is converted first, and the carbohydrate fraction is a by-product to be converted separately (Abu-Omar et al. 2021; Schutyser et al. 2018). This chapter considers processes leading to liquid fuels and renewable chemicals, especially phenolics. Many of the monomeric phenolics that can be produced via thermo-catalytic conversion of lignin are shown in Fig. 7.1. Much of the research focused on the thermochemical conversion of lignin for these products falls into one of two categories. The first is pyrolysis, which involves the rapid heating of the lignin to high temperatures (400–700 °C) under an inert atmosphere to produce biochar, bio-oil (organic liquids and water), and noncondensable gases (Ha et al. 2019; Leng et al. 2022; Fan et al. 2017; Schutyser et al. 2018; Liu et al. 2020; Cao et al. 2019). The second is solvent phase depolymerization also called solvent liquefaction or solvolysis (Ha et al. 2019; Patil et al. 2020; Ragauskas et al. 2014; Rinaldi et al. 2016; Schutyser et al. 2018; Ahmed et al. 2020; Kumar et al. 2021; Cao et al. 2019; Bourbiaux et al. 2021). There are many different types of processes that can occur in the solution phase including reductive and oxidative depolymerization, and their classification depends on catalysts and conditions used. This chapter does not cover gasification in depth as the technologies used for lignin are similar to other materials and have been well described (Sansaniwal et al. 2017; Lee et al. 2021; Braghiroli et al. 2019; Cao et al. 2021; Guo et al. 2021; Hu et al. 2020; San Miguel et al. 2012; Wang et al. 2021a, b). Additionally, carbonization, (Ragauskas et al. 2014; Elkasabi and Mullen 2021; Stanzione et al. 2016) (Snowdon et al. 2014) a slower pyrolysis-type process, that is useful for the production of solid carbon materials such as biochar, carbon black, coke, carbon fiber, and graphite has also been covered extensively in the literature and will not be covered in this chapter.

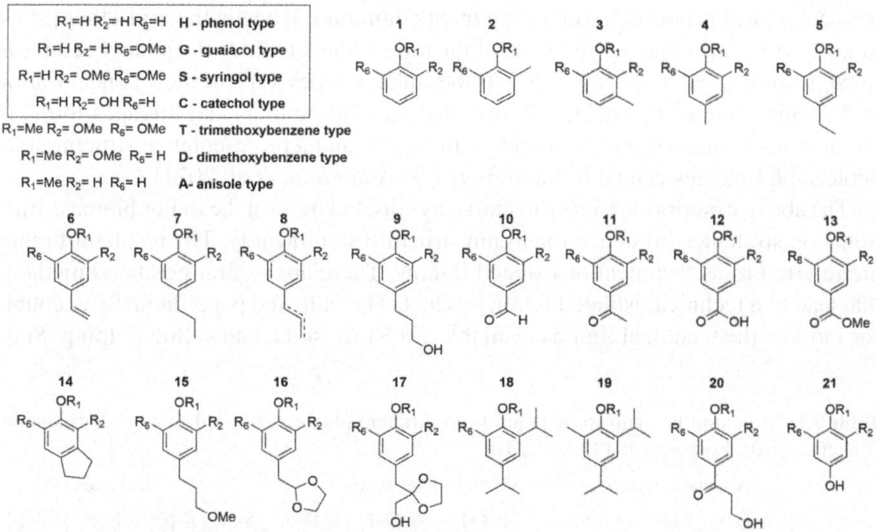

Fig. 7.1 Selected phenolic monomers produced from thermochemical conversion of lignin. Structure abbreviations found throughout the text and tables of this chapter are defined. True phenolic types are shown in box, H, G, S, and C

7.2 Lignin Structure and Types

Lignin is situated between cellulose and hemicellulose in plant cell walls and provides plants structural integrity and defense against natural biological and chemical threats. It is biosynthesized via radical coupling reactions of three major monolignols, sinapyl, coniferyl, and *p*-coumaryl alcohols. These compounds differ based on the number of methoxy groups on the aromatic ring, as shown in Fig. 7.2. They give rise to syringyl (S), guaiacyl (G), and *p* coumaryl (H) lignin units. Other lesser concentrated lignin units include the ferulate subunit, most highly concentrated in herbaceous plants, and the caffeyl (C) unit found in certain species (Ralph et al. 2019). Hardwoods contain predominantly S-lignin units, softwoods nearly exclusively G-lignin units, and grasses contain a mixture of all three. These units are

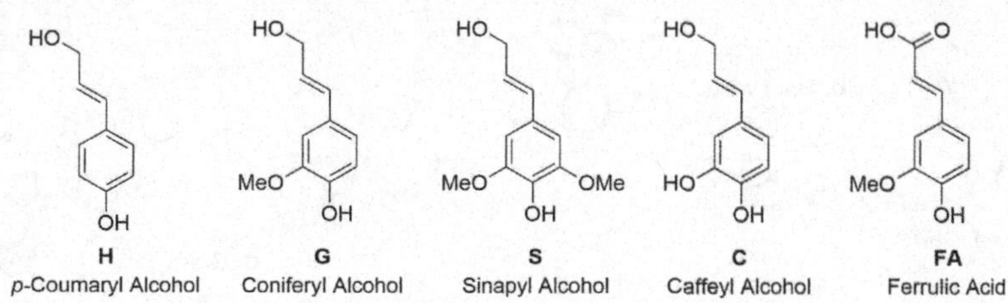

Fig. 7.2 Lignin monomers

linked by various boding motifs. The most common linkage is the aryl-ether β-O-4 linkage which accounts for 40–60% of the total. Others include the phenyl coumarin (β-5), resinol (β-β), β-1, and 5–5. All the linkage types vary in their concentration in the native lignins by species. Ranges for each of the units and linkages found in major types of species can be found in Table 7.1, and a representative structure that depicts the linkages can be found in Fig. 7.3 (Abu-Omar et al. 2021).

The above descriptions apply to native lignins, but most of the major biomass fractionation strategies influence the lignin structure significantly. The resultant lignins are referred to as technical or isolated lignins. There are several practiced methods that lead to a technical isolated lignin product. The pulp and paper industry accounts for most of the technical lignins available via Kraft, soda, and sulfite pulping. Soda

Table 7.1 Structural breakdown of lignin from different plants. Compiled from various sources (Liu et al. 2020; Lourenço and Pereira 2018)

	Aromatic unit type (%)			C-O linkages (%)				C-C linkages (%)		
	H	G	S	β-O4	α-O-4	4-O-5	5-5	β-β	β-5	β-1
Native										
Hardwood	0–8	25–50	46–75	50–65	<1	6–7	<1	3–12	3–11	1–7
Softwood	<5	>95	0	43–50	5–7	4	5–7	2–6	9–12	1–9
Grass	5–33	33–80	20–54	74–84				1–7	5–11	

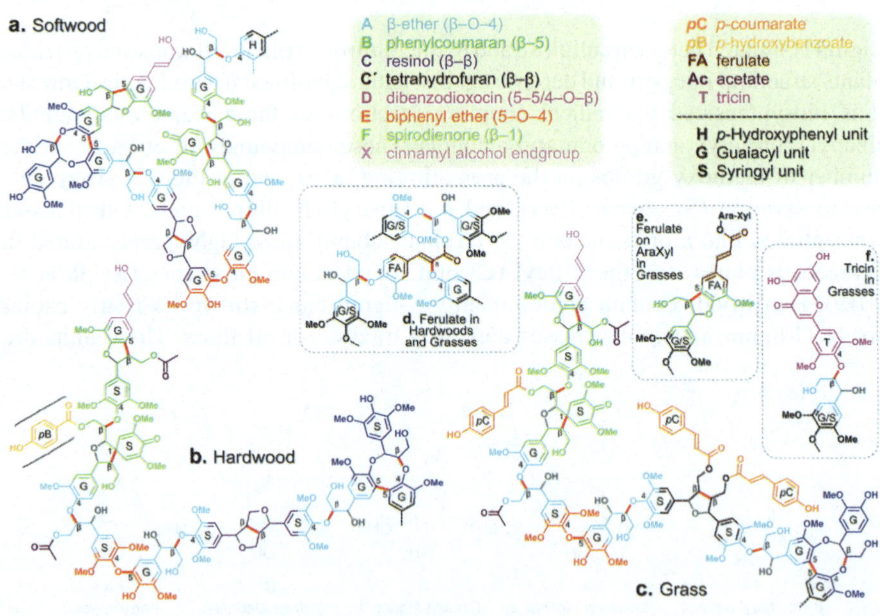

Fig. 7.3 Representative 11-mer lignin structures depicting major linkages. Reproduced with permission from the Royal Society of Chemistry (Abu-Omar et al. 2021)

and Kraft use alkaline (aqueous NaOH) conditions at temperatures of 140–170 °C to decouple lignin from carbohydrates. A key difference between Kraft and soda pulping is that Kraft pulping uses sodium hydrosulfide, which introduces sulfur to the side chain of the lignins (Liu et al. 2020; Vishtal and Kraslawski 2011; Santos et al. 2013). However, both processes result in a significant reaction of the β-O-4 linkages, resulting in condensed lignin structures where the aromatic groups are mainly linked via C-C bonds. Another major lignin isolation procedure is the organosolv process (Liu et al. 2020; Vishtal and Kraslawski 2011). There are variations in this process using various organic solvents (methanol, ethanol, acetone, THF (tetrahydrofuran), etc.) or mixtures of organic solvents and water. After acidic workup, lignins can be isolated. Again, this process results in a reduction in the number of β-O-4 structures in favor of condensed mostly C-C linked structures, with the degree of condensation depending on the severity of the conditions used. Hydrolysis of lignin comes from dilute acid and/or enzymatic treatment of biomass usually as a pretreatment of the biomass for fermentation of sugars from cellulose (e.g., cellulosic ethanol production) (Cao et al. 2012; Moxley et al. 2012). Other less common methods for lignin isolation include the use of ionic liquids (Gillet et al. 2017) or deep eutectic solvents (DES) (Zhou et al. 2022). Table 7.2 illustrates the decrease in natural bonding linkages from native to technical lignins. As we will see in the following sections, these condensed and C-C-containing linkages are much more recalcitrant to thermal and catalytic breakdown, with the bond dissociation energies increasing from about 75 kcal/mol for β-O-4 linkages to 90–120 kcal/mol for the condensed C-C structures (Rinaldi et al. 2016).

7.3 Pyrolysis of Technical Lignins

Pyrolysis is the heating of organic materials in an inert atmosphere causing the breaking of chemical bonds usually resulting in the production of smaller molecules. For biomass, most of these molecules can be condensed into a liquid called bio-oil. Biomass fast pyrolysis is typically conducted at temperatures of 450–650 °C, with 500–550 °C considered the temperature that will maximize bio-oil yield. Lignin decomposes over a wider temperature range than does cellulose; therefore, variation in the optimum temperature for pyrolytic depolymerization can occur. Upon pyrolysis, the input of thermal energy induces bond breaking that can occur via homolytic (free radical), heterolytic (ionic), or concerted mechanisms (Zhou et al. 2016a, b; Leng et al. 2022; Kawamoto 2017). For pyrolysis of lignin, it is unclear which is the dominant reaction, but most researchers invoke radical mechanisms to explain their observations. The complexity of the lignin structure means that many pathways are in competition and that the major mechanism may change with specific lignin structures, reaction conditions, or reactor types. There are many more operational mechanisms active than can be discussed here, but two radical mechanisms for the thermal breaking of the β-O-4 linkage are shown in Fig. 7.4 (Kawamoto 2017). Bond

Table 7.2 Characteristics of technical lignins. Compiled from various sources (Liu et al. 2020; Vishtal and Kraslawski 2011; Poveda-Giraldo et al. 2021; Cao et al. 2012)

Type	Biomass separation conditions	Notes	Remaining native linkages (per 100 aromatic rings)			MW (g/mol)
			β-O4	β-β	β-5	
Kraft (or Alkali)	NaOH/Na_2S 150–180 °C	High S content, condensed structures. Methoxy group loss	3–6	2–13	1–5	1,500–5,000 (up to 25,000)[a]
Soda	NaOH, 90–150 C	Low S content, condensed structures	3–6	1–6	~0	1,000–3,000 (up to 15,000)[a]
Organosolv	Various organic solvents 90–210 °C	Structural changes vary with severity	0–4.3	0–1	2–5	500–5,000
Acid hydrolysis	H_2SO_4, 120 °C	Decrease in aryl-ether, decrease in S/G ratio	40–50 (Hardwood)			5,000–10,000
Enzymatic hydrolysis	Enzyme, 20–45 °C					4,500–9,500
Milled	Mechanical ball milling followed by extraction with 1,4-dioxane and water	Most like native structure, low isolation yield	34–60			11,000–40,000

[a] Typical range of MW is shown. The highest value observed is in parentheses

breaking by thermal homolytic cleavage is controlled by the bond dissociation energies of the various bonds, so as temperatures increase, more types of bonds can be broken. Therefore, as pyrolysis temperatures increase, the number and complexity of substituents on the aromatic rings decrease. At lower pyrolysis temperatures (~300–400 °C), guaiacols and syringols with unsaturated side chains can be produced, but as the temperature increases, less oxygenated alkyl phenols become more prominent (Fig. 7.5).

On a more general level, there is debate about how the product mixture, which includes monomeric phenols, oligomeric phenols (also called pyrolytic lignin), char, and permanent gases, is formed. One proposal was that primary products include char, gases, and monomeric phenols, and oligomeric products are formed via secondary repolymerization reactions of the monomers. Another possible mechanism, referred

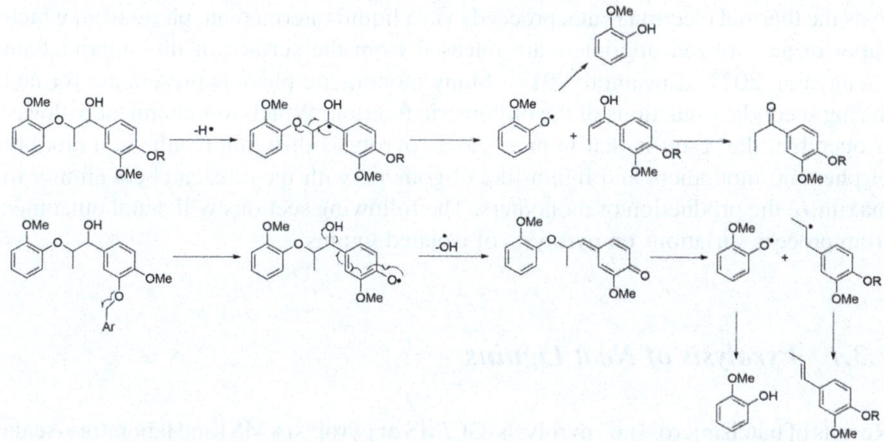

Fig. 7.4 Radical mechanisms for the pyrolytic cleavage of β-O4 linkages in lignin. Adapted with permission from Springer (Kawamoto 2017)

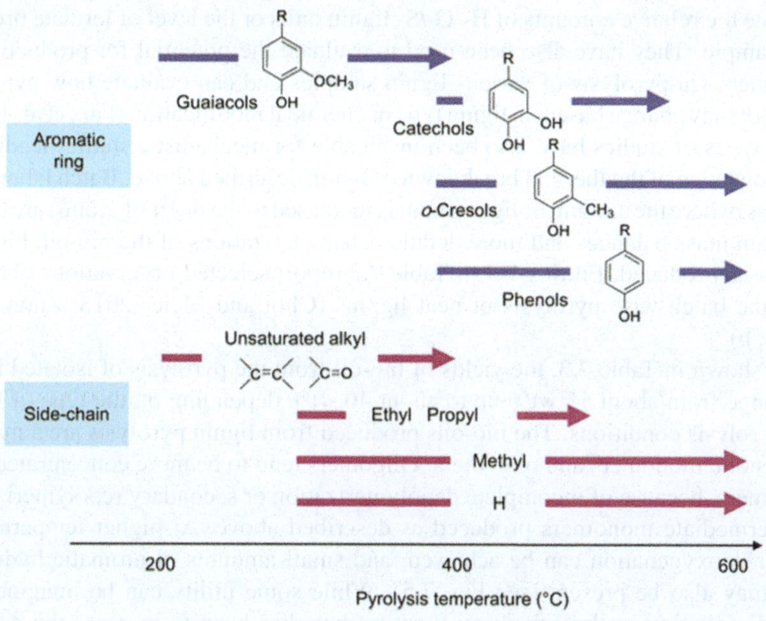

Fig. 7.5 Effect of Pyrolysis Temperature on Product Profile from Fast Pyrolysis of Lignin. Reproduced with permission from Springer (Kawamoto 2017)

to as the thermal ejection route, proceeds via a liquid intermediate phase from which vapor or aerosolized oligomers are released from the surface of this intermediate (Leng et al. 2022; Kawamoto 2017). Many monomeric phenols present are formed during secondary reactions of the oligomeric fraction. Whichever chemical pathway is operable, the result is that in most cases, pyrolysis of lignin results in a mixture of phenolic monomers and lignin-like oligomers, with most researchers aiming to maximize the production of monomers. The following sections will detail outcomes from process variations on pyrolysis of isolated lignins.

7.3.1 Pyrolysis of Neat Lignins

Results of batch micro- (i.e., pyrolysis-GC/MS or pyrolysis-MS) and laboratory-scale pyrolysis of lignin are well reported in the literature. Micro or analytical pyrolysis systems such as pyrolysis-GC/MS systems have been used to characterize the lignin structure (Faix et al. 1987). Based on the observable monomeric products, one can estimate the relative amounts of H-/G-/S- lignin units or the level of ferulate present, for example. They have also been used to evaluate the potential for production of monomers via pyrolysis of various lignin samples and can evaluate how pyrolysis behavior may change based on lignin type or chemical modification (Faix et al. 1987). These types of studies have also been invaluable for mechanistic studies leading to the knowledge of the thermal breakdown of lignin described above. Batch laboratory systems, where the amount of lignin input is increased to the order of grams, are useful to obtain mass balances and more detailed characterizations of the bio-oil, biochar, and gases produced. Entries 1-9 in Table 7.2 report selected observations obtained from the batch-wise pyrolysis of neat lignins (Choi and Meier 2013; Zhou et al. 2016a, b).

As shown in Table 7.3, the yields of bio-oil from the pyrolysis of isolated lignin can range from about 15 wt% up to about 40 wt% depending on the type of lignin and pyrolysis conditions. The bio-oils produced from lignin pyrolysis are a mixture of phenolic monomers and oligomers. Oligomers tend to be more concentrated than monomers, because of incomplete depolymerization or secondary repolymerization of intermediate monomers produced as described above. At higher temperatures, further deoxygenation can be achieved, and small amounts of aromatic hydrocarbons may also be present (see Fig. 7.5). While some utility can be imagined for phenolic oligomers, the aim of most researchers has been to increase the production of monomers. Zhou et al. compared organosolv lignin and milled wood lignin (MWL) from oak (hardwood), pine (softwood), and corn stover in a microreactor (Zhou et al. 2016a, b). The bio-oil yields obtained were in a range from 43–55 wt%, but in the MWLs, all performed slightly better than their organosolv counterparts in terms of yield. The corn stover lignins yielded a significantly higher proportion of lignin monomers (~16 wt%) than did the wood-derived lignins.

Moving from batch-wise to continuously fed reactor systems has proved difficult for the pyrolysis of technical lignins. The reason for this is that the physical

Table 7.3 Selected results for pyrolysis of neat technical lignins

	Lignin	Reactor type	Py. Temp (°C)	Bio-oil yield (wt%)	Monomer yield (wt%)	Major monomers[a]	Char yield (wt%)	Notes	References
1	Kraft	Batch fixed bed (40 mg)	470	30.0	~9.6	G1 (3.3), G2, G6, G8	50.1		Choi and Meier (2013)
2	Kraft	Batch fixed bed (40 mg)	560	38.1	–	–	47.1		Choi and Meier (2013)
3	MWL red oak	Batch–micro	500	48	8.6 C%[b]	A1, S1, S8, G4, G1			Zhou et al. (2016a, b)
4	Organosolv red oak	Batch–micro	500	45	6.3 C%[b]	A1, S1, S8, G4, G1			Zhou et al. (2016a, b)
5	MWL pine	Batch–micro	500	45	9.5 C%[b]	G8, G4, G1, G10,			Zhou et al. (2016a, b)
6	Organosolv pine	Batch–micro	500	43	8.3 C%[b]	G8, G4, G1			Zhou et al. (2016a, b)
7	MWL corn stover	Batch–micro	500	55	16.3 C%[b]	H6 (6.8)[b], G6 (2.8)[b], S1, A1, G1, H1			Zhou et al. (2016a, b)
8	Organosolv corn stover	Batch–micro	500	50	16.3 C%[b]	H6 (5.6)[b], G6 (3.7)[b], S1, A1, G1, H1			Zhou et al. (2016a, b)
9	Soda	Continuous—fluidized beds (Multiple labs)	500-530	~16		G1, G4, G5, H1, H4, C1	~41	Plugging due to melting and agglomeration of bed materials	Nowakowski et al. (2010)

(continued)

Table 7.3 (continued)

	Lignin	Reactor type	Py. Temp (°C)	Bio-oil yield (wt%)	Monomer yield (wt%)	Major monomers[a]	Char yield (wt%)	Notes	References
9	Organosolv	Continuous—mechanically agitated fluid bed	500	~10	~2.5	Cs major, Hs[c]	~75	Process shutdown due to plugging	Ghysels et al. (2020)
10	Kraft (pine)	Continuous—fluid bed (0.4 g/min)	550	23			41	Slow feed rate to prevent plugging	Beis et al. (2010)
11	Kraft (Lignobost)	Continuous—fluid bed	550	22			29		Beis et al. (2010)
12	Acetolcell (Organosolv)	Continuous—fluid bed	550	16			63		Beis et al. (2010)
13	Hydrolysis	Continuous—mechanically stirred FB	500	40	23.1		34	Bed agglomeration, downstream deposits	Pienihäkkinen et al. (2021)
14	Hydrolysis	Continuous—circulating FB	525	37			43	Downstream deposits, but No bed agglomeration,	Pienihäkkinen et al. (2021)
15	Hydrolysis	Continuous—centrifuge	550	47		S1, G6, G1, G4, G5, G6, S11	27	Plugged at cyclone outlet	Trinh et al. (2013)

[a] Refer to Fig. 7.1 for definition of key compounds listed in order of yield, value in parenthesis are notable high yields of an individual compound, [b] Carbon yield (%), and [c] Hs, Cs, Gs and Ss are products that are generally phenols, catechols, guaiacols and syringols, respectively, as opposed to a specific compound

properties of isolated lignins have made the process of pyrolyzing lignin using a continuously fed fluidized bed reactor challenging for scientists and engineers, especially compared with the relative ease in which non-fractionated biomass materials are pyrolyzed. A diagram of a typical continuous fluidized bed fast pyrolysis system used for biomass pyrolysis is shown in Fig. 7.6. The lower melting temperature of isolated lignins results in clogged feeding tubes, agglomerates with bed media, and accumulation of unreacted feeds all of which eventually lead to process shutdowns (Fig. 7.7) (Nowakowski et al. 2010; Pienihäkkinen et al. 2021; Trinh et al. 2013; Ghysels et al. 2020; Beis et al. 2010). These problems were well described by Nowakowski et al., who distributed both isolated soda lignin and a residue from ethanol production containing up to 50% cellulosic material in addition to lignin

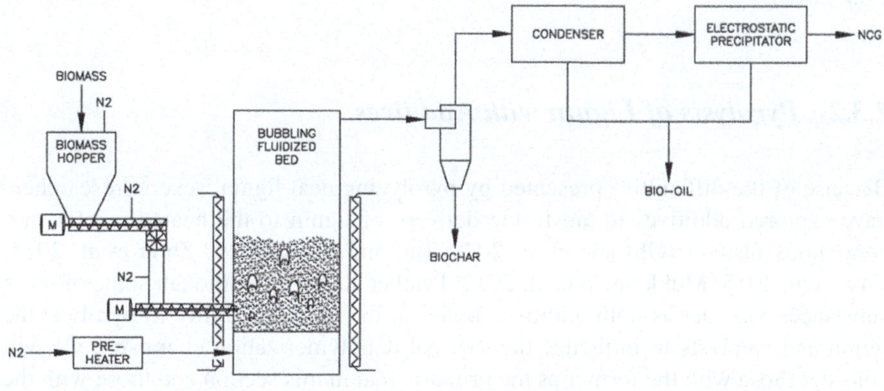

Fig. 7.6 Example schematic of a continuously fed biomass fast pyrolysis reactor system

Fig. 7.7 Clogged process tubes and char-sand agglomerates, the typical results of feeding technical lignins continuously to fluidized bed reactors. Reproduced with permission from Elsevier (Nowakowski et al. 2010)

to fourteen laboratories for fluidized bed pyrolysis (Nowakowski et al. 2010). Each of the laboratories reported extreme difficulties with processing the isolated lignin while the lignin diluted with cellulosic material was easily pyrolyzed. Pienihakkinen et al. reported being able to overcome bed agglomeration problems by incorporating reactor modifications (Pienihäkkinen et al. 2021). In one instance, they added a mechanical mixer to a bubbling fluidized bed reactor, and in another instance, a highly agitated circulating fluidized bed reactor was used. However, although they were able to eliminate agglomeration, their process was still under shutdown by the buildup of deposits in cyclones and gas lines. Other types of reactors have not been immune to these problems either. Trinh et al. have reported the use of a centrifuge reactor for the pyrolysis of hydrolyzed lignin and also observed the rapid formation of deposits and plugging of the biochar and bio-oil product collection units (Trinh et al. 2013).

7.3.2 Pyrolysis of Lignin with Additives

Because of the difficulties presented by pyrolyzing neat lignin, several researchers have explored additives to aid in the delivery of lignin to the heated reactor in a continuous fashion (Ghysels et al. 2020; Bai and Shuai 2019; Zhou et al. 2015; Case et al. 2015; Mukkamala et al. 2012; Patel et al. 2020). Although some of these substances may act as both additives to aid in the physical ability to pyrolyze the lignin and catalysts to influence the thermal depolymerization chemistry, we will consider those with the former as the primary goal in this section and those with the latter as the main goal in the following section.

Table 7.4 presents selected results of lignin pyrolysis with an additive. The most common additives tested include alkaline earth metal hydroxides and various clays. The clay acts as a diluent, making the lignin become less prone to melting. Furthermore, when pelletized as done by De Wild et al., the particles become more robust and easier to feed (de Wild et al. 2012, 2017). That group reported success in feeding lignins with 30 wt% palygorskite-family clay as extrudates using a cooled feeding screw which slowed the melting of lignin allowing feeding without material accumulating in the feed screw or agglomeration of the bed material (sand). Kraft, soda, and organosolv lignins were successfully pyrolyzed; however, the monomer yields were modest, and the average molecular weights remained high (>1000 g/mol) compared with some other results. These oils were also subjected to hydrotreatment which increased the yield of monomeric phenols. Ghysels et al. also reported the use of a palygorskite clay (attapulgite) for pyrolysis of organosolv lignin (Ghysels et al. 2020). Their loading was higher (50%) than that used by de Wild, and their mixtures were not extruded. They obtained a similar yield of monomeric phenols but reported a molecular weight distribution weighted lower than that reported by de Wild. Furthermore, they also tested $Ca(OH)_2$ and NaCOOH as additives. $Ca(OH)_2$ provided a higher concentration of lignin monomers in the liquid with high selectivity for alkyl phenols. NaCOOH had a lower concentration of monomers overall but with higher

Table 7.4 Selected results for Continuous Pyrolysis of Technical Lignins with Additives (non-catalytic)

	Lignin	Additive	Reactor type[a]	Temp (°C)	Bio-oil yield (wt%)	Bio-oil O (Wt%)	Monomer yield (wt%)	Major monomers[b]	Char yield (wt%)	Notes	References
1	Soda	Clay	Bubbling FB	450	42		5.5	Hs, Cs	43	Clay prevented melting and agglomeration	de Wild et al. (2017)
2	Organosolv	Clay	Bubbling FB	450	41		4.5	Hs, Cs, Gs, Ss	45		de Wild et al. (2017)
3	Kraft	Clay	Bubbling FB	450	34		4.9	Gs, Cs, Hs	47		de Wild et al. (2017)
4	Kraft	Clay	Bubbling FB	550	48		8.1	Gs, Cs, Hs	37		de Wild et al. (2017)
5	Organosolv	Attapulgite (50%)	Mech. Agit. FB	500	~29	35.1	~6.2	Cs, Gs, Hs	~32		Ghysels et al. (2020)
6	Organosolv	$Ca(OH)_2$ (17%)	Mech. Agit. FB	500	~15	26.0	~4.0	Hs, Cs, Gs,		Small coke particles get through to liquid	Ghysels et al. (2020)
7	Organosolv lignin	NaCOOH (53%)	Mech. Agit. FB	500	~17	32.8	~3.2	Hs, Gs, Cs			Ghysels et al. (2020)
8	Organosolv	$Ca(OH)_2$	FB	500	25	44.5	6	H6, G6, H5, H1, G1	~32		Zhou et al. (2015), Bai and Shuai (2019)

[a] FB = fluidized bed and [b] Refer to Fig. 7.1 for definition of key compounds listed in order of yield. Hs, Cs, Gs and Ss are products that are generally phenols, catechols, guaiacols and syringols, respectively, as opposed to a specific compound.

selectivity for methoxy phenols. $Ca(OH)_2$ is converted to CaO during pyrolysis, and it was suggested that the Ca^{2+} can catalyze deoxygenation reactions leading to the formation of polyaromatic, while the role of NaCOOH was to quench radicals leading to retention of methoxy groups. DeSisto has also similarly reported the use of $Ca(OH)_2$, $Ca(COOH)_2$, and $Mg(COOH)_2$ with comparable results (Case et al. 2015; Mukkamala et al. 2012; Patel et al. 2020). Their work indicates that the divalent formate salts of Ca and Mg enhance the yield of liquid products in contrast to the findings of Ghysels et al. for monovalent sodium formate.

7.3.3 Catalytic Pyrolysis of Lignin

Catalytic fast pyrolysis (CFP) processes for biomass are often aimed at reducing the oxygen content of the bio-oils to reduce the level of deoxygenation required during hydrotreatment to produce hydrocarbon fuels (Table 7.5). Alternatively, catalytic pyrolysis can be aimed at generating a certain chemical or chemical class to use as a renewable chemical source. Catalytic cracking of pyrolysis vapors over zeolites, particularly the ZSM-5 zeolites, has been the most practiced form of biomass catalytic pyrolysis. ZSM-5-type zeolites promote deoxygenation and selectively produce aromatic hydrocarbons from a wide variety of substrate molecules at pyrolysis temperatures. Catalytic pyrolysis of lignocellulosic biomass over ZSM-5-type zeolites can produce aromatic hydrocarbons with carbon yields ranging from 20–35% and from isolated cellulose, and the carbon yields of aromatic hydrocarbons can reach up to 40% (Mullen 2019). However, when the same processes are used with isolated lignins as the feedstocks, the yields of aromatic hydrocarbons are much more modest, with most researchers reporting about 10% carbon yield of aromatic hydrocarbons. The major drawback of these catalysts is their susceptibility to deactivation by the formation of coke, a problem which for biomass has been directly attributable to the lignin content of the biomass. Another drawback is the possibility of irreversible poisoning by alkali metals contained in biomass (Mullen and Boateng 2013). Despite these drawbacks, zeolites, particularly ZSM-5 types, are still the most tested catalyst for catalytic pyrolysis of isolated lignins. Catalytic cracking over zeolites is also strongly dependent on the total number of the acid sites per substrate, a function of the Si/Al ratio of the material, the catalyst substrate ratio, and/or the substrate residence time (space velocity). Like non-catalytic pyrolysis, nearly all the work on lignin pyrolysis has been limited to micro or small laboratory-scale batch reactors. In one of the few examples of catalytic pyrolysis using a continuous reactor, Bond et al. reported very low yields of aromatic hydrocarbons (2.2 wt%) and very high coke formation (69%) of CFP of hydrolysis lignin over HZSM-5 (Bond et al. 2014).

ZSM-5 is a microporous catalyst, and its pore structure limits the molecules which can enter the confined space around its active site. This shape selectivity is responsible for the selective formation of aromatic hydrocarbons, but the large size of many lignin-derived pyrolysis vapors is why its effectiveness is limited for the production of aromatics from lignin. ZSM-5 will accommodate substrate molecules with kinetic

Table 7.5 Selected results for catalytic pyrolysis of technical lignins

	Lignin	Catalyst	Reactor type	Atmosphere	C/L (w/w)[a]	Py. Temp (°C)	Monomer type[b]	Monomer yield	References
1	Soda	HZSM-5	Micro	He	15	550	Aromatic hydrocarbons	8.2 C% [d]	Mullen and Boateng (2010)
2	Hydrolysis	None	Drop tube	N_2	N/A	450	Phenolics S1 major	6.8 wt%	Paysepar et al. (2020a, b)
3	Hydrolysis	Zeolite X4	Drop tube	N_2	1	450	Phenolics syringol (S1) Major	7.4 wt%	Paysepar et al. (2020a, b)
4	Hydrolysis	HZSM-5	Fluidized bed	N_2	0.2 h^{-1} WHSV [e]	600	Aromatic hydrocarbons	2.2 C% [d]	Bond et al. (2014)
5	Hydrolysis	HZSM-5	Micro	He	20	550	Aromatic hydrocarbons	9.9 C% [d]	Li et al. (2014)
6	Hydrolysis	Meso-HZSM-5[c]	Micro	He	20	550	Aromatic hydrocarbons	13.2 C% [d]	Li et al. (2014)
7	Kraft	None	Batch—fixed bed	N_2	N/A	500	Phenolics Gs major	11 wt%	Shafaghat et al. (2017)
8	Kraft	Al_2O_3	Batch—Fixed bed	N_2	2	500	Phenolics Hs, Cs, Gs	~11 wt%	Shafaghat et al. (2017)
9	Kraft	HZSM-5	Batch—fixed bed	N_2	1	500	Phenolics Hs	~10 wt%	Shafaghat et al. (2017)
10	Kraft	Hβ Zeolite	Batch – fixed bed	N_2	1	500	Phenolics Hs, Gs	~12 wt%	Shafaghat et al. (2017)

(continued)

Table 7.5 (continued)

	Lignin	Catalyst	Reactor type	Atmosphere	C/L (w/w)[a]	Py. Temp (°C)	Monomer type[b]	Monomer yield	References
11	Kraft	HY Zeolite	Batch—fixed bed	N_2	1	500	Phenolics Hs	~10 wt%	Shafaghat et al. (2017)
12	Kraft	MgO	Batch—fixed bed	N_2	1	500	Phenolics Hs, Gs	~10 wt%	Shafaghat et al. (2017)
13	Milled—corn stover	HZSM-5	Micro	He	20	600	Aromatic hydrocarbons	11.0 C% [d]	Xue et al. (2020)
14	Milled—corn stover	HZSM-5	Micro	H_2	20	600	Aromatic hydrocarbons	20.5 C% [d]	Xue et al. (2020)
15	Milled—corn stover	MoO_3ZSM-5	Micro	He	20	600	Aromatic hydrocarbons	9.7 C% [d]	Xue et al. (2020)
16	Milled—corn stover	MoO_3ZSM-5	Micro	H_2	20	600	Aromatic hydrocarbons	27.1 C% [d]	Xue et al. (2020)
17	Milled—corn stover	NiZSM-5	Micro	He	20	600	Aromatic hydrocarbons	10.0 C% [d]	Xue et al. (2020)
18	Milled—corn stover	NiZSM-5	Micro	H_2	20	600	Aromatic hydrocarbons	19.2 C% [d]	Xue et al. (2020)
19	Organosolv—corn stover	MoO_3	Micro	H_2/He (1/1)	~166	500	≥C4 Hydrocarbons	35.7 C% [d]	Nolte et al. (2017)

[a] C/L catalyst/lignin ratio, [b] Refer to Fig. 7.1 for definition of key compounds listed in order of yield. Hs, Cs, Gs and Ss are products that are generally phenols, catechols, guaiacols and syringols, respectively, as opposed to a specific compound., [c] Meso-HZSM-5 prepared by NaOH (aq) treatment of HZSM-5, [d] Carbon yield (%), and [e] WHSV = weight hourly space velocity

diameters of up to about 5 Å, and some cracking of oligomeric species can happen on the surface of the catalysts but even monomeric guaiacols and syringols have kinetic diameters of about 8 Å and up. Therefore, there has been interest in using mesoporous materials, such as those based on Al-SBA-15 (Wang et al. 2021a, b), or larger pore zeolites such as zeolite Y or β. Others have considered a variety of metal oxides as a catalyst. While these materials have generally been less useful than ZSM-5 for deoxygenation of lignocellulosic pyrolysis vapors, for lignin where the best approach may be depolymerization with limited deoxygenation, these catalysts may have more interest. Shafaghat et al. tested several catalysts at relatively low catalyst-to-lignin ratios (2/1) including alumina, zeolites Y and B, HZSM-5, and MgO (Table 7.5, entries 7-12) (Shafaghat et al. 2017). While most of the catalysts decreased the overall liquid yield, they generally also exhibited a decrease in oligomers which is attributed to cracking reactions. A corresponding increase in monomers was also observed, particularly the presence of demethoxylated alkyl phenols. Some completely deoxygenated compounds were also present, but at low loadings, the process was more selective for phenolics.

There have also been efforts to introduce mesoporosity into microporous zeolites to overcome the diffusion problem. This type of hybrid pore structure is often referred to as "hierarchical" or the materials called "hierarchical zeolites" (Mardiana et al. 2022). This type of pore structure can be achieved by partially desilylating ZSM-5 by treatment with NaOH or can be introduced during the zeolite synthesis via templating. Much of the literature on testing these types of materials has been for catalytic pyrolysis of lignocellulosic biomass, but Li et al. have reported testing mesopore containing HZSM-5 for catalytic pyrolysis of isolated lignin and have shown 33% increase in the yield of aromatic hydrocarbons compared with the parent HZSM-5 (Li et al. 2014).

Catalysts with metal functionality have been used in reactive atmospheres (low pressure H_2) to enhance the production of hydrocarbons from biomass. These reactive catalytic processes have been mostly applied to whole biomass, but there are a few examples of small-scale studies with isolated lignins. MoO_3-based catalysts are most commonly used for this. In this case, the role of the hydrogen is thought to be to create an oxygen vacancy on the metal surface by reacting with metal-bound oxygen to generate water. An oxygen atom is then abstracted from the biomass-derived substrate to fill the vacancy, deoxygenating the biomass carbon structure. Nolte et al. have reported using MoO_3 for reactive catalytic pyrolysis of organosolv lignin using 0.5 atm partial pressure of H_2 and achieved 35.7 carbon yield (C%) of hydrocarbons higher than C4 (Nolte et al. 2017). Xue et al. have used bifunctional catalysts, MoO_3 or Ni supported on ZSM-5 for the conversion of milled corn stover lignin to aromatic hydrocarbons (Xue et al. 2020). For both supported catalysts and ZSM-5 alone, the yield of aromatics was enhanced when performed under an H_2 atmosphere, with a maximum yield of 27.1 C% of aromatic hydrocarbons using the MoO_3ZSM-5 catalyst. Although the enhanced yields under H_2 suggest that it helps mitigate catalyst deactivation via coke deposition, more research is needed to access catalyst lifetimes for these types of processes.

7.4 Solvent Phase Lignin Depolymerization

Solvent-assisted thermal depolymerization pathways have long been considered as an alternative method to pyrolysis to produce bio-oil. Hydrothermal (water is the solvent) liquefaction of biomass has received significant attention over the years as an alternative to pyrolysis (Kumar et al. 2018). While its ability to accommodate wet feedstocks can be an advantage over pyrolysis, particularly for resources like sludges, animal wastes, and algae, its drawbacks include the need for pressurized equipment. While hydrothermal liquefaction has been tested for isolated lignins with some success at supercritical conditions, more often, organic solvents are utilized for lignin depolymerization. The solvent has many roles in lignin depolymerization chemistry. One role is as a diluent to decrease the concentration of intermediates to limit their repolymerization rate. Solvents can also act as hydrogen donors in reductive depolymerization routes, where the solvent can also quench radicals and limit repolymerization reactions. Solvents in both sub- and supercritical conditions have been considered, as have ionic liquids. Many types of catalysts have been considered for solvent liquefaction including both homogeneous and heterogeneous catalysts. Simple homogeneous catalysts such as bases (NaOH), acids (H_2SO_4), and metal salts are often used to increase the solubility of lignin and induce the breaking of C-O bonds at lower temperatures than required for non-catalytic liquefaction. Heterogeneous catalysts (e.g., zeolite and transition metals) are often used to induce other types of reactions such as decarbonylation, demethoxylation, and hydrogenation (in the presence of H_2 or transferred from an organic H-donor) which can increase the yield of phenolic monomers. The literature on solvent liquefaction is extensive, with a wide variety of combinations of solvents, additives, catalysts, temperatures, and atmospheres being explored. These can generally be classified as acid/base reactions, reductive catalytic depolymerization, and oxidative catalytic depolymerization.

7.4.1 Non-catalytic Solvent Liquefaction

Although usually performed with a catalyst, non-catalytic solvent-assisted depolymerization is possible. These reactions are usually performed on the harsher end of the reaction condition spectrum, to make up for the lack of catalyst. For technical lignins, yields of monomers in the absence of catalysts are generally limited to less than 10 wt% (Patil et al. 2020); however, non-catalytic conditions can still be effective for depolymerization leading to lower molecular weight oligomers. Experiments without catalysts can also give insights into the role of solvent in depolymerization chemistry. Table 7.6 presents selected results for the non-catalytic lignin solvent liquefaction.

Table 7.6 Selected results from non-catalytic solvent phase depolymerization of technical lignins

	Lignin	Solvent	Temp (°C)	Lignin/Solvent (m/v)[a]	Reaction time (min)	Bio-oil yield (wt%)	Bio-oil MW (Da)	Monomer yield (wt%)	Major monomers[b]	Char yield (wt%)	References
1	Kraft	H_2O	130	1/6	60	10.6	1105		G1	80.1	Zhou (2014)
2	Kraft	H_2O	230	1/6	60	10.0	717		G1	77.8	Zhou (2014)
3	Kraft	H_2O	280	1/6	60	9.2	620		G1, S11 S10, G11, G10	74.2	Zhou (2014)
4	Organosolv	H_2O	300	1/20	40	~7		~2.5	S1, G1		Toledano et al. (2012)
5	Alkali	H_2O	265	1/60	360	22.3			G1 (12 wt%), C1, G4	46	Onwudili and Williams (2014)
6	Alkali	H_2O-FA[c] (60/1)	265	1/61	360	33.1			C1 (18 wt%), G1, G7	1.1	Onwudili and Williams (2014)
7	Kraft	H_2O-FA	390	1/7.42 (w/w)	240	75.5			H2-4, H14, H5, H7	4.3	Løhre et al. (2016)
8	Alkali	EtOH	350	1/20 (w/w)	30			9.6	G1, G5, G9	40.5	Riaz et al. (2018)
9	Alkali	EtOH-FA[c] (5/1)	350	1/32 (w/w)	30			19.4	G7, G5, G1	2.5	Riaz et al. (2018)
10	Alkali	EtOH-FA[c] (5/1)	350	1/32 (w/w)	60			36.7	G7 (4.9 wt%, G7 (4.5 wt%), G5 (4.4 wt%), G9, C5	1.3	Riaz et al. (2018)
11	Alkali	EtOH-FA[c] (5/1)	350	1/32 (w/w)	300			25.6	G5 (4.7 wt%), C5 (4.0 wt%)	3.3	Riaz et al. (2018)

(continued)

Table 7.6 (continued)

	Lignin	Solvent	Temp (°C)	Lignin/Solvent (m/v)[a]	Reaction time (min)	Bio-oil yield (wt%)	Bio-oil MW (Da)	Monomer yield (wt%)	Major monomers[b]	Char yield (wt%)	References
12	Alkali	EtOH-FA[c] (5/1)	350	1/22 (w/w)	60			30.0	G5 (4.8 wt%), G7, G9, C5 (2.7 wt%)	4.7	Riaz et al. (2018)
13	Soda	MeOH	300	1/20	240	24		4		46	Huang et al. (2014)
14	Soda	EtOH	300	1/20	240	23		5	G5, S1, A5, S5, H5	40	Huang et al. (2014)
15	Organosolv	Tetralin	400	1/10	15		258	~11	H5, H1, H6		Kim et al. (2014)
16	Organosolv	*i*PrOH	400	1/10	15		315	~10	H5, H1		Kim et al. (2014)
17	Organosolv	Naphthalene	400	1/10	15		476	~7.5	H5, H1, G1, G5, H6		Kim et al. (2014)
18	Organosolv	Acetone	300	1/20	40	37.6	3451[d]		S1, G1	~3	Erdocia et al. (2015)
19	Organosolv	MeOH	300	1/20	40	~33	3224[d]		S1, G1	~5	Erdocia et al. (2015)
20	Organosolv	EtOH	300	1/20	40	~30	2086[d]		S1, G1	~3.5	Erdocia et al. (2015)

[a] g/mL unless otherwise noted, [b] Refer to Fig. 7.1 for definition of key compounds listed in order of yield, value in parenthesis is notable high yield of individual compound, [c] FA = formic acid, and [d] Lignin MW = 15,088 Da

Simple hydrothermal treatment of technical lignins has not proved an effective method for the production of monomers or any liquid products (Zhou 2014; Onwudili and Williams 2014; Toledano et al. 2012). Zhou et al. reported hydrothermal treatment of Kraft lignin with a maximum bio-oil yield of 10.6% at 130 °C, with slightly lower yields at higher temperatures (Zhou 2014). The major monomer produced was guaiacol. Similarly, Toledano reported a 7% yield of bio-oil and a 2.5 wt% yield of monomers from hydrothermal treatment of soda lignin (Toledano et al. 2012).

Organic solvents have proven slightly more effective, owing at least in part to better initial solubility of the lignins. Alcohols are the most common solvent used, particularly ethanol. The use of hydrogen donor solvents such as tetralin, isopropanol, and formic acid has proven effective. Kim et al. compared the conversion of lignin in tetralin, isopropanol, and naphthalene and achieved an 11 wt% yield of monomers in tetralin at 400 °C (Kim et al. 2014). This was higher than found in naphthalene, the product of dehydrogenation of tetralin. The average molecular weight of the bio-oil was also lower when it was produced in tetralin. The authors suggested that using the hydrogen donating solvent suppressed repolymerization reactions.

The addition of formic acid has proved effective for increasing monomer production. Although formic acid is a hydrogen donor and could be considered an acidic catalyst, in the absence of another catalyst, we are classifying its use as a non-catalytic process and discussing it in this section. The addition of formic acid to water greatly increased the yield of bio-oil compared with water alone. In one report, 75.5% bio-oil yield was reported from Kraft lignin (Onwudili and Williams 2014). Riaz et al. thoroughly investigated the role formic acid plays in solvent liquefaction in ethanol (Riaz et al. 2018). Compared with neat ethanol (9.6 wt%), they found a large increase in monomer production when a 5/1 ethanol to the formic acid solvent system (36.7 wt%) was at 350 °C. They also found a significant decrease in the consumption of ethanol via its thermal conversion to other liquid products.

7.4.2 Base-Catalyzed Depolymerization

Solvent depolymerization of lignin can be catalyzed by both acids and bases (Fernández-Rodríguez et al. 2017). These reactions are usually performed in the absence of a dedicated oxidizing or reducing agent. Oxidative and reductive depolymerization methods will be covered in the following sections. Selected base-catalyzed lignin depolymerization results are shown in Table 7.7. Both homogeneous and heterogeneous bases have been tested as simple catalysts for lignin depolymerization. One advantage of using a soluble strong base (e.g., NaOH) as a liquefaction catalyst is the improved solubility in water and water/alcohol mixtures allowing for the use of the most environmentally sustainable medium. These processes are similar to some lignin isolation techniques as the alkaline aqueous solutions aid in breaking lignin-carbohydrate bonds in addition to the destruction of lignin-lignin ether bonds. Furthermore, these processes are more effective on native lignins where there are many remaining β-O4 linkages remaining than on technical lignins (Katahira et al. 2016).

Table 7.7 Selected results of base-catalyzed technical lignin depolymerization

	Lignin (MW, g/mol)[a]	Solvent	Temp (°C)	Catalyst (Conc./Loading)	Lignin/solvent[b]	Reaction time	Bio-oil yield (wt%)	Bio-oil MW (g/mol)	Monomer yield (wt%)	Major monomers[c]	Solid yield (wt%)	References
1	Soda	H_2O	300	NaOH (4wt%)	1/20	80	~15		6.32	Cs	~31	Fernández-Rodríguez et al. (2017)
2	Organosolv	H_2O	300	NaOH (4wt%)	1/20	80	~22		8.27	Cs	~41	Fernández-Rodríguez et al. (2017)
3	Enzymatic hydrolysis[e] (~9,000)	H_2O	300	NaOH (4 wt%)	1/10	40	70	~650		C1, C5	25	Katahira et al. (2016)
4	Enzymatic hydrolysis (~9,000)	H_2O	270	NaOH (4wt%)	1/10	40	60	~800		C1, S11. S1, C5	31	Katahira et al. (2016)
5	Enzymatic hydrolysis (~9,000)	H_2O	330	NaOH (4wt%)	1/10	40	69	~700		C5, C1	21	Katahira et al. (2016)
6	Kraft	H_2O	300	NaOH (4 wt%)	1/10	40	23				61	Katahira et al. (2016)
7	Organosolv	H_2O	300	NaOH (4 wt%)	1/20	40	~20		~3.7	C1, C4	43	Toledano et al. (2012)
8	Organosolv	H_2O	300	KOH	1/20	40	~12.5		~2.7	C1, C4	42	Toledano et al. (2012)
9	Organosolv	H_2O	300	LiOH	1/20	40	~12.7		~2.2	C1, C4	42	Toledano et al. (2012)
10	Organosolv	H_2O	300	K_2CO_3	1/20	40	~17		~3.8	C1, C4	35	Toledano et al. (2012)
11	Organosolv	H_2O	300	$Ca(OH)_2$	1/20	40	~11		~0.8	G1	40	Toledano et al. (2012)
12	Kraft	H_2O	270	NaOH	1/10	LHSV[d] = 1.4	78.8		4.5	G1		Beauchet et al. (2012)

(continued)

Table 7.7 (continued)

	Lignin (MW, g/mol)[a]	Solvent	Temp (°C)	Catalyst (Conc./Loading)	Lignin/ solvent[b]	Reaction time	Bio-oil yield (wt%)	Bio-oil MW (g/mol)	Monomer yield (wt%)	Major monomers[c]	Solid yield (wt%)	References
13	Kraft	H_2O	315	NaOH	1/10	LHSV [d]= 1.4	68.4		8.4	C1, C4, C5		Beauchet et al. (2012)
14	Organosolv (1,636)	H_2O	190	NaOH	1/10	60		1,578	5.2	H1, S10		Cornejo et al. (2020)
15	Organosolv (1,636)	$EtOH/H_2O$	250	NaOH	1/10	60			7.5	S1, Hs		Cornejo et al. (2020)
16	Organosolv (1,636)	EtOH	190	NaOH	1/10	60		948	2.1	H1		Cornejo et al. (2020)
17	Organosolv (1,636)	EtOH	270	NaOH	1/10	60		1,515	1.8	H1		Cornejo et al. (2020)
18	Ionic liquid (1,360)	MeOH	250	MgO (10 mol%)	1/80	30			9.3	G1	37.5	Long et al. (2014)
19	Ionic liquid (1,360)	EtOH	250	MgO (10 mol%)	1/80	30			8.5	G1	37.2	Long et al. (2014)
20	Ionic liquid (1,360)	$EtOH/H_2O$	250	MgO (10 mol%)	1/80	30			11.3	G1	32.2	Long et al. (2014)
21	Ionic liquid (1,360)	THF	250	MgO (10 mol%)	1/80	30		543	12.8	G1	38.5	Long et al. (2014)

(continued)

Table 7.7 (continued)

	Lignin (MW, g/mol)[a]	Solvent	Temp (°C)	Catalyst (Conc./Loading)	Lignin/ solvent[b]	Reaction time	Bio-oil yield (wt%)	Bio-oil MW (g/mol)	Monomer yield (wt%)	Major monomers[c]	Solid yield (wt%)	References
22	Ionic liquid (1,360)	H_2O	250	MgO (10 mol%)	1/80	30		884	2.2	G1		Long et al. (2014)
23	Kraft	$EtOH/H_2O$ 1/2	250	NaX (100 mass%)	1/60	60	51		18	G11, G10, G4		Chaudhary and Dhepe (2017)
24	Dilute acid (Corn stover)	H_2O	275	HT [f]	1/150	60			~9	H1, G1, S1		Kruger et al. (2016)
25	Dilute acid (Corn stover)	3-methyl-3-pentanol	275	HT [f]	1/150	60			~7	H6 (~5 wt%)		Kruger et al. (2016)
26	Organosolv	Dimethyl carbonate	180	$CsCO_3$(10 wt%)	1/50	8				T13 (4.0 wt%), D13 (0.9 wt%)		Dabral et al. (2018)
27	Kraft	DCM	180	$CsCO_3$(10 wt%)	1/50	8				D13 (1.5 wt%)		Dabral et al. (2018)

[a] M_w from GPC/SEC analysis, [b] g/mL unless otherwise noted, [c] Refer to Fig. 7.1 for definitions of key compounds listed in order of yield, values in parenthesis are notable high yield of individual compounds, [d] LHSV = liquid hourly space velocity, [e] enzymatic hydrolysis residual was 52.8% lignin, and [f] HT = hydrotalcite

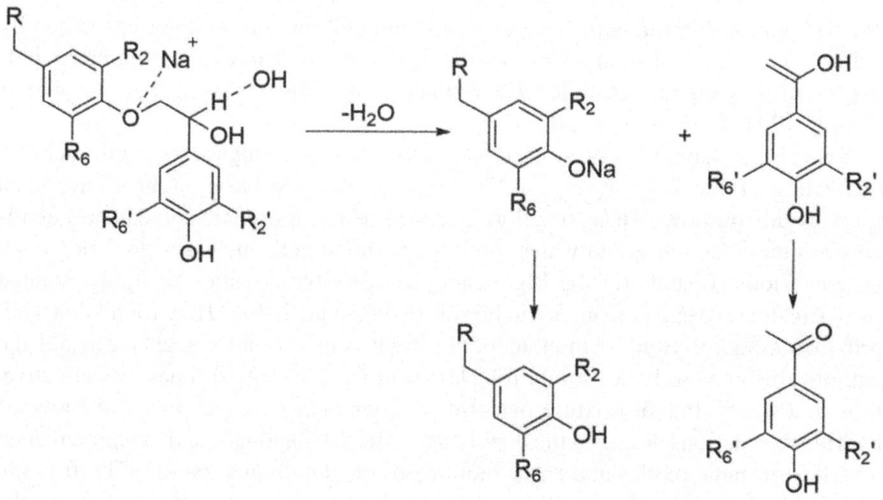

Fig. 7.8 Mechanisms for NaOH-catalyzed cleavage of β-O4 linkages in lignin

This is demonstrated in the work by Katahira, where they studied NaOH (aq) mediated depolymerization of several lignins (selected results in Table 7.7, entries 3-6) where they found fewer structural changes correlated with higher oil yield (Katahira et al. 2016). The mechanism of the breaking of the β-O4 linkage is shown in Fig. 7.8. Both the Na^+ and OH^- ions participate in polarizing the C-O bond and reducing the energy for its ionic cleavage with the OH^- accepting a proton from the α-carbon. However, the resulting phenolate species are still quite reactive, making these processes susceptible to repolymerization reactions much like other lignin depolymerization methods.

The selectivity of the monomer production in NaOH-catalyzed depolymerization is very sensitive to temperature and solvent selection. High temperatures (>300 °C) favor the production of catechol while when lower temperatures are employed the selectivity shifts to methoxyphenols (Schutyser et al. 2018; Fernández-Rodríguez et al. 2017). For example, Beauchet et al. reported an increasing yield of monomers but a decreasing yield of oil when the temperature of NaOH-catalyzed depolymerization of Kraft lignin was increased from 270 to 315 °C (Beauchet et al. 2012). At 270 °C, the most concentrated monomers included guaiacol and vanillin while at 315 °C catechol was the dominant species present. The selectivity towards catechol was also observed for other bases. According to Toledano et al., the presence of several bases (Table 7.7, entries 7-11) increased the oil yield and shifted the monomer selectivity from guaiacols and syringols to catechol compared with non-catalytic depolymerization in water (Table 7.6, entry 4) at 300 °C (Toledano et al. 2012). The monomer yield was increased over the non-catalytic case for the most effective bases tested (NaOH, KOH) but was only about 8.4 wt%. However, coke yield was higher without a catalyst, and most of the bases showed a significant shift towards lower molecular products (monomers, dimers, and trimers) compared with no catalyst. $Ca(OH)_2$ was not an effective catalyst for yield or oil quality. Performing

NaOH-catalyzed lignin depolymerization in ethanol instead of water has generally led to lower yields of monomers, but there is some evidence that monomer yields may be slightly higher in a mixed water/ethanol solvent mixture than in neat solvents (Thring 1994; Cornejo et al. 2020).

Solid bases have also been used as heterogeneous catalysts for lignin depolymerization (Table 7.7, entries, 23-27); metal oxides and basic zeolites have been used for this purpose. These reactions perform better in organic solvents as heterogeneous bases do not aid in water solubility as homogeneous bases do. Long et al. tested various solvents for the MgO-catalyzed depolymerization of lignin isolated from sugarcane bagasse using ionic liquids (Long et al. 2014). They found that THF provided a higher yield of monomers (12.8 wt%) than other solvents, a result the authors attributed to better solubility. Neat water was about 10 times less effective, though a water/ethanol mixture performed better than neat alcohols, indicative of hydrolysis reactions aiding in the depolymerization. Chaudhary and Dhepe screened a variety of metal oxides and basic zeolites for the depolymerization of kraft lignin in EtOH/water. Generally, they found the zeolites to perform better and found the highest yield of bio-oil (51%) and phenolic monomers (18 wt%) using Na-zeolite X (Chaudhary and Dhepe 2017). The major monomeric products were guaiacols in this case. In some cases, the choice of solvent can affect not only the yield but also the selectivity for base-catalyzed lignin depolymerization. Over nitrate-loaded hydrotalcites, Kruger et al. found high selectivity for 4-vinylphenol in 3-methyl-3-petananol; in contrast, phenol and guaiacol were the major monomers formed in water (Kruger et al. 2016). With these catalysts, yields of monomers were maximized at about 9 wt% from acid hydrolysis lignin from corn stover.

As seen from the above discussion, similar to pyrolysis or non-catalytic solvent liquefaction that although the products of base-catalyzed depolymerization exhibit drastically reduced molecular weight ranges, the yield of monomers is still generally limited by repolymerization reactions. Some attempts to reduce these reactions by providing an alternate quench mechanism for reactive intermediates using a "capping" reagent have been made. Barrett et al. (2016) reported using dimethyl carbonate (DMC) in combination with a copper-doped porous metal catalyst for depolymerization of organosolv lignin. The average molecular weights of the bio-oils produced with and without DMC were similar; however, O-methylated phenolics were the major monomers detected by GC. Similarly, Dabral et al. (2018) have reported using DMC to produce depolymerized lignins containing methoxy capped monomers; in this case, using a $CsCO_3$ as the base catalyst, respectively. They found up to 4.0 wt% yield of **T13** from organosolv lignin. Utilization of boric acid was also reported to prevent repolymerization reactions by the formation of borate esters at the phenolic hydroxy group. The use of boric acid in combination with NaOH leads to an increase in bio-oil yield from about 22 wt% to 59 wt%, and the bio-oil was mainly comprising monomers, dimers, and trimers (Roberts et al. 2011). Recently, Gan and Pan have reported phenolation in combination with NaOH-catalyzed depolymerization of kraft lignin, observing phenolation in both the oligomeric (51%) and small molecule (13%) portions of the products (Gan and Pan 2019).

7.4.3 Acid-Catalyzed Depolymerization

Acid-catalyzed solvent depolymerization, much like base-catalyzed, is very effective for breaking down the β-O4 linkages. However, unlike the use of aqueous bases, the addition of acid does not significantly improve the solubility of lignins in water. Like non-catalytic and base-catalyzed methods, the product selectivity is very sensitive to the temperature and solvents used, as well as the properties of the acids. Many different types of acids have been tested as catalysts, including both solid (including zeolites) (Wanmolee et al. 2018; Deepa and Dhepe 2014a, b) and soluble Bronsted acids as well as homogeneous Lewis acids. As shown in Table 7.8, there is a large spread in the reported yields of monomers from acid-catalyzed depolymerization methods. Some of this may be attributable to the observation that acid-catalyzed depolymerization efficiency seems to be very correlated to the amount of remaining β-aryl-ether linkages in the substrate lignin (Schutyser et al. 2018; Deuss et al. 2017). Acid is very effective at promoting the cleavage of these bonds. Additionally, acid catalysis can also promote significant reactions of solvent, which in some cases can be exploited as capping reagents to prevent repolymerization reactions but can also mean solvent cannot be completely recycled adding expense to the process.

Deepa and Dhepe have screened a variety of solid acid catalysts and also compared them with the use of strong mineral acids (Deepa and Dhepe 2014a, b). They have found very high yields of THF solubles (highly concentrated with phenolic monomers) with zeolites and other alumina-silica materials, maximizing yields with HUSY and HZSM-5 zeolites, and Al_2O_3-SiO_2 at near 60 wt% for conversion of commercially available dealkaline lignin. It is possible that some ethanol-derived compounds are included in this, but they also found a high selectivity to vanillin **(G10)** (~9 wt% yield). With HCl and H_2SO_4, the yields of the same THF soluble fractions were 29 and 39 wt%, respectively, but the oil composition had a higher proportion of dimers, trimers, and oligomers compared with those from the solid acids (Deepa and Dhepe 2014a). The yield was also reduced when organosolv lignin was used in place of the dealkaline lignin, to about 32 wt%. However, when Wang et al. used HZSM-5 for the depolymerization of organosolv lignin under similar conditions, they only obtained a monomer yield of 12.5 wt% (Wang et al. 2018a, b, c). Guvenatam et al. have screened several Lewis acids based on the acetate, chloride, and triflate salts of Fe, Cu, Co, and Al for the depolymerization of soda lignin in ethanol/water at 400 °C (Güvenatam et al. 2016a, b). They found the highest phenolic monomer yield at 12.6 wt% using copper acetate, but produced bio-oil with overall lower molecular weights from the metal triflates. However, in all cases, significant conversion of the ethanol solvent was observed and interfered with determining the actual depolymerization rate of the lignin by GPC. Additionally, ethanol acted as a reactant in alkylation reactions that capped reactive intermediates, preventing some char formation.

Duess et al. have also had success utilizing a capping agent to suppress repolymerization reactions during acid-catalyzed depolymerization (Deuss et al. 2017, 2015, 2016). In their early work, they reported that cyclic acetals (**H16**, **G16, S16, H17,**

Table 7.8 Selected results of acid-catalyzed solvent phase lignin depolymerization

	Lignin (MW, g/mol)[a]	Solvent	Temp (°C)	Catalyst (Loading)	Lignin/Solvent[b]	Reaction time (min)	Bio-oil yield (wt%)	Bio-oil MW (g/mol)[a]	Monomer yield (wt%)	Major monomers[c]	References
1	Organosolv (4280)	EtOH/H_2O	350	HZSM-5	1/100	240		1,620 (residue)	12.5	H1 (3.5 wt%), G1, G5, G4, S1	Wang et al. (2018a, b, c)
2	Dealkaline	MeOH/H_2O	250	HUSY	1/60	30			~60	G10, G12, G13	Deepa and Dhepe (2014b)
3	Dealkaline	MeOH/H_2O		HZSM-5	1/60	30			~55	G10, G12, G13	Deepa and Dhepe (2014b)
4	Dealkaline	MeOH/H_2O	250	SiO2/Al_2O_3	1/60	30			~59	G10 (18 wt%), G12, G13	Deepa and Dhepe (2014b)
5	Dealkaline	MeOH/H_2O	250	HCl	1/60	30	29				Deepa and Dhepe (2014b)
6	Dealkaline	MeOH/H_2O	250	H2S04	1/60	30	39				Deepa and Dhepe (2014b)
7	Organosolv	MeOH/H_2O	250	SiO_2/Al_2O_3	1/60	30			~32		Deepa and Dhepe (2014a)
8	Organosolv	MeOH	300	H_2SO_4 (1% w/w)	1/57	60	~55		1.8	H5, G5, G1, H1. S1	Wanmolee et al. (2018)
9	Organosolv	MeOH	300	HUSY (1% w/w)	1/57	60			2.5	H5, G5, G1, S1	Wanmolee et al. (2018)

(continued)

Table 7.8 (continued)

	Lignin (MW, g/mol)[a]	Solvent	Temp (°C)	Catalyst (Loading)	Lignin/Solvent[b]	Reaction time (min)	Bio-oil yield (wt%)	Bio-oil MW (g/mol)[a]	Monomer yield (wt%)	Major monomers[c]	References
10	Organosolv	MeOH	300	HUSY (50) (1% w/w)	1/57	60			1.6	H5, G5, G1, H1, S1	Wanmolee et al. (2018)
11	Organosolv	MeOH	300	HZSM-5 (1% w/w)	1/57	60			1.7	H5, G5, G1, H1, S1	Wanmolee et al. (2018)
12	Organosolv	MeOH	300	SiO_2(1% w/w)	1/57	60			1.4	H5, G5, G1, H1, S1	Wanmolee et al. (2018)
13	Organosolv	MeOH	300	HUSY (5) (10% w/w)	1/57	60	~52		3.6	H5, G5, G1, H1,	Wanmolee et al. (2018)
14	Organosolv	MIBK	300	HUSY (5) (10% w/w)	1/57	60	~44		4.9	H5, G5, G1, H1	Wanmolee et al. (2018)
15	Organosolv	MIBK	350	HUSY (5) (10% w/w)	1/57	60	~60		10.6	H5, G5, G1, H1	Wanmolee et al. (2018)
16	Soda	EtOH	400	$Fe(OAc)_2$	1/43	240	198[d]	404	7.3	Hs, Gs	Güvenatam et al. (2016b)
17	Soda	EtOH	400	$Cu(OAc)_2$	1/43	240	308[d]	417	12.6	Hs, Gs	Güvenatam et al. (2016b)
18	Soda	EtOH	400	$FeCl_2$	1/43	240	185[d]	345	9.3	Hs, Gs	Güvenatam et al. (2016b)
19	Soda	EtOH	400	$CuCl_2$	1/43	240	201[d]	300	10.0	Hs, Gs	Güvenatam et al. (2016b)
20	Soda	EtOH	400	$Cu(OTf)_2$	1/43	240	245[d]	268	5.3	Hs, Gs	(Güvenatam et al. 2016b)
21	Soda	EtOH	400	$Ni(OTf)_2$	1/43	240	362[d]	285	10.6	Hs, Gs	Güvenatam et al. (2016b)

(continued)

Table 7.8 (continued)

	Lignin (MW, g/mol)[a]	Solvent	Temp (°C)	Catalyst (Loading)	Lignin/Solvent[b]	Reaction time (min)	Bio-oil yield (wt%)	Bio-oil MW (g/mol)[a]	Monomer yield (wt%)	Major monomers[c]	References
22	Soda	EtOH	400	$Al(OTf)_3$	1/43	240	435[d]	257	8.7	Hs, Gs	Güvenatam et al. (2016b)
23	Kraft	EtOH	300	$ZnCl_2$	1/40	120	75.8		38.1	G1 (11 wt%), C1 (5.4 wt%)	Zhang et al. (2014)
24	Organosolv	1,4-dioxane + ethylene glycol	140	HOTf	1/20		91		6.4	S16 (2.6 wt%), G16 (2.3 wt%), H16 (1.7 wt%), G17, S17	Deuss et al. (2015)
25	Organosolv—Walnut (2,333)	1,4-dioxane + ethylene glycol	140	$Fe(OTf)_3$ 10 wt%	1/20	15			38.2	G16, S16, H16	Deuss et al. (2017)
26	Organosolv—Pine (3,089)	1,4-dioxane + ethylene glycol	140	$Fe(OTf)_3$ 10 wt%	1/20	15			17.3	G16 (16.5 wt%)	Deuss et al. (2017)
27	Kraft (986-1,400)	1,4-dioxane + ethylene glycol	140	$Fe(OTf)_3$ 10 wt%	1/20	15			0.5-2.6	G16, H16	Deuss et al. (2017)
28	Soda (777)	1,4-dioxane + ethylene glycol	140	$Fe(OTf)_3$ 10 wt%	1/20	15			2.4	H16, G16	Deuss et al. (2017)

[a] M_w from GPC/SEC analysis, [b] g/mL unless otherwise noted, [c] Refer to Fig. 7.1 for definitions of key compounds listed in order of yield, value in parenthesis is notable high yield of individual compound, and [d] Also contains solvent-derived liquid products

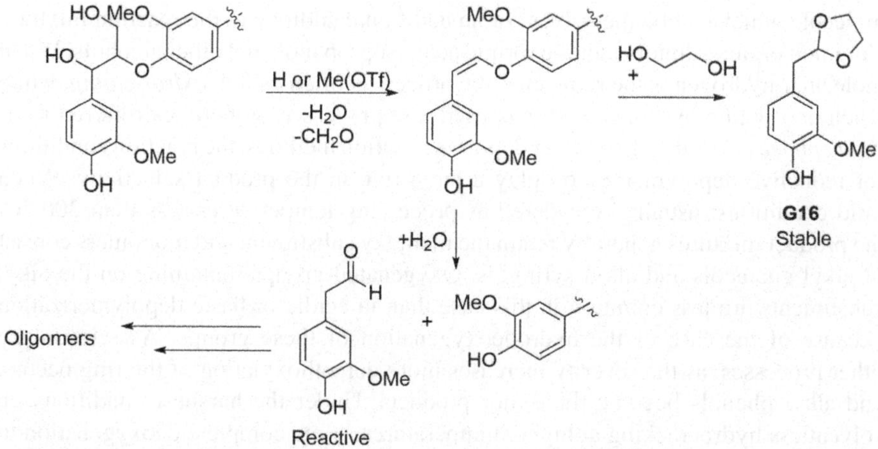

Fig. 7.9 Production of cyclic acetals from acid-catalyzed depolymerization of lignin with intermediate trapping by ethylene glycol. Adapted with permission from The Royal Society of Chemistry (Deuss et al. 2016)

G17, and **S17**) can be produced in high yield from β-O4 model dimers using triflic acid catalysis in the presence of various diols (Deuss et al. 2015). The pathway for the formation of the acetals is trapping of an intermediate aldehyde from breaking of the β-aryl-ether bond, forming the acetal on one end of the bond, and releasing the phenolic monomer from the other (Fig. 7.9). When they applied this methodology to organosolv lignin depolymerization in 1,4-dioxane at 140 °C, they found the yield of monomers increased from 2.0 wt% without ethylene glycol to 6.4 wt% in its presence. Furthermore, the yield of soluble bio-oil-containing monomers and other lower molecular weight species increased from 10 to 27% while the yield of insoluble was reduced from 44 to only 9% (Deuss et al. 2015). They went on to optimize the system by using $Fe(OTf)_3$ as the optimum catalyst, obtaining 19 wt% monomer yields from walnut shell organosolv lignin (Deuss et al. 2016). With this catalyst, they were able to obtain up to 35.5 wt% of monomeric acetals from a β-O4 rich organosolv lignin isolated from walnut shells (Deuss et al. 2017). In another case, they were able to obtain 16.5 wt% yield of a single acetal compound (**G16**) from an organosolv lignin from pine wood. However, in technical lignins with lower levels of β-O4 linkages, such as commercially available kraft or soda lignins, the yields were much lower, and <10 wt% of the acetals were formed.

7.4.4 Reductive Depolymerization

Solution phase lignin depolymerization can also be performed in a reductive manner, which includes a metal catalyst and a hydrogen source. The hydrogen source can be H_2, or it can be obtained via a transfer hydrogen route from an organic donor molecule which can be the solvent or an additional additive to the reaction mixture. Common organic donors include formic acid, isopropanol, and other alcohols. When molecular hydrogen is the reductant, the process is often called *hydrotreating*, while when an organic hydrogen donor is used, the process is sometimes referred to as *liquid phase reforming*. Like other depolymerization methods, the reaction conditions for reductive depolymerization play a large role in the product selectivity. When mild conditions, usually considered as processing temperatures less than 300 °C, the product mixtures generally retain the methoxy substituent and monomers consist of alkyl guaiacols and alkyl syringols. Oxygenated groups remaining on the alkyl substituents are less common in this case than in acidic or basic depolymerization because of the ease of the hydrodeoxygenation of these groups. As seen in the other processes, as the severity increases more demethoxylation of the ring occurs, and alkyl phenols become the major products. Under the harshest conditions, in solventless hydrocracking at higher temperatures, some complete deoxygenation to aromatic hydrocarbons and polyaromatics can occur. These processes are more akin to hydrodeoxygenation of biomass pyrolysis oils but act directly on the lignin rather than first producing the bio-oil intermediate. Under both mild and harsh conditions, repolymerization reactions hinder the overall yield of monomers.

7.4.4.1 Hydrotreating

Mild hydrotreating of isolated lignins is often done using traditional hydrogenation catalysts such as Ru, Pd, Pt, or Ni usually supported on carbon, alumina, or silica-based materials (Shuai et al. 2016; Tymchyshyn et al. 2020; Shu et al. 2015, 2018; Van den Bosch et al. 2015a, b; Torr et al. 2011; Bouxin et al. 2015; Lancefield et al. 2016; McVeigh et al. 2016; de Albuquerque 2020; Zhao et al. 2019; Kim et al. 2015). Solvents are often alcohols or water/alcohol mixtures because these provide good solubility for the lignins. Table 7.9 provides selected results of hydrotreating various isolated lignins. Kim et al. screened carbon-supported Pt, Pd, Ru, and Ni for the depolymerization of soda lignin in alcohols at 350 °C under 3 MPa H_2 pressure (Kim et al. 2015). The yield of phenolic monomers was only slightly increased by the presence of the catalysts compared with the non-catalytic control; 4-ethylphenol (**H5**) and 4-ethylguaiacol (**G5**) were the most concentrated monomers in the bio-oils. However, the catalysts did increase the overall yield of bio-oil and decreased the production of biochar. Those bio-oils produced in ethanol had lower oxygen content compared to those produced in methanol or isopropanol. The average molecular weight of the bio-oils was reported to range from about 650–900 g/mol, compared with about 3,600 g/mol for the raw lignin. Zhao et al. have also reported high selectivity towards ethyl phenol, in this case for the hydrotreating of organosolv lignin over Ni catalysts (Zhao et al. 2019). They screened supports for the Ni catalyst including MgO, Al_2O_3, and SiO_2, finding a 20% Ni loading on MgO most effective for the production of monomeric phenols and producing the bio-oil with the lowest overall weight average molecular weight (M_w). Monomeric phenols were produced in 13.5 wt% yields, with a yield of ethyl phenol (**H5**) of 5.1 wt%. Similarly, Ma et al.

Table 7.9 Selected results of reductive solvent phase depolymerization of lignin with H_2

	Lignin (MW, g/mol)[a]	Catalyst	Temp (°C)	Solvent	H_2 pressure (bar)	Reaction time (min)	Bio-oil yield (wt%)/O content (wt%)	Bio-oil MW (g/mol)[a]	Monomer yield (wt%)	Major monomers[b]	References
1	Soda (wheat straw) (3,698)	Pt/C	350	EtOH	30	40	77.4/18.5	901	10.9	G5, S1, H5, S11, G4, S4	Kim et al. (2015)
2	Soda (wheat straw)	Pd/C	350	EtOH	30	40	76.7/19.2	958	11.0	G5, S1, H5, S11, G4, S4	Kim et al. (2015)
3	Soda (wheat straw)	Ru/C	350	EtOH	30	40	76.3/17.7	828	11.1	G5, S1, H5, S11, G4, S4	Kim et al. (2015)
4	Soda (wheat straw)	Ni/C	350	EtOH	30	40	60.2/19.5	655	10.0	G5, S1, H5, S11, G4, S4	Kim et al. (2015)
5	Hydrolysis (hardwood)	Ru/C	300	Acetone	50	60	~82/27.1				Tymchyshyn et al. (2020)
6	Hydrolysis (hardwood)	Ru/C	340	Acetone	50	60	~90/25.0				Tymchyshyn et al. (2020)
7	Hydrolysis (hardwood)	MoRu/AC	300	Acetone	50	60	~84/28.5	190			Tymchyshyn et al. (2020)
8	Hydrolysis (hardwood)	MoRu/AC	340	Acetone	50	60	~88/23.9	154			Yan et al. (2021; Tymchyshyn et al. (2020)
9	Organosolv (poplar)	MoS_2/AC	300	MeCy[c]	3	600			6.3	H7, H4, AHCs[d]	Ji et al. (2020)
10	Acid hydrolysis (Wood)	Pd/C	195	Dioxane/Water	34.5	195	89			G9 (~20 wt%), G7	Torr et al. (2011)

(continued)

Table 7.9 (continued)

	Lignin (MW, g/mol)[a]	Catalyst	Temp (°C)	Solvent	H_2pressure (bar)	Reaction time (min)	Bio-oil yield (wt%)/O content (wt%)	Bio-oil MW (g/mol)[a]	Monomer yield (wt%)	Major monomers[b]	References
11	Alkali (7,512)	Pd/C	260	MeOH	40	300	58.5		6.8	Gs, Hs	Shu et al. (2015, 2018)
12	Alkali (7,512)	Pd/C+ $AlCl_3$	260	MeOH	40	300	91.4		26.0	Gs, Hs	Shu et al. (2015, 2018)
13	Alkali (7,512)	Pd/C+ $CrCl_3$	260	MeOH	40	300	80.7/25.89	655	28.5	Gs, Hs	Shu et al. (2015, 2018)
14	Organosolv (Sugarcane bagasse)	Ni/Zr-P	260	iPrOH	20	240			15.1	H 5 (6.1 wt%)	Ma et al. (2019)
15	Not reported	Ni/MgO	270	iPrOH	30	240			15.0	H5 (6.3 wt%)	Zhao et al. (2019)
16	Kraft	Ru/C	225	EtOH/H_2O	30	90			3.7	G7, G5, G4, G1	Zakzeski et al. (2012)
17	Kraft	Pd/C	225	EtOH/H_2O	30	90			4.8	G7, G5, G4, G1	Zakzeski et al. (2012)
18	Kraft	Pt/Al_2O_3	225	EtOH/H_2O	30	90			6.2	G7, G5, G4, G1	Zakzeski et al. (2012)
19	Kraft	Ni/SiO_2	225	EtOH/H_2O	30	90			5.3	G10, G1	Zakzeski et al. (2012)
20	NH_3—poplar	Pt/Al_2O_3	300	MeOH/H_2O	20	120			14.0	S7, S1, S8	Bouxin et al. (2015)

(continued)

Table 7.9 (continued)

	Lignin (MW, g/mol)[a]	Catalyst	Temp (°C)	Solvent	H_2 pressure (bar)	Reaction time (min)	Bio-oil yield (wt%)/O content (wt%)	Bio-oil MW (g/mol)[a]	Monomer yield (wt%)	Major monomers[b]	References
21	Organosolv—poplar	Pt/Al_2O_3	300	$MeOH/H_2O$	20	120			6.6	S7	Bouxin et al. (2015)
22	Soda—wheat straw	Pt/Al_2O_3	300	$MeOH/H_2O$	20	120			5.7	S1, G5	Bouxin et al. (2015)
23	Kraft (4,973)	Pt/Al_2O_3	300	EtOH/Water	20	180		1,390	~9	G1 (3.8 wt%), G5, G7, G15	de Albuquerque (2020)
21	Kraft (4,973)	Pt/Al_2O_3	300	Acetone/Water	20	180		968	~9	G1 (5.2 wt%), G4	de Albuquerque (2020)
22	NH_3—Poplar	Pt/Al_2O_3	300	MeOH	20	300			~45%	S7 (6.5 wt%), G7	McVeigh et al. (2016)
23	NH_3—Poplar	Pt/Al_2O_3	300	MeOH/Water	20	300			16.4	S7	McVeigh et al. (2016)
24	NH_3—Poplar	Ir/Al_2O_3	300	MeOH/Water	20	300			9.3	S1, H1	Bouxin et al. (2015); McVeigh et al. (2016)
25	NH_3—Poplar	Rh/Al_2O_3	300	MeOH/Water	20	300			19.9	S7	McVeigh et al. (2016)

(continued)

Table 7.9 (continued)

	Lignin (MW, g/mol)[a]	Catalyst	Temp (°C)	Solvent	H_2pressure (bar)	Reaction time (min)	Bio-oil yield (wt%)/O content (wt%)	Bio-oil MW (g/mol)[a]	Monomer yield (wt%)	Major monomers[b]	References
26	Organosolv (Candlenut shells)	Cu-PMO (50 wt%)	180	MeOH	40	840			54.8	C9 (36.1 wt%)	Barta et al. (2014)
27	Organosolv (Candlenut shells)	Cu-PMO (50 wt%)	180	MeOH	40	1200			63.7	C9 (43.3 wt%)	Barta et al. (2014)
28	Formaldehyde protected—poplar	Ru/C	250	1,4-Dioxane	40	900			78 mol%	S7 (~20 mol%, S5 (~17 wt%) S9 (~15 wt%)	Shuai et al. (2016)
29	Non-formaldehyde protected—poplar	Ru/C	250	1,4-Dioxane	40	900			~22 mol%	S5 (~14 mol%)	Shuai et al. (2016)
30	Hydrolysis (Sorghum Straw) (14,661)	Ni-Al/SBA-15	280	EtOH	10	90	79.9	1,238	17.8	Hs major	Chen et al. (2017)
31	Hydrolysis (Sorghum Straw) (14,661)	Ni-Al/SBA-15	280	EtOH	10	90	69.0	1,695	13.8	Hs major	Chen et al. (2017)
32	Hydrolysis (Sorghum Straw) (14,661)	Ni-Al/MCM-41	280	EtOH	10	90	61.6	1,798	11.0	Hs major	Chen et al. (2017)
33	Hydrolysis (Sorghum Straw) (14,661)	Ni-Al/SBA-15	280	EtOH	10	240	~72	1,695	21.9	Hs major	Chen et al. (2017)

(continued)

Table 7.9 (continued)

	Lignin (MW, g/mol)[a]	Catalyst	Temp (°C)	Solvent	H_2 pressure (bar)	Reaction time (min)	Bio-oil yield (wt%)/O content (wt%)	Bio-oil MW (g/mol)[a]	Monomer yield (wt%)	Major monomers[b]	References
34	Organosolv—Beech	Ni/SiO_2	250	Hexadecane	20	360			23	Hydrocarbons, alcohols	Kasakov et al. (2015)
35	Organosolv—Beech	Ni/HZSM-5	250	Hexadecane	20	360			28	Hydrocarbons	Kasakov et al. (2015)
36	Organosolv—Beech	Ni/Hβ	250	Hexadecane	20	360			35	Hydrocarbons	Kasakov et al. (2015)
37	Organosolv—Beech	Ni/HZSM-5	320	Hexadecane	20	360			70	Hydrocarbons	Kasakov et al. (2015)
38	Corncob lignin	Ni/SiO_2	300	Dodecane	60	120			~32	Hydrocarbons (13.4% yield C8) to EB (13.3 wt%)[e]	Luo et al. (2020)
39	Corncob lignin	Ni/HZSM-5 (300)	300	Dodecane	60	120			~32	Hydrocarbons (11.2% yield C8)	Luo et al. (2020)

[a] M_w from GPC/SEC analysis, [b] Refer to Fig. 7.1 for definitions of key compounds listed in order of yield, value in parenthesis is notable high yield of individual compound,[c] MeCy = methylcyclohexane, and [d] AHC = aromatic hydrocarbon, and [e] After distillation, C8 isolated hydrocarbons dehydrogenation to ethylbenzene via reaction over $Pt\text{-}Sn/Al_2O_3$

have also reported the use of a Ni/ZrPO4 catalyst for depolymerization of organosolv lignin with monomer yields up to 15 wt%, with high selectivity for ethyl phenol (**H5**) (Ma et al. 2019). Zaho et al. have suggested that the nickel catalysts help promote decarboxylation of *p*-coumaric acids or esters that are intermediate depolymerization products (Zhao et al. 2019).

Bouxin et al. have reported testing Pt/Al_2O_3 catalyst for hydrotreating lignins (Bouxin et al. 2015; Lancefield et al. 2016; McVeigh et al. 2016; de Albuquerque 2020). Their work has found a correlation between the quantity of β-O4 linkages remaining in the lignin and the yield of monomers produced from the process, as has been the case for other depolymerization processes. The correlation is not perfect, however, and other factors must have an influence. They found a higher yield of monomers from lignin isolated from poplar (14 wt%) and/or wheat straw by an ammonia process (9.7 wt%) which preserved a large fraction of the aryl-ether linkages than from organosolv (6.6 wt%) or soda lignin (5.7 wt%) in reactions done in a 1/1 methanol/water mix under 20 bar H_2 at 300 °C (Bouxin et al. 2015). In the case of the ammonia lignins, the monomeric products contained higher concentrations of alkylated phenolics, while those from soda and organosolv lignins were more selective for non-alkylated products. However, in another study, they found their highest yields (23 wt%) from an oak organosolv lignin that had lower β-O4 content than the ammonia poplar lignin (Lancefield et al. 2016). They also screened other metal catalysts including Rh/Al_2O_3 and Ir/Al_2O_3 and found that Rh was more active on a weight basis, but Pt was more efficient on a molar basis (McVeigh et al. 2016). In the case of Ir, the activity of the catalyst was only slightly higher than the Al_2O_3 support, and the product selectivity was shifted towards demethoxylated and dealkylated products. Similarly, others have found correlations between the method of isolation and yield of monomers in the hydrotreating of lignins, with those milder methods that produce structures with aryl-ether content similar to native lignins performing better than those leading to a more condensed structure (Van den Bosch et al. 2015a, b; Torr et al. 2011).

On the other hand, Shu et al. have reported a methodology that has been successful at producing a relatively high yield of phenolic monomers from kraft lignin in their process which uses Lewis acids in combination with Pd/C for hydrotreating (Shu et al. 2015, 2018). They screened several metal chlorides as Lewis acids and found that the use of $CrCl_3$ and $AlCl_3$ produced the highest yields of phenolic monomers. When the optimized conditions (280 °C, 5 h, MeOH, and 4 MPa H_2) were used, 35.4 wt% yield of monomers was reported. Interestingly, monomer yield using only $CrCl_3$ in the absence of Pd was still at 24 wt%, whereas yield with Pd/C but without the Lewis acid was only 6.8 wt%. Their more recent work suggests that the selectivity can be controlled by choice of the metal chloride, with $CrCl_3$ providing selectivity towards phenols and $ZnCl_2$ providing selectivity towards guaiacols, and this was confirmed via testing on a β-O4 model dimer (Shu et al. 2018).

Remarkably high selectivity for catechols from the hydrotreatment of candlenut-derived organosolv lignin over Cu supported on porous metal oxides was observed by Barta et al. (2014). Whereas most other hydrotreatments of lignin at mild conditions produce guaiacols or phenols, this particular lignin catalyst combination yields 63.7

wt% yield of only four catechols. This narrow product distribution has allowed for the isolation of the compounds via standard column chromatography and isolation of 4-propanolcatechol (**C9**) in 43.3% yield.

As mentioned above, there is a strong correlation between the aryl-ether linkage content of the lignin materials used and the yield of phenolic monomers and other lower molecular weight products for most of the methodologies summarized in this section. Shuai et al. have reported a lignin isolation procedure using formaldehyde to protect the β-O4 during acid-catalyzed lignin removal that preserves the β-O4 linkages and prevents the formation of condensed structures (Shuai et al. 2016). The yields of monomers from subsequent hydrotreating of the lignins isolated by this method over Ru/C in THF are much higher than when the lignin is isolated without formaldehyde Figure 7.

Some studies have added a hydrogenation-activating metal to an acidic supporting catalyst, with the aim of coupling depolymerization, hydrogenolysis, and hydrogenation in one step. These are sometimes called *bifunctional* catalysts or the process *bifunctional hydroprocessing* (Kasakov et al. 2015; Luo et al. 2020; Chen et al. 2017). These processes can convert lignin directly into cycloalkanes resulting from both the deoxygenation and saturation of the lignin-derived phenolics. Table 7.9, entries 34-39 report selected results for hydrotreating over these types of catalysts. Kasakov et al. have reported using Ni on acidic zeolites (HZSM-5 and HBEA) to produce mono and bicyclic alkanes (Kasakov et al. 2015). The reactions were performed in *n*-hexadecane with a hydrogen pressure of 20 bar, and the optimum temperature found was 320 °C. The selectivity of the product was influenced by the properties of the silicate support with Ni/SiO_2 providing a mixture of cyclic alkanes and alcohols, while the acidic zeolite supports provided exclusively hydrocarbons demonstrating the importance of the acidic support in achieving complete deoxygenation. Ni/HBEA provided the highest yield of hydrocarbons (70%). Luo et al. have also reported the use of silicate-supported Ni catalyst for the production of cyclic alkanes (Luo et al. 2020). When silicate-1 was used as the support, 33 wt% of the lignin was converted to cyclic alkanes with a high selectivity towards ethylcyclohexane. They then demonstrate that the C8 compound could be isolated by distillation and nearly quantitatively dehydrogenated to ethylbenzene using a Pt-Sn/Al_2O_3 catalyst. They proposed a scheme where this valuable chemical is produced while the remaining cyclic alkanes are used in gasoline and jet fuel formulations. Other reports for the production of cyclic alkanes via bifunctional catalysts have focused on Ru and Pd on zeolites and other aluminosilicate materials, with yields of cycloalkanes from lignin ranging from about 20 wt% to about 40 wt% (Schutyser et al. 2018).

7.4.5 *Reductive Depolymerization by Transfer Hydrogenation*

Transfer hydrogenation of liquid phase reforming processes obtains hydrogen from the solvent or other organic additive and avoids the need for adding hydrogen gas. The reactions are still often pressurized with inert gas so that they are run at supercritical or near supercritical solvent conditions. Catalysts for these reactions include the same classes of catalysts as used for mild hydrotreating, and the solvents and reaction conditions are similar. In general, higher average yields of monomers are reported than for mild hydrotreating, but a drawback is the necessary consumption of solvent. Table 7.10 provides some selected results for depolymerization via liquid phase reforming or transfer hydrogenation.

Zakzeski et al. compared the reductive depolymerization via hydrotreating and liquid phase reforming (Zakzeski et al. 2012). In the liquid phase reforming route,hydrogen was transferred from ethanol and was catalyzed by Pt/Al_2O_3 with acidic and basic co-catalysts tested while the hydrotreating was tested with catalysts including Ru/C, Pd/C, Ni/SiO_2, and a direct comparison with the Pt/Al_2O_3 catalyst. They found higher monomer yields were obtained in the liquid phase reforming process, which was carried out in a water-ethanol mixture at 225 °C (12.8–17.6) with H_2SO_4 performing the best as a co-catalyst. For hydrotreating under the same conditions, the best monomer yield was 6.2 wt%, also obtained with Pt/Al_2O_3.

Catalysts based on Cu, Ni, and Mo (Ma et al. 2014; Ma et al. 2015a, b) have been among the most effective catalysts for reductive depolymerization with transfer hydrogenation from alcohols. Hung et al. have reported the use of Cu-Mg-Al mixed metal oxides as an effective catalyst for depolymerization of soda lignin in ethanol (Huang et al. 2014, 2015, 2017a, b, 2021). They reported monomer yields up to 86 wt% from Kraft lignin, 60 wt% from soda lignin, and 62 wt% from organosolv lignin at 380 °C (Huang et al. 2015). Pt and Ni catalysts on the same Mg-Al oxide support were less effective catalysts (Huang et al. 2014). They reported a strong temperature dependence on the chemistry observed; at lower temperatures (<300 °C), recondensation of reactive intermediates is dominant, whereas at temperatures between 300–380 °C, hydrogenation of non-ring carbon-carbon bonds and capping and alkylation reactions from ethanol are active (Huang et al. 2017a, b). Furthermore, more thermal cracking of bonds that are more recalcitrant than the aryl-ether bonds begins at this temperature. However, at even higher temperatures (380–420 °C), char forming reactions become dominant.

Cheng et al. have reported the use of Ni-Cu/C for depolymerization of organosolv lignin in an ethanol/isopropanol mixture at 270 °C (Cheng et al. 2020). They obtained a maximum monomer yield consisting mostly of alkyl guaiacols and alkyl syringols. The bimetallic Ni-Cu catalyst proved better at transferring hydrogen from the solvents than Ni or Cu alone, while Ni-W was not an effective catalyst. Biswas recently reported the use of a bimetallic Ni-Co catalyst supported on biochar-derived activated carbon for lignin depolymerization in methanol. Although monomer yields were not quantified, high selectivity towards vanillin production was observed (Biswas et al. 2021). Chen et al. have reported a trimetallic W-Ni-Mo catalyst on sepiolite as the support which was selective for the production of guaiacol (**G1**) and ethyoxyphenol (**OEt-H1**) which could be produced in 4.2 wt% and 6.2 wt%, respectively, from Kraft lignin at 280 °C in ethanol (Chen et al. 2021a, b). Ma et al. have reported various Mo-based catalysts for the depolymerization of Kraft lignin in ethanol, with monomer yields reaching up to 33.3 wt% (Ma et al. 2014; Ma et al. 2015a, b).

Table 7.10 Selected results for solvent phase catalytic reductive depolymerization of lignin via transfer hydrogenation

	Lignin (MW, g/mol)[a]	Catalyst	Temp (°C)	Solvent	Pressure (bar)[b]	Reaction time (min)	Bio-oil yield (wt%)/O content (wt%)	Bio-oil MW (g/mol)[a]	Monomer yield (wt%)	Major monomers[c]	References
1	Kraft (4,973)	Pt/Al_2O_3 + H_2SO_4	300	EtOH/Water	58	180			17.6	G1, C5, G10	Zakzeski et al. (2012)
2	Kraft (4,973)	Pt/Al_2O_3 + $H_3PW_{12}O_{40}$	300	EtOH/Water	58	180			13.2	G1, O-EtG4[d]	Zakzeski et al. (2012)
3	Kraft (4,973)	Pt/Al_2O_3 + NaOH	300	EtOH/Water	58	180			12.8	OEt-G4[d]	Zakzeski et al. (2012)
4	Alkali (10,000)	Ru/C	290	EtOH	62	240	39.4	3,770			Zhou et al. (2018)
5	Alkali (10,000)	Ni/ZSM-5	290	EtOH	62	240	39.5	3,680			Zhou et al. (2018)
6	Alkali (10,000)	CuNiAl-HT[e]	290	EtOH	62	240	62.2	3,250			Zhou et al. (2018)
7	Alkali (10,000)	CuNiAl-HT[e-]	290	EtOH-PhOH	62	240	81.8/36.7	1,880			Zhou et al. (2018)
8	Alkali	Ni-Co/AC[f]	280	Water		15	25/25.8			G10	Biswas et al. (2021)
9	Alkali	Ni-CO/AC[f]	280	EtOH		15	70/24.2			G10	Biswas et al. (2021)

(continued)

Table 7.10 (continued)

	Lignin (MW, g/mol)[a]	Catalyst	Temp (°C)	Solvent	Pressure (bar)[b]	Reaction time (min)	Bio-oil yield (wt%)/O content (wt%)	Bio-oil MW (g/mol)[a]	Monomer yield (wt%)	Major monomers[c]	References
10	Soda	$CuMgAlO_x$	300	EtOH		120			8		Huang et al. (2014)
11	Soda	$CuMgAlO_x$	300	EtOH		480			23	G4, G5, H4	Huang et al. (2014)
12	Soda	$CuMgAlO_x$	340	EtOH		240			30	H5, Hydrocarbons, Alcohols	Huang et al. (2017a, b)
13	Soda	$CuMgAlO_x$	380	EtOH		480			60[g]	Hydrocarbons, Hs	Huang et al. (2015)
14	Organosolv	$CuMgAlO_x$	380	EtOH		480			62[g]		Huang et al. (2015)
15	Kraft	$CuMgAlO_x$	380	EtOH		480			86[g]		Huang et al. (2015)
16	Soda	TiN	300	EtOH	10				12	H18	Chen et al. (2016)
17	Soda	TiN	340	EtOH	10				19	H18, H19	Chen et al. (2016)

(continued)

Table 7.10 (continued)

	Lignin (MW, g/mol)[a]	Catalyst	Temp (°C)	Solvent	Pressure (bar)[b]	Reaction time (min)	Bio-oil yield (wt%)/O content (wt%)	Bio-oil MW (g/mol)[a]	Monomer yield (wt%)	Major monomers[c]	References
18	Kraft	Mo/SEP[g]	280	EtOH	5	240	65/12.9		~61	Hs, Gs, OEt-H1 (2.5 wt)	Chen et al. (2021a, b)
19	Kraft	Ni-Mo/SEP[g]	280	EtOH	5	240	36/14.3			Hs, Gs, OEt-H1 (~3.7 wt%)	Chen et al. (2021a, b)
20	Kraft	W-Mo/SEP[g]	280	EtOH	5	240	35/14.5			Hs, Gs, OEt-H1 (~3.7 wt%)	Chen et al. (2021a, b)
21	Kraft	W-Ni-Mo/SEP[g]	280	EtOH	5	240	39/14.5			Hs, Gs, OEt-H1 (~5 wt%)	Chen et al. (2021a, b)
22	Organosolv—poplar	Ni/C	270	EtOH/iPrOH	1	240			51.3	S8 (23.3 wt%), G8 (10.9 wt%). S9, G9	Cheng et al. (2020)
23	Organosolv—poplar	Cu/C	270	EtOH/iPrOH	1	240			46.9	S8 (24.1 wt%), G8 (13.6 wt%), S9	Cheng et al. (2020)
24	Organosolv—poplar	NiCu/C	270	EtOH/iPrOH	1	240			63.4	S8 (27.2 wt%), G8 (11.5 wt%), S9, G9	Cheng et al. (2020)
25	Organosolv—poplar	NiW/C	270	EtOH/iPrOH	1	240			<1	–	Cheng et al. (2020)

(continued)

Table 7.10 (continued)

	Lignin (MW, g/mol)[a]	Catalyst	Temp (°C)	Solvent	Pressure (bar)[b]	Reaction time (min)	Bio-oil yield (wt%)/O content (wt%)	Bio-oil MW (g/mol)[a]	Monomer yield (wt%)	Major monomers[c]	References
26	Organosolv—switchgrass	Pt/C	350	EtOH	1	20		469			Cheng et al. (2020)
27	Organosolv—switchgrass	Pt/C	350	EtOH/FA	1	20		419		G5 (3.5 wt%), G7 (2.9 wt%)	Cheng et al. (2020)
28	Organosolv—switchgrass	Pt/C	350	EtOH/FA	1	4		512	21	G5 (7.8 wt%), G7 (4.9 wt%), G1	Cheng et al. (2020)
29	Alkali	Pd/C	265	H_2O/FA	1	240	31.6		26	C1 (13 wt%)	Onwudili and Williams (2014)
30	Alkali—wheat	Pd/C	280	MeOH/FA	1	120			10.9		Zhu et al. (2016)
31	Alkali—wheat (benzyl methylated)	Pd/C	280	MeOH/FA	1	120			16.8		Zhu et al. (2016)

[a] M_w from GPC/SEC analysis, [b] Pressure at ambient temperature,[c] Refer to Fig 7.1 for defintions of key compounds listed in order of yield, value in parenthesis is notable high yield of individual compound, [d] Et-G4 = 4-methy-2-methoxyethyoxybenzene AHC = aromatic hydrocarbon, [e] hydrotalcite, [f] AC = biochar-derived activated carbon, and [g] SEP = sepiolite

After alcohol solvents, the most studied hydrogen donor molecule is formic acid. Xu et al. used Pt/C in combination with formic acid as the hydrogen donor in the depolymerization of organosolv lignin from switchgrass (Xu et al. 2012). Performing the reactions in ethanol at 350 °C, they reported a maximum monomer yield of 21 wt% with a reaction time of 4 h. The monomers were alkyl phenols and guaiacols. At longer reaction times (20 h), monomer yields were increased but the average molecular weight of the entire bio-oil was higher, indicating that repolymerization reactions were likely occurring. Other hydrogenation active metals tested for transfer hydrogenation from formic acid include Pd/C (Zhu et al. 2016; Onwudili and Williams 2014) and Ru/C (Kloekhorst et al. 2015). Zhu et al. reported monomer yields of 7-11 wt% from alkali lignins in methanol in formic acid over Pd/C at 280 °C. Premethylation of the benzyl hydroxy groups increased those yields to 10.3–16.8 wt% (Zhu et al. 2016). Under similar conditions, but using water as the solvent, Onwudili and Williams reported 26 wt% of monomer production from kraft lignin with catechol as the major monomeric product (Onwudili and Williams 2014). At higher temperatures and using Ru/C as the catalyst in a 1/1 mixture of isopropanol and formic acid, Kloekhorst et al. reported both depolymerization and hydrodeoxygenation of organosolv lignin yielding a bio-oil product at 71 wt% yield (Kloekhorst et al. 2015). The monomer yield was greater than 30 wt% with alkyl phenols, catechols, and aromatic hydrocarbons produced. Other effective hydrogen donor additives have included tetralin and glycerol.

7.4.6 Oxidative Depolymerization

As the name suggests, oxidative lignin depolymerization is usually induced by a catalyst in the presence of an oxidant (Table 7.11). Usually, this oxidant is O_2, provided as pure molecular oxygen or air. Other oxidants have been used in lignin oxidation for characterization purposes. Lignin oxidation can occur in acidic or basic aqueous solutions or in organic solvents. In contrast to reductive depolymerization, the products of oxidative lignin depolymerization tend to completely retain their oxygen functionalities or even contain further oxidized carbon atoms. When stronger oxidation conditions are used (e.g., high O_2 pressure and/or longer reaction times), further oxidation can occur to open the aromatic ring and produce non-aromatic carboxylic acids and dicarboxylic acids. However, under most reported conditions, aromatic aldehydes such as vanillin (**G10**) and carboxylic acids are commonly the most concentrated monomers in the produced bio-oil mixture. In fact, vanillin production is one of the few value-added commercial products ever produced from lignin, produced via oxidative conversion of sulfite pulping lignin as early as 1936 (Fache et al. 2015). At one time, this process supplied the majority of the world's vanillin, but today that share has dropped to 15%, and most vanillin is synthesized from fossil sources (Kumar et al. 2021). However, around 3000 tons/year are still commercially produced from lignin. Because of this past commercial success, there are many studies on oxidative cleavage of lignin model compounds in attempt to understand

Table 7.11 Selected results for oxidative depolymerization of lignin

	Lignin	Catalyst	Temp (°C)	Solvent	Atmosphere (P, bar)[a]	Reaction time (min)	Bio-oil yield (wt%)	Monomer yield (wt%)	Major monomers[b]	References
1	Kraft	None	130	NaOH(aq)	O_2 (3), N_2 (9)	35			G10 (10.8 wt%)	Fargues et al. (1996)
2	Hydrolysis—poplar	None	170	NaOH (aq)		10	53.4	10.0	S10 (5.1 wt%), G10, S11, G11	Xiang and Lee (2001)
3	Hydrolysis—poplar	$CuSO_4$	170	NaOH (aq)		10	62.7	13.8	S10 (10.3 wt), G10, S11, G11	Xiang and Lee (2001)
4	Hydrolysis—poplar	$FeCl_3$	170	NaOH (aq)		10	58.9	11.7	S10 (5.3 wt), G10, S11, G11	Xiang and Lee (2001)
5	Hydrolysis—poplar	CuSO4 + $FeCl_3$	170	NaOH (aq)		10	70.3	16.7	S10 (8.8 wt%), G10 (4.7 wt%), S11, G11	Xiang and Lee (2001)
6	Hydrolysis—poplar	CuSO4 + $FeCl_3$	170	NaOH (aq)		20	81.4	11.5 (78)[d]	S10 (8.8 wt%), G10, (4.7 wt%), S11, G11, Acetic acid (31.7 wt%), Aliphatic acids	Xiang and Lee (2001)

(continued)

Table 7.11 (continued)

	Lignin	Catalyst	Temp (°C)	Solvent	Atmosphere (P, bar)[a]	Reaction time (min)	Bio-oil yield (wt%)	Monomer yield (wt%)	Major monomers[b]	References
7	Enzymatic hydrolysis—cornstalk	$LaCoO_3$	120	NaOH (aq)	O_2 (5), N_2 (15)	50	~37	~15	S10 (10 wt%), G10 (4.3 wt%), H10	Deng et al. (2009)
8	Enzymatic hydrolysis—cornstalk	$LaFe_{0.8}Cu_{0.2}O_3$	120	NaOH (aq)	O_2 (5), N_2 (15)	30	~32	~18	S10 (12 wt%), G10 (4.3 wt%), H10	Deng et al. (2009)
9	Alkali	$MoPO^c/CeO_2$		NaOH (aq)	O_2 (5)	180		12.5	G10 (9 wt%)	Rawat et al. (2020)
10	Kraft	$H_3PMo_{12}O_{40}$	170	$MeOH/H_2O$	O_2(10.8)	20	60	7	G10, G11	Voitl and Rudolf von Rohr (2008)
11	Hydrolysis	$H_3PMO_{10}V_2O_{40}$	190	$MeOH/H_2O$	O_2 (25)	10	36.0			Zhao et al. (2013)
12	Alkali	$H_3PMO_{10}V_2O_{40}$	190	$MeOH/H_2O$	O_2 (25)	10	42.5			Zhao et al. (2013)
13	Pyrolytic—rice husk	$H_3PMO_{10}V_2O_{40}$	190	$MeOH/H_2O$	O_2 (25)	10	65.2			Zhao et al. (2013)
14	Organosolv	Pd/CeO_2	225	MeOH	O_2(1)	1440	40		G10 (5.2 wt%)	Deng et al. (2015)
15	Organosolv	Co/Mn/Zr/Br	180	Acetic Acid	Air (70)	120		10.9	S12, G12, S10, G10	Partenheimer (2009)

(continued)

Table 7.11 (continued)

	Lignin	Catalyst	Temp (°C)	Solvent	Atmosphere (P, bar)[a]	Reaction time (min)	Bio-oil yield (wt%)	Monomer yield (wt%)	Major monomers[b]	References
16	Organosolv—eucalyptus	Co(Salen)/GO[e]	80	MeCN/THF	Air (1)	1440		~9	G10 (3wt%), G12 (2.5 wt%)	Zhou and Lu (2016)
17	Organosolv—birch	DDQ/*t*BuONO	80	MOE[f]/Et_2O	O_2 (1)	840			S20 (5 wt% - isolated) after Zn red	Lancefield et al. (2015)
18	Kraft—pine	Au/LHD[h]	120	DMF	O_2 (1)		10 (20)[i]	4 (9)[i]	G10, G12	Song et al. (2018)
19	GVL Organosolv—maple	Au/LHD[h]	120	DMF	O_2 (1)		20 (56)[i]	8 (40)[i]	S10 (11.2 wt%)[i], S20, FA, S12	Song et al. (2018)
20	Kraft—softwood	Al-V-Mo	377	NaOH (aq)	Ar/O_2	Cont. FB			Lactic Acid (6%)	Lotfi et al. (2015)
21	Hydrolysis—corn stover	Peracetic acid	60	H_2O	Air (1)	100		17	G12, G22, H12, H22	Ma et al. (2016)
22	Hydrolysis—corn stover	Peracetic Acid/Nb_2O_5	60	H_2O	Air (1)	100		47	G12, G22, H12, H22	Ma et al. (2016)

[a] Pressure at ambient temperature, [b] Refer to Fig. 7.1 for definitions key compounds listed in order of yield, value in parenthesis is notable high yield of individual compound,[c] MoPO = molybdenum pyrophosphate, [d] First value phenolic only, valued in parenthesis includes aliphatic acids, [e] GO = graphene oxide, [f] ACN = acetonitrile, [g] MOE = 2-methoxyethanol, [h] LHD = layered double hydroxide (Li-Al), and [i] Values before and after hydrolysis with NaOH, Yield of S10 after hydrolysis

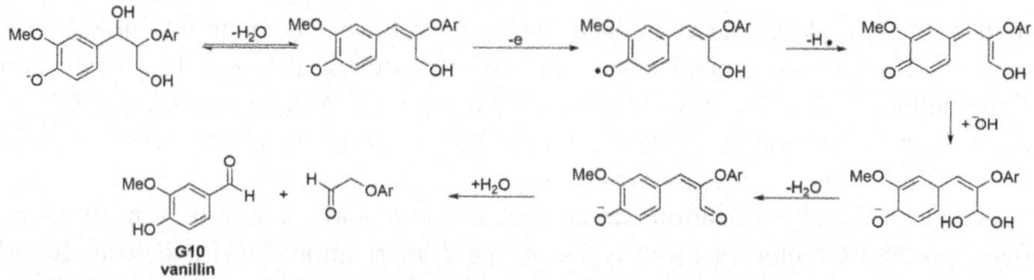

Fig. 7.10 Possible mechanism for oxidative depolymerization of lignin β-O4 linkage yielding vanillin. Adapted with permission from Springer (Tarabanko et al. 2004)

and improve the process. This section, however, focuses on the results of the studies on technical lignins.

The commercial production of vanillin is accomplished via the aerobic oxidation of lignin in a basic (NaOH) solution, and therefore, these reaction conditions have received significant attention in the literature. The studied reaction conditions vary in oxygen pressure from 2–14 bar and 120–190 °C, with the oxygen supplied by an O_2 or air atmosphere. The mechanism is not completely understood but is thought to be initiated by electron abstractions from a phenoxy anion to form phenoxy radicals (Tarabanko et al. 2004). The mechanisms based on both reactions of various reactive positions with O_2 or additional radical reactions and nucleophilic -OH attacks that lead to bond cleavage have been proposed. The -OH attack mechanism (Fig. 7.10) is supported by an observation of higher rates at very high pH (>13) (Tarabanko et al. 2004). Fargues et al. studied the effect of oxygen partial pressure and temperature and found an optimized yield of 10.8 wt% of vanillin at 133 °C at 2.8 bar of O_2 (9.7 bar total pressure) from Kraft lignin (Fargues et al. 1996). Higher temperature or higher oxygen partial pressure led to the oxidative breakdown of the vanillin, reducing the yield.

The addition of metal catalysts to the basic aqueous solution can enhance the production of monomers. Most catalysts tested have shown moderate increases in monomer yield over the process described above. The most common catalyst used is $CuSO_4$, sometimes in combination with $FeCl_3$, but other metals such as Pd, Mn, and Mo have also been tested. For example, Xiang and Lee studied the depolymerization of poplar hydrolysate lignin to form vanillin (**G10**) and syringaldehyde (**S10**) as the major monomers (Xiang and Lee 2001). They found an increase in the yield of monomers from 10.0 wt% non-catalytically to 13.8 wt% with the addition of $CuSO_4$ and 16.2 wt% with the addition of $CuSO_4$ and $FeCl_3$ (Table 7.11, entries 2-6). The Cu is thought to aid in the formation of the phenoxy radical via acceptance of an electron, while Fe is thought to activate and deliver dissolved oxygen. Perovskite-type mixed metal oxides ($LaMnO_3$) have also been well studied as a catalyst for alkaline oxidation of lignin (Tarabanko et al. 2004). For enzymatic hydrolysis of lignin from corn stover, an increase in the yield of vanillin and particularly syringaldehyde was noted when the perovskites were added. Doping the catalysts with Cu and Fe ($LaFe_{0.8}Cu_{0.2}O_3$) provided an even higher yield of the aldehydes, with the maximum yield of the two aldehydes combined reaching 16.1 wt%, compared with 8.7 wt% without a catalyst. Recently, Rawat et al. have reported producing a 12.5 wt% yield of monomers, with a 9 wt% yield of vanillin using a MoPO/CeO_2 catalyst at 150 °C under 5 bar O_2 (Rawat et al. 2020). Under these conditions, 55 wt% yield of ethyl acetate soluble bio-oil was also obtained.

Oxidative depolymerization can also occur in an acidic solution. Vanillin is the major product targeted in these types of depolymerization. Voitl and Rohr found $H_3PMo_{12}O_{40}$ as a more effective acid than H_2SO_4 for the oxidative depolymerization of Kraft lignin under O_2 in a mixture of methanol and water for the production of vanillin and methylvanillate (**G13**) (Voitl and Rudolf von Rohr 2018). The yields were higher in the presence of methanol than in neat water, and methylvanillate was not detected in water alone. The authors suggest that methanol plays a role in preventing repolymerization reactions by quenching radical intermediates and suggest a variety of mechanisms that could explain the formation of methylvanillate from vanillin and methanol.

Aromatic acids can also be produced via oxidative depolymerization of lignin. Partenheimer reported producing 11 wt% of a monomeric mixture consisting of predominately vanillic (**G12**) and syringic (**S12**) acid in concentrated acetic acid at 180 °C over a Co/Mn/Zr/Br catalyst system (Partenheimer 2009). Gonclaves and Schuchardt similarly used a Co/Br catalyst in acetic acid at 210 °C to produce a 5% yield of vanillin and vanillic acid. In an example using an organic oxidant, Ma et al. have reported the use of peracetic acid for depolymerization. In water at only 60 °C, they achieved 18 wt% and 22 wt% yield of monomers from Kraft lignin and enzymatic hydrolysis-derived lignin, respectively. The addition of Nb_2O_5 as a catalyst further increased those yields to 35 and 47 wt%. In this case, the major monomers produced were 4-hydroxy-2-methoxyphenol (**G21**), *p*-hydroxybenzoic acid (**H12**), vanillic acid (**G12**), syringic acid (**S12**), and 3,4-dihydroxybenzoic acid (**C12**).

Oxidative lignin depolymerization can also take place in organic solvents or neutral aqueous solutions. These procedures have the advantage of not using corrosive basic or acidic solutions. These reactions usually occur with metal catalysts under O_2 or air atmosphere. Metals used included Pd, Cu-V, Co, Co-Fe, and Au. Deng et al. have reported oxidatively depolymerizing organosolv lignin in methanol at 185 °C under 1 bar O_2 using a Pd/CeO_2 catalyst. They obtained a bio-oil with monomers at 8.5 wt% yield, with the major products being vanillin, 4-hydroxybenzaldehyde (**H10**), and guaiacol (Deng et al. 2015). Zhu and Lu have reported the use of Co(salen) supported on graphene oxide (Zhou and Lu 2016). They used mild conditions, only 80 °C under air in methanol, and obtained about 8.5 wt% of monomers, with vanillin and vanillic acid as the major monomeric components. The graphene oxide support had a considerable effect as the yields were better than zeolite or mesoporous alumina supports. Gold nanoparticles have been reported by Song et al. to be effective catalysts for aerobic oxidation of the benzylic alcohol position in lignin (Song et al. 2018). These reactions take place in dimethylformamide (DMF) at 120 °C under 1 bar O_2 for 24 h. After oxidation of the benzylic group, the lignin is easily hydrolyzed using a NaOH solution to afford a bio-oil-containing monomers. In the case of γ-valerolactone (GVL) extracted lignin, the yield of monomers was 40 wt%, with acetovanillone (**G11**), ferulic acid, and vanillic acid as the most concentrated components. From Kraft lignin, however, the yield of monomers was only 9 wt% with vanillin as the major component. The authors attributed the difference to the availability of the benzylic position to be oxidized in the GVL lignin, while benzylic alcohols were of lesser concentration in the more processed Kraft lignin, limiting the effectiveness of this method for that technical lignin. Zhao has reported triazine frameworks as metal-free catalysts for lignin oxidation (Zhao et al. 2018). Although neither the liquid nor monomer yields were reported, they did report a breakdown of 80% of the β-O-4 linkages in organosolv lignin after reaction in methanol at 180 °C under O_2.

7.4.7 *"Lignin-First" in situ Depolymerization*

A common theme of the above sections has been that most thermal and catalytic lignin depolymerization methods are more effective on lignins that retain more of the aryl-ether β-O-4 linkages from their native structures than those that contain more condensed structures. We have also seen that another major factor contributing to low monomer yields is repolymerization reactions. An approach that has gained significant traction in recent years is to combine the lignin fractionation and depolymerization into a single step; this allows for operation on native lignin with a maximized fraction of β-O4 linkages to break. This approach has been dubbed an "early stage" or "lignin-first" in contrast to most biorefinery schemes which would be considered "carbohydrate-first" because lignin is removed while cellulose is depolymerized to release sugars (Fig. 7.11) (Liu et al. 2020). In the lignin-first approach, the products are a lignin oil, rich in phenolic monomers, and a carbohydrate fraction sometimes called the pulp that can be converted to other products separately (Renders et al. 2019; Abu-Omar et al. 2021; Liu et al. 2020).

The depolymerization methods used are the same as many of the above-described solvent phase methods. The large majority of lignin-first methods use reductive depolymerization methods (in this case often referred to as catalytic reductive fractionation), either with H_2 gas or via hydrogen donation from an alcohol solvent or cellulosic portions of the biomass; although acid (Kramarenko et al. 2021) or base-catalyzed and oxidative (Hafezisefat et al. 2020) methods have also been reported. The conditions used are on the mild side to minimize side reactions of the carbohydrate residue. The one-pot lignin-first approach can be considered as a three-step process, (1) solvent extraction of lignin from the biomass, (2) depolymerization of the lignin via catalytic or solvolysis mechanisms, and (3) stabilization of reactive intermediates, usually via hydrogenation (Renders et al. 2019). The nature of the process, where contact between the solid biomass and the catalyst is unnecessary, lends itself to performance in semi-continuous flow-through reactors, and the catalytic stabilization step can occur in the same or a separate unit operation from the solvation (Kumaniaev et al. 2017; Renders et al. 2019; Kramarenko et al. 2021).

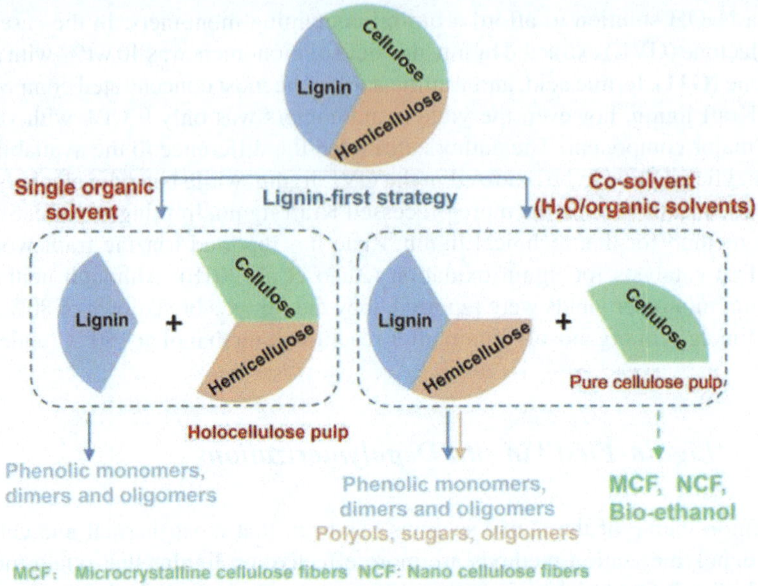

Fig. 7.11 General schematic for variations of lignin-first biorefineries. Reproduced with permission of Wiley (Liu et al. 2020)

However, flow-through reactors tend to increase solvent consumption, and longer residence times of the lignin lead to familiar concerns about repolymerization reactions (Renders et al. 2019). Table 7.12 reports selected lignin-first results. Yields are reported on a lignin basis in Table 7.12 and in the following paragraphs.

As a result of the combination of operating on a high concentration of relatively labile β-O-4 linkages and rapid stabilization of monomers, the yields of monomers produced via lignin-first reductive depolymerization are generally higher (~25-60 wt% under optimized conditions) than those found from isolated lignin, with the variation found between biomass sources, catalysts, and reaction conditions. The monomeric product distribution is also narrower than found for conversion from isolated lignin. Most products are propyl, propenyl, or propanol-substituted methoxyphenols (e.g., **H/G/S7-9**). The usually mild conditions used lead to retention of most of the methoxy groups. Most reports have used traditional hydrogenation catalysts such as heterogeneous supported Pd, Pt, or Ni materials. Catalyst and conditions play a crucial role in monomer selectivity. Under an H_2 atmosphere, Pt/C, Ru/C, and Rh/C have each been reported to be selective for propyl substituted guaiacols and syringols, whereas under similar conditions, the well-studied Pd/C or Ni/Al_2O_3 catalysts have proven not to remove the γ-OH group leading to a high selectivity for propanol guaiacols and syringols. For example, Van den Bosch et al. (2015a, b) reported a similar yield of monomer from birch wood using Ru/C (48 wt%) or Pd/C (47 wt%) catalysts at 250 °C in methanol under 3 MPa H_2, but using Ru/C, a 75% selectivity towards propyl substituted guaiacols and syringols was observed while

Table 7.12 Selected Results of "Lignin-First" Simultaneous Biomass Fractionation and Depolymerization

	Biomass	Process	Catalyst	Temp (°C)	Solvent	Atmosphere (P, bar)[a]	Reaction time (min)	Delignification (wt%)	Monomer yield (wt%)[b]	Major monomers[c]	References
1	Corn stover	Red	$Ni/AlPO_4$	235	MeOH	H_2 (20)	180	50	35.7	G7 (7.5 wt%), H5, H21, G21	Zhang et al. (2021)
2	Birch	Red	Pd/C	250	MeOH	H_2(30)	180	90	49 C% [e]	S/G9	Van den Bosch et al. (2015a, b)
3	Birch	Red	Ru/C	250	MeOH	H_2(30)	180	85	49 C% [e]	S/G7	Van den Bosch et al. (2015a, b)
4	Birch wood	Red	Ru/C	250	MeOH	H_2 (30)	240	68	52 C% [e]	G7, S7	Van den Bosch et al. (2015a, b)
5	Birch	Red	Pd/C	200	MeOH	H_2 (30)	60		25	S7, G7	Huang et al. (2016)
6	Birch	Red	Pd/C + $Yb(OTf)_3$	200	MeOH	H_2 (30)	60		43	S7, S9	Huang et al. (2016)
7	Birch	Red	Pd/C + $Yb(OTf)_3$	200	MeOH	H_2 (30)	120		46	S7, S9	Huang et al. (2016)
8	Oak	Red	Pd/C + $b(OTf)_3$	200	MeOH	H_2 (30)	60		45		Huang et al. (2016)

(continued)

Table 7.12 (continued)

	Biomass	Process	Catalyst	Temp (°C)	Solvent	Atmosphere (P, bar)[a]	Reaction time (min)	Delignification (wt%)	Monomer yield (wt%)[b]	Major monomers[c]	References
9	Pine	Red	Pd/C + $Yb(OTf)_3$	200	MeOH	H_2 (30)	60		25		Huang et al. (2016)
10	Wheat Chaff	Red	Pd/C + $Yb(OTf)_3$	200	MeOH	H_2 (30)	60		5		Huang et al. (2016)
11	Birch	Red	Ni-Al_2O_3	250	MeOH	H_2 (30)	180	87	44 C% [e]	G/S9 (66C% [e]), G/S7	Van den Bosch et al. (2017)
12	Poplar	Red	Pd/C	225	MeOH	H_2 (34)	720		59	S9 (35 wt%), G9	Klein et al. (2016)
13	Poplar	Red	Pd/C + Zn(OAc) 1/0.1	225	MeOH	H_2 (34)	720		61	S9 (24wt%), S7 (16 wt%). G9, G7	Klein et al. (2016)
14	Poplar	Red	Pd/C + Zn(OAc) 1/1	225	MeOH	H_2 (34)	720		43	S9 (24wt%), S7 (28 wt%), G7	Klein et al. (2016)
15	Poplar	Red	Ni-W_2C/AC	235	H_2O	H_2 (60)	240		32.4	S9, G7, G9, G7	Li et al. (2012)
16	Ash	Red	Ni-W_2C/AC	235	H_2O	H_2 (60)	240		40.5	S7 (14.8), S9, G7, G9	Li et al. (2012)

(continued)

Table 7.12 (continued)

	Biomass	Process	Catalyst	Temp (°C)	Solvent	Atmosphere (P, bar)[a]	Reaction time (min)	Delignification (wt%)	Monomer yield (wt%)[b]	Major monomers[c]	References
17	Pine	Red	Ni-W_2C/AC	235	H_2O	H_2 (60)	240		10.1	G7, G9	Li et al. (2012)
18	Birch	Red – trans	Pd/C	195	EtOH/H_2O	Ar (1)	900		31	S8 (19 wt%)	Galkin et al. (2016)
19	Birch	Red – trans	Pd/C	210	EtOH/H_2O	Ar (1)	900		34	S7 (20 wt%), S8 (11 wt%)	Galkin et al. (2016)
20	*Quercus suber* bark	Red – trans	Pd/C + NaOH	200	MeOH/H_2O	H_2	120			G5 (12wt% isolated after distillation)	Kumaniaev and Samec (2018)
21	*Quercus suber* bark	Red – trans	Pd/C	200	MeOH/H_2O	H_2	120			G8	Kumaniaev and Samec (2018)
22	*Quercus suber* bark	Red – trans	Pd/C + H_3PO_4	200	MeOH/H_2O	H_2	120			G9, G8	Kumaniaev and Samec (2018)
23	Birch	Red – trans (flow)	Pd/C	180	MeOH/H_2O	N_2	30		37	G/S9 (18.2 wt%), GS/7 (10.6 wt%), G/S15	Kumaniaev et al. (2017)
24	Birch	Acid (Flow)	Hβ	220	EtOH/H_2O	N_2 (50)	180		16.5	S8 (7.2 wt%), G11, S10	Kramarenko et al. (2021)

(continued)

Table 7.12 (continued)

	Biomass	Process	Catalyst	Temp (°C)	Solvent	Atmosphere (P, bar)[a]	Reaction time (min)	Delignification (wt%)	Monomer yield (wt%)[b]	Major monomers[c]	References
25	Birch	Acid (Flow)	Hβ + Oxalic Acid	220	$EtOH/H_2O$	N_2 (50)	180		20	S8, G11, S10	Kramarenko et al. (2021)
26	Pine	Acid	H_2SO_4	140	DMC + EG	N_2 (1)	40			G16 (9 wt%)	De Santi et al. (2020)

[a] Pressure at ambient conditions, [b] Yield Based on Lignin Contained in Biomass, [c] Refer to Fig. 7.1 for definitions key compounds listed in order of yield, value in parenthesis is notable high yield of individual compound, [d] APO = aluminum phosphate, and [e] Carbon yield (%)

Pd/C provided 91% selectivity for the propanol substituted compounds (Van den Bosch et al. 2015a, b). In contrast, without H_2 but instead using hydrogen transfer from ethanol, Galkin et al. reported 35% yield from birch with high selectivity for 4-propylsyringol (**S7**) using Pd/C at 210 °C. The selectivity can also be tuned with additives (Galkin et al. 2016). For example, the addition of Lewis Acids with Pd/C can shift the selectivity back towards the elimination of the alkyl hydroxy group even under H_2. Klein et al. found $ZnCl_2$ in addition to Pd/C most selective for the propylmethoxyphenols (Klein et al. 2016). Huang et al. found metal triflates in combination with Pd/C to produce methoxypropylguaiacol (**G15**) and syringols under H_2 in methanol, with the Lewis Acid catalyzing the methylation of the hydroxy group (Huang et al. 2017a, b; Huang et al. 2017a, b; Huang et al. 2016). $Al(OTf)_3$ and $Yb(OTf)_3$ were the most effective additives providing monomers in 45 and 43 wt%, respectively. This process simultaneously produced methylated sugars from the carbohydrate fraction of the biomass.

The identity of the biomass has a larger influence on the product yield and distribution in lignin-first approaches than from isolated lignin. For isolated lignin, the most important structural differences are influenced by the isolation procedure. With that step removed, the influence of biological structures between species becomes more noticeable. For reductive catalytic fractionation, phenolic monomer yields are often highest from hardwood, followed by herbaceous species, and softwoods provide the lowest yield of monomers. This is directly related to the number of β-O-4 linkages. (S)-lignin which has its highest concentration in hardwoods is known to be correlated with the amount of β-O-4s. For example, Van den Bosch et al. reported monomer yields of 50 wt%, 44 wt%, 27 wt%, and 21 wt% for Ru/C-catalyzed reductive fractionation from birch, poplar, miscanthus, and mixed softwoods, respectively (Van den Bosch et al. 2015a, b). Similarly, Galkin et al. found higher yields from hardwood than softwood for the Pd/C catalyzed process (Galkin et al. 2016).

In addition to reductive fractionation, other processes developed for the conversion of isolated lignin have similarly found success when adapted to a lignin-first approach. For example, De Santi et al. (2020) have reported the production of cyclic acetals from softwood by using ethylene glycol to trap reactive intermediates, similar to their process for isolated lignin (De Santi et al. 2020). The maximum yield of one guaiacyl-derived acetal (**G16**) is 9 wt% from pinewood using H_2SO_4 as the catalyst in dimethyl carbonate at 140 °C. In addition to that acetal, the ethylene glycol is well incorporated into other lignin structures via this process as well, and a high level of depolymerization is achieved. In another acid-catalyzed process, Kramarenko et al. have reported in situ depolymerization of lignin from birch, spruce, and walnut shells, maximizing monomer yield at 17 wt% from birch (Kramarenko et al. 2021). Hafezisefat et al. (2020) have reported a lignin-first non-catalytic oxidative method that produced 10.5 wt% of phenolic aldehydes and acids from oak wood in perfluorodecalin under O_2 (Hafezisefat et al. 2020). The high solubility of oxygen in fluorocarbons is noted as the reason a catalyst is unnecessary.

7.5 Downstream Processing: Separations and Upgrading

As we have seen in the above sections, the primary conversion of lignin has made many advances towards bio-oils containing high concentrations of monomeric compounds. However, in most cases, the concentration of any one compound is low, and the required separations for isolation only make sense for the highest value products. Furthermore, even with significant advances in monomer production, nearly all the reported bio-oil mixtures still also contain higher molecular weight species. Therefore, there is a significant need to upgrade the phenolic mixtures to more fungible products via novel separation technologies or further conversion (upgrading) that reduces the chemical complexity to fit the needs of potential end products.

Methods for separations of the phenolic bio-oils include liquid-liquid extraction (Wang et al. 2018a, b, c; Sun et al. 2020) and distillation (Sun et al. 2020; Koelewijn et al. 2018; Chan et al. 2020; Mante et al. 2018). Liquid-liquid extractions are useful for the separation of monomers from heavier portions on the bio-oil but are of limited value for obtaining high purities of any single compound, although some limited concentrations can be achieved based on differences in pKa of the phenolic hydroxy proton and careful use of varying the pH of the aqueous phase (Elkasabi et al. 2020). Distillation of phenolic fractions has also been reported as useful, particularly if the compounds isolated are of high value. Koelewijn et al. have reported distillation of hexane-extracted lignin-derived bio-oils rich in 4-propylguaiacol (**G7**) and 4-propylsyringol (**S7**) (Koelewijn et al. 2018). Their distillation conditions allowed for recovery of the 4-propylsyringol in > 98% purity and the 4-propylguaiacol in 67% purity, with high recovery yields (>94%) for both compounds.

The most reported form of upgrading for bio-oils from lignocellulosic biomass is hydrotreating (Schutyser et al. 2018; Shu et al. 2020; Kim et al. 2019), and bio-oils from lignin are no exception to this. Hydrotreating of bio-oils from pyrolysis or liquefaction of lignocellulosic biomass has been studied for decades and extensively reported (Elliott 2007). We will only consider briefly the specific hydrotreating of lignin-derived bio-oils here. Hydrotreating ranges a spectrum from complete deoxygenation and saturation leading to cycloalkanes to selective removal of oxygen from side chains providing alkyl aromatics or selective hydrogenation of the rings leading to cyclohexanols. Lignin oligomers can also be cracked to monoaromatics or cycloalkanes during hydrotreating. Severe hydrotreating can also be applied directly to technical lignins to directly produce hydrocarbons without the intermediate bio-oil. Most hydrotreating of lignin or lignin bio-oil is ultimately aimed at the production of hydrocarbon biofuels. Selectivity for the types of aromatics produced can be controlled by the utilization of different catalysts and conditions (Schutyser et al. 2018; Shu et al. 2020; Kim et al. 2019). Aromatic hydrocarbons can be produced via direct hydrogenolysis of ArC-O bonds. Catalysts used for this purpose include widely used petroleum refining hydrotreating catalysts such as sulfided CoMo and NiMo catalysts (Shu et al. 2020; Schutyser et al. 2018). Other catalysts active for this type of transformation include MoO_3 (Li et al. 2021) and bimetallics such as PtMo,

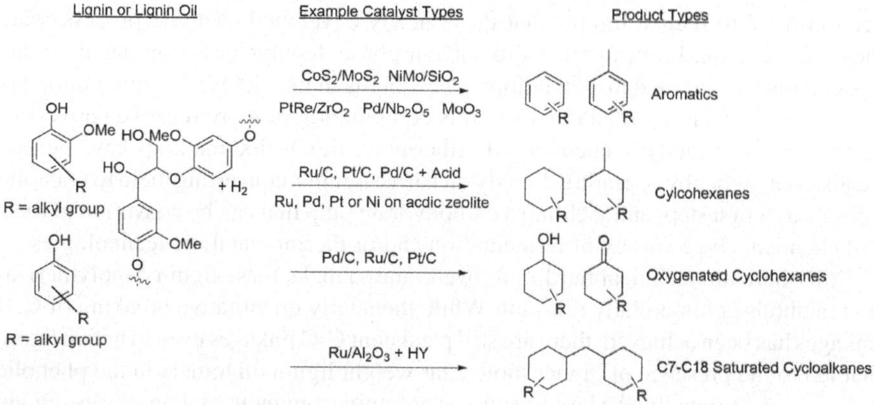

Fig. 7.12 Summary of catalysts and products form hydrotreating of lignin or depolymerized lignin liquids

PtRe, RuLa, RuZr, PdNb, and PdZr (Agarwal et al. 2019; Ohta et al. 2014; Li et al. 2017; Yohe et al. 2016). Cyclohexanes can be produced from lignin-derived phenolics by highly hydrogenation-active metal catalysts (e.g., Pd/C, Pt/C, Ru/C, Ni/C) in combination with an acidic moiety that can dehydrate the cyclohexanol intermediates (Wang et al. 2017; Agarwal et al. 2019; Rover et al. 2015; Lin et al. 2020; de Wild et al. 2017). In the absence of the acidic co-catalyst, a mixture of cyclohexanes and cylcohexanones is often produced (Teles et al. 2021). The use of Ru/Al_2O_3 and acidic Y-zeolite has been used to access a mixture of C7-C18 cycloalkanes, a range suitable for use in jet fuel, via one-pot depolymerization, HDO, and coupling of monomers from alkali lignin (Wang et al. 2018a, b, c; Shen et al. 2018). A summary of hydrotreating chemistry is provided in Fig. 7.12. A key upcoming challenge will be developing higher value, non-fuel uses for depolymerized lignin, and separations of the mixed phenolics and subsequent chemical conversion will be at the forefront of lignin conversion research.

7.6 Summary and Outlook

Efficient conversion and utilization of lignin have been a long-standing impediment to the development of environmentally and economically sustainable biorefineries. As has been demonstrated in this chapter, the last decade has seen tremendous progress in the development of catalytic methods that have greatly increased the chemical efficiency of lignin depolymerization. New chemistries have focused on breaking C-O aryl-ether (β-O-4) linkages that make up most native lignin structures. At the same time, there has been a recognition that technical lignins have largely altered structures with few of these types of bonds. Mild isolation methods or methods that use protective chemistry to prevent the formation of condensed C-C linked aromatic units have been utilized to realize lignins that these newly developed catalytic processes are most effective on. From pyrolysis to solvent phase depolymerization under acidic, basic, reductive, or oxidative conditions, new highs in the yields of lignin monomers have recently been reported thanks to this combination of more reactive native-like lignins and new catalytic chemistry. Furthermore, lignin-first methods have pushed yields even higher by operating directly on native lignins, converting them to phenolic bio-oils as a first step, and yielding a carbohydrate pulp that can be converted to fuels and chemicals by a variety of fermentation and/or thermo-catalytic technologies.

Still, there are technical hurdles to overcome to make these lignin depolymerization methods commercially relevant. While the nearly quantitative breaking of C-O linkages has been achieved, there are still persistent C-C linkages even in native lignin that led to the presence of higher molecular weight lignin oligomers in the phenolic mixtures. Additionally, the largest sources of lignin from industrial processes remain technical lignins, which come largely from pulp producers and carbohydrate-first biorefineries, nearly 50 million tons worldwide annually. There are still few processes that efficiently convert these types of materials, so methods for depolymerizing C-C linked in condensed lignins remain a significant breakthrough still to be achieved.

References

Abu-Omar MM, Barta K, Beckham GT, Luterbacher JS, Ralph J, Rinaldi R, Román-Leshkov Y, Samec JSM, Sels BF, Wang F (2021) Guidelines for performing lignin-first biorefining. Energy Environ Sci 14:262–292

Agarwal A, Park S-J, Park J-H (2019) Upgrading of kraft lignin-derived bio-oil over hierarchical and nonhierarchical Ni and/or Zn/HZSM5 catalysts. Ind Eng Chem Res 58:22791–22803

Ahmed R, Liu G, Yousaf B, Abbas Q, Ullah H, Ali MU (2020) Recent advances in carbon-based renewable adsorbent for selective carbon dioxide capture and separation-a review. J Clean Prod 242:118409

Bai X, Zhou, S. (2019) Pyrolysis of Lignin US Patent 10,190,053B2

Barrett JA, Gao Y, Bernt CM, Chui M, Tran AT, Foston MB, Ford PC (2016) Enhancing Aromatic production from reductive lignin disassembly: in situ O-Methylation of phenolic intermediates. ACS Sustain Chem Eng 4:6877–6886

Barta K, Warner GR, Beach ES, Anastas PT (2014) Depolymerization of organosolv lignin to aromatic compounds over Cu-doped porous metal oxides. Green Chem 16:191–196

Beauchet R, Monteil-Rivera F, Lavoie JM (2012) Conversion of lignin to aromatic-based chemicals (L-chems) and biofuels (L-fuels). Bioresour Technol 121:328–334

Beis SH, Mukkamala S, Hill N, Joseph J, Baker C, Jensen B, Stemmler EA, Wheeler MC, Frederick BG, van Heiningen ARP, Berg AG, deSisto WJ (2010) Fast pyrolysis of lignins. BioResources 5:1408–1424

Biswas B, Kumar A, Kaur R, Krishna BB, Bhaskar T (2021) Catalytic hydrothermal liquefaction of alkali lignin over activated bio-char supported bimetallic catalyst. Bioresour Technol 337:125439

Bond JQ, Upadhye AA, Olcay H, Tompsett GA, Jae J, Xing R, Alonso DM, Wang D, Zhang T, Kumar R, Foster A, Sen SM, Maravelias CT, Malina R, Barrett SRH, Lobo R, Wyman CE, Dumesic JA, Huber GW (2014) Production of renewable jet fuel range alkanes and commodity chemicals from integrated catalytic processing of biomass. Energy Environ Sci 7:1500–1523

Bourbiaux D, Pu J, Rataboul F, Djakovitch L, Geantet C, Laurenti D (2021) Reductive or oxidative catalytic lignin depolymerization: an overview of recent advances. Catal Today 373:24–37

Bouxin FP, McVeigh A, Tran F, Westwood NJ, Jarvis MC, Jackson SD (2015) Catalytic depolymerisation of isolated lignins to fine chemicals using a Pt/alumina catalyst: part 1—impact of the lignin structure. Green Chem 17:1235–1242

Braghiroli FL, Bouafif H, Neculita CM, Koubaa A (2019) Influence of pyro-gasification and activation conditions on the porosity of activated biochars: a literature review. Waste Biomass Valoriz 11:5079–5098

Cao S, Pu Y, Studer M, Wyman C, Ragauskas AJ (2012) Chemical transformations of Populus trichocarpa during dilute acid pretreatment. RSC Adv 2:10925–10936

Cao Y, Chen SS, Zhang S, Ok YS, Matsagar BM, Wu KC, Tsang DCW (2019) Advances in lignin valorization towards bio-based chemicals and fuels: lignin biorefinery. Bioresour Technol 291:121878

Cao C, Xie Y, Li L, Wei W, Jin H, Wang S, Li W (2021) Supercritical water gasification of lignin and cellulose catalyzed with Co-precipitated CeO2–ZrO2. Energy Fuels 35:6030–6039

Case PA, Truong C, Wheeler MC, DeSisto WJ (2015) Calcium-catalyzed pyrolysis of lignocellulosic biomass components. Bioresour Technol 192:247–252

Chan YH, Loh SK, Chin BLF, Yiin CL, How BS, Cheah KW, Wong MK, Loy ACM, Gwee YL, Lo SLY, Yusup S, Lam SS (2020) Fractionation and extraction of bio-oil for production of greener fuel and value-added chemicals: Recent advances and future prospects. Chem Eng J 397:125406

Chaudhary R, Dhepe PL (2017) Solid base catalyzed depolymerization of lignin into low molecular weight products. Green Chem 19:778–788

Chen L, Koranyi TI, Hensen EJ (2016) Transition metal (Ti, Mo, Nb, W) nitride catalysts for lignin depolymerisation. Chem Commun (camb) 52:9375–9378

Chen P, Zhang Q, Shu R, Xu Y, Ma L, Wang T (2017) Catalytic depolymerization of the hydrolyzed lignin over mesoporous catalysts. Bioresour Technol 226:125–131

Chen M, Shi J, Wang Y, Tang Z, Yang Z, Wang J, Zhang H (2021a) Conversion of Kraft lignin to phenol monomers and liquid fuel over trimetallic catalyst W-Ni-Mo/sepiolite under supercritical ethanol. Fuel 303:121332

Chen M, Zhang J, Wang Y, Tang Z, Shi J, Wang C, Yang Z, Wang J, Zhang H (2021b) Lignin catalytic depolymerization for liquid fuel and phenols by using Mo/sepiolite catalysts calcined at different temperature. J Environ Chem Eng 9:105348

Cheng C, Li P, Yu W, Shen D, Jiang X, Gu S (2020) Nonprecious metal/bimetallic catalytic hydrogenolysis of lignin in a mixed-solvent system. ACS Sustain Chem Eng 8:16217–16228

Choi HS, Meier D (2013) Fast pyrolysis of Kraft lignin—vapor cracking over various fixed-bed catalysts. J Anal Appl Pyrolysis 100:207–212

Cornejo A, Bimbela F, Moreira R, Hablich K, Garcia-Yoldi I, Maisterra M, Portugal A, Gandia LM, Martinez Merino V (2020) Production of aromatic compounds by catalytic depolymerization of technical and downstream biorefinery lignins. Biomolecules 10:1338

Dabral S, Engel J, Mottweiler J, Spoehrle SSM, Lahive CW, Bolm C (2018) Mechanistic studies of base-catalysed lignin depolymerisation in dimethyl carbonate. Green Chem 20:170–182

de Albuquerque M, Fragoso D, Bouxin FP, Montgomery JRD, Westwood NJ, Jackson SD (2020) Catalytic depolymerisation of isolated lignin to fine chemicals: depolymerisation of Kraft lignin. Bioresource Technol Reports 9:100400

de Wild PJ, Huijgen WJJ, Heeres HJ (2012) Pyrolysis of wheat straw-derived organosolv lignin. J Anal Appl Pyrolysis 93:95–103

de Wild PJ, Huijgen WJ, Kloekhorst A, Chowdari RK, Heeres HJ (2017) Biobased alkylphenols from lignins via a two-step pyrolysis-Hydrodeoxygenation approach. Bioresour Technol 229:160–168

De Santi A, Galkin MV, Lahive CW, Deuss PJ, Barta K (2020) Lignin-first fractionation of softwood lignocellulose using a mild dimethyl carbonate and ethylene glycol organosolv process. Chemsuschem 13:4468–4477

Deepa AK, Dhepe PL (2014a) Lignin depolymerization into aromatic monomers over solid acid catalysts. ACS Catal 5:365–379

Deepa AK, Dhepe PL (2014b) Solid acid catalyzed depolymerization of lignin into value added aromatic monomers. RSC Adv 4:12625

Deng H, Lin L, Sun Y, Pang C, Zhuang J, Ouyang P, Li J, Liu S (2009) Activity and stability oif perovskite-type Oxide LaCoO3 catalyst in lignin wet oxidation to aromatic aldehydes process. Energy Fuels 23:19–24
Deng W, Zhang H, Wu X, Li R, Zhang Q, Wang Y (2015) Oxidative conversion of lignin and lignin model compounds catalyzed by CeO2-supported Pd nanoparticles. Green Chem 17:5009–5018
Deuss PJ, Scott M, Tran F, Westwood NJ, de Vries JG, Barta K (2015) Aromatic monomers by in situ conversion of reactive intermediates in the acid-catalyzed depolymerization of lignin. J Am Chem Soc 137:7456–7467
Deuss PJ, Lahive CW, Lancefield CS, Westwood NJ, Kamer PC, Barta K, de Vries JG (2016) Metal triflates for the production of aromatics from lignin. Chemsuschem 9:2974–2981
Deuss PJ, Lancefield CS, Narani A, de Vries JG, Westwood NJ, Barta K (2017) Phenolic acetals from lignins of varying compositions via iron(iii) triflate catalysed depolymerisation. Green Chem 19:2774–2782
Elkasabi Y, Mullen CA (2021) Progress on biobased industrial carbons as thermochemical biorefinery coproducts. Energy Fuels 35:5627–5642
Elkasabi Y, Mullen CA, Boateng AA (2020) Continuous extraction of phenol and cresols from advanced pyrolysis oils. SN Appl Sci 2:491
Elliott DC (2007) Historical developments in hydroprocessing Bio-oils. Energy Fuels 21:1792–1815
Erdocia X, Prado R, Fernández-Rodríguez J, Labidi J (2015) Depolymerization of different organosolv lignins in supercritical methanol, ethanol, and acetone to produce phenolic monomers. ACS Sustain Chem Eng 4:1373–1380
Fache M, Boutevin B, Caillol S (2015) Vanillin production from lignin and its use as a renewable chemical. ACS Sustain Chem Eng 4:35–46
Faix O, Meier D, Grobe I (1987) Studies on isolated lignins and lignins in wood biomass by pyrolysis-gas chromatography-mass spectometry and off-line pyrolysis-gas chromatography with flame ionization detection. J Anal Appl Pyrolysis 11:403–416
Fan L, Zhang Y, Liu S, Zhou N, Chen P, Cheng Y, Addy M, Lu Q, Omar MM, Liu Y, Wang Y, Dai L, Anderson E, Peng P, Lei H, Ruan R (2017) Bio-oil from fast pyrolysis of lignin: effects of process and upgrading parameters. Bioresour Technol 241:1118–1126
Fargues C, Mathias A, Rodrigues A (1996) Kinetics of vanillin production from kraft lignin oxidation. Ind Eng Chem Res 25:28–36
Fernández-Rodríguez J, Erdocia X, Sánchez C, González Alriols M, Labidi J (2017) Lignin depolymerization for phenolic monomers production by sustainable processes. J Energy Chem 26:622–631
Galkin MV, Smit AT, Subbotina E, Artemenko KA, Bergquist J, Huijgen WJ, Samec JS (2016) Hydrogen-free catalytic fractionation of woody biomass. Chemsuschem 9:3280–3287
Gan L, Pan X (2019) Phenol-enhanced depolymerization and activation of kraft lignin in alkaline medium. Ind Eng Chem Res 58:7794–7800
Ghysels S, Dubuisson B, Pala M, Rohrbach L, Van den Bulcke J, Heeres HJ, Ronsse F (2020) Improving fast pyrolysis of lignin using three additives with different modes of action. Green Chem 22:6471–6488
Gillet S, Aguedo M, Petitjean L, Morais ARC, da Costa Lopes AM, Łukasik RM, Anastas PT (2017) Lignin transformations for high value applications: towards targeted modifications using green chemistry. Green Chem 19:4200–4233
Gosselink RJA, de Jong E, Guran B, Abächerli A (2004) Co-ordination network for lignin—standardisation, production and applications adapted to market requirements (EUROLIGNIN). Ind Crops Prod 20:121–129
Guo W, Zhang B, Zhang R, Zhang J, Li Y, Wu Z, Ma J, Yang B (2021) Thermodynamics and kinetics analysis from liquid chemical looping gasification of lignin with bismuth-based oxygen carrier. Fuel Process Technol 219:106888
Güvenatam B, Heeres EHJ, Pidko EA, Hensen EJM (2016a) Lewis-acid catalyzed depolymerization of Protobind lignin in supercritical water and ethanol. Catal Today 259:460–466

Güvenatam B, Heeres EHJ, Pidko EA, Hensen EJM (2016b) Lewis acid-catalyzed depolymerization of soda lignin in supercritical ethanol/water mixtures. Catal Today 269:9–20

Ha J-M, Hwang K-R, Kim Y-M, Jae J, Kim KH, Lee HW, Kim J-Y, Park Y-K (2019) Recent progress in the thermal and catalytic conversion of lignin. Renew Sustain Energy Rev 111:422–441

Hafezisefat P, Lindstrom JK, Brown RC, Qi L (2020) Non-catalytic oxidative depolymerization of lignin in perfluorodecalin to produce phenolic monomers. Green Chem 22:6567–6578

Hu Y, Gong M, Xing X, Wang H, Zeng Y, Xu CC (2020) Supercritical water gasification of biomass model compounds: a review. Renew Sustain Energy Rev 118:109529

Huang X, Koranyi TI, Boot MD, Hensen EJ (2014) Catalytic depolymerization of lignin in supercritical ethanol. Chemsuschem 7:2276–2288

Huang X, Korányi TI, Boot MD, Hensen EJM (2015) Ethanol as capping agent and formaldehyde scavenger for efficient depolymerization of lignin to aromatics. Green Chem 17:4941–4950

Huang X, Zhu J, Koranyi TI, Boot MD, Hensen EJ (2016) Effective release of lignin fragments from lignocellulose by Lewis acid metal Triflates in the lignin-first approach. Chemsuschem 9:3262–3267

Huang X, Atay C, Zhu J, Palstra SWL, Koranyi TI, Boot MD, Hensen EJM (2017a) Catalytic depolymerization of lignin and woody biomass in supercritical ethanol: influence of reaction temperature and feedstock. ACS Sustain Chem Eng 5:10864–10874

Huang X, Morales Gonzalez OM, Zhu J, Korányi TI, Boot MD, Hensen EJM (2017b) Reductive fractionation of woody biomass into lignin monomers and cellulose by tandem metal triflate and Pd/C catalysis. Green Chem 19:175–187

Huang WH, Lee DJ, Huang C (2021) Modification on biochars for applications: a research update. Bioresour Technol 319:124100

Ji N, Diao X, Li X, Jia Z, Zhao Y, Lu X, Song C, Liu Q, Li C (2020) Toward Alkylphenols production: lignin depolymerization coupling with Methoxy removal over supported MoS2 catalyst. Ind Eng Chem Res 59:17287–17299

Kasakov S, Shi H, Camaioni DM, Zhao C, Baráth E, Jentys A, Lercher JA (2015) Reductive deconstruction of organosolv lignin catalyzed by zeolite supported nickel nanoparticles. Green Chem 17:5079–5090

Katahira R, Mittal A, McKinney K, Chen X, Tucker MP, Johnson DK, Beckham GT (2016) Base-catalyzed depolymerization of biorefinery lignins. ACS Sustain Chem Eng 4:1474–1486

Kawamoto H (2017) Lignin pyrolysis reactions. J Wood Sci 63:117–132

Kim KH, Brown RC, Kieffer M, Bai X (2014) Hydrogen-donor-assisted solvent liquefaction of lignin to short-chain Alkylphenols using a micro reactor/gas chromatography system. Energy Fuels 28:6429–6437

Kim J-Y, Park J, Kim U-J, Choi JW (2015) Conversion of lignin to phenol-rich oil fraction under supercritical alcohols in the presence of metal catalysts. Energy Fuels 29:5154–5163

Kim S, Kwon EE, Kim YT, Jung S, Kim HJ, Huber GW, Lee J (2019) Recent advances in hydrodeoxygenation of biomass-derived oxygenates over heterogeneous catalysts. Green Chem 21:3715–3743

Klein I, Marcum C, Kenttämaa H, Abu-Omar MM (2016) Mechanistic investigation of the Zn/Pd/C catalyzed cleavage and hydrodeoxygenation of lignin. Green Chem 18:2399–2405

Kloekhorst A, Shen Y, Yie Y, Fang M, Heeres HJ (2015) Catalytic hydrodeoxygenation and hydrocracking of Alcell ® lignin in alcohol/formic acid mixtures using a Ru/C catalyst. Biomass Bioenergy 80:147–161

Koelewijn SF, Cooreman C, Renders T, Andecochea Saiz C, Van den Bosch S, Schutyser W, De Leger W, Smet M, Van Puyvelde P, Witters H, Van der Bruggen B, Sels BF (2018) Promising bulk production of a potentially benign bisphenol a replacement from a hardwood lignin platform. Green Chem 20:1050–1058

Kramarenko A, Etit D, Laudadio G, D'Angelo FN (2021) Beta-zeolite-assisted lignin-first fractionation in a flow-through reactor*. Chemsuschem 14:3838–3849

Kruger JS, Cleveland NS, Zhang S, Katahira R, Black BA, Chupka GM, Lammens T, Hamilton PG, Biddy MJ, Beckham GT (2016) Lignin depolymerization with nitrate-intercalated hydrotalcite catalysts. ACS Catal 6:1316–1328
Kumaniaev I, Samec JSM (2018) Valorization of Quercus suber bark toward hydrocarbon Bio-Oil and 4-ethylguaiacol. ACS Sustain Chem Eng 6:5737–5742
Kumaniaev I, Subbotina E, Sävmarker J, Larhed M, Galkin MV, Samec JSM (2017) Lignin depolymerization to monophenolic compounds in a flow-through system. Green Chem 19:5767–5771
Kumar M, Olajire Oyedun A, Kumar A (2018) A review on the current status of various hydrothermal technologies on biomass feedstock. Renew Sustain Energy Rev 81:1742–1770
Kumar A, Biswas B, Kaur R, Krishna BB, Bhaskar T (2021) Hydrothermal oxidative valorisation of lignin into functional chemicals: a review. Bioresour Technol:126016
Lancefield CS, Ojo OS, Tran F, Westwood NJ (2015) Isolation of functionalized phenolic monomers through selective oxidation and C-O bond cleavage of the beta-O-4 linkages in lignin. Angew Chem Int Ed Engl 54:258–262
Lancefield CS, Rashid GMM, Bouxin F, Wasak A, Tu W-C, Hallett J, Zein S, Rodríguez J, Jackson SD, Westwood NJ, Bugg TDH (2016) Investigation of the Chemocatalytic and Biocatalytic valorization of a range of different lignin preparations: the importance of β-O-4 content. ACS Sustain Chem Eng 4:6921–6930
Langholtz MH, Stokes BJ, Eaton LM (2016) 2016 Billion-ton report: advancing domestic resources for a thriving bioeconomy, Volume 1: economic availability of feedstocks. In edited by Energy U S D o. Oak Ridge, TN
Lee CS, Conradie AV, Lester E (2021) Review of supercritical water gasification with lignocellulosic real biomass as the feedstocks: process parameters, biomass composition, catalyst development, reactor design and its challenges. Chem Eng J 415:128837
Leng E, Guo Y, Chen J, Liu S, E J, Xue Y (2022) A comprehensive review on lignin pyrolysis: mechanism, modeling and the effects of inherent metals in biomass. Fuel, 309
Li C, Zheng M, Wang A, Zhang T (2012) One-pot catalytic hydrocracking of raw woody biomass into chemicals over supported carbide catalysts: simultaneous conversion of cellulose, hemicellulose and lignin. Energy Environ Sci 5:6383–6390
Li J, Li X, Zhou G, Wang W, Wang C, Komarneni S, Wang Y (2014) Catalytic fast pyrolysis of biomass with mesoporous ZSM-5 zeolites prepared by desilication with NaOH solutions. Appl Catal A 470:115–122
Li B, Li L, Zhao C (2017) A highly stable Ru/LaCO3OH catalyst consisting of support-coated Ru nanoparticles in aqueous-phase hydrogenolysis reactions. Green Chem 19:5412–5421
Li N, Zhang X, Zhang Q, Chen L, Li Y, Wang C, Ma L (2021) Mo-based catalyst for chemical looping deoxygenation of phenolic compounds to aromatic hydrocarbons. Fuel Process Technol 221:106936
Lin B, Li R, Shu R, Wang C, Yuan Z, Liu Y, Chen Y (2020) Synergistic effect of highly dispersed Ru and moderate acid site on the hydrodeoxygenation of phenolic compounds and raw bio-oil. J Energy Inst 93:847–856
Liu X, Bouxin FP, Fan J, Budarin VL, Hu C, Clark JH (2020) Recent advances in the catalytic depolymerization of lignin towards phenolic chemicals: a review. Chemsuschem 13:4296–4317
Løhre C, Barth T, Kleinert M (2016) The effect of solvent and input material pretreatment on product yield and composition of bio-oils from lignin solvolysis. J Anal Appl Pyrolysis 119:208–216
Long J, Zhang Q, Wang T, Zhang X, Xu Y, Ma L (2014) An efficient and economical process for lignin depolymerization in biomass-derived solvent tetrahydrofuran. Bioresour Technol 154:10–17
Lotfi S, Boffito DC, Patience GS (2015) Gas-Phase Partial oxidation of lignin to carboxylic acids over vanadium pyrophosphate and aluminum-vanadium-molybdenum. Chemsuschem 8:3424–3432
Lourenço A, Pereira H (2018) Compositional variability of lignin in biomass. In Lignin-Trends and Applications

Luo Z, Qin S, Chen S, Hui Y, Zhao C (2020) Selective conversion of lignin to ethylbenzene. Green Chem 22:1842–1850

Ma R, Hao W, Ma X, Tian Y, Li Y (2014) Catalytic ethanolysis of Kraft lignin into high-value small-molecular chemicals over a nanostructured alpha-molybdenum carbide catalyst. Angew Chem Int Ed Engl 53:7310–7315

Ma X, Cui K, Hao W, Ma R, Tian Y, Li Y (2015a) Alumina supported molybdenum catalyst for lignin valorization: effect of reduction temperature. Bioresour Technol 192:17–22

Ma X, Ma R, Hao W, Chen M, Yan F, Cui K, Tian Y, Li Y (2015b) Common pathways in ethanolysis of kraft lignin to platform chemicals over molybdenum-based catalysts. ACS Catal 5:4803–4813

Ma R, Guo M, Lin KT, Hebert VR, Zhang J, Wolcott MP, Quintero M, Ramasamy KK, Chen X, Zhang X (2016) Peracetic acid depolymerization of biorefinery lignin for production of selective monomeric phenolic compounds. Chemistry 22:10884–10891

Ma H, Li H, Zhao W, Li L, Liu S, Long J, Li X (2019) Selective depolymerization of lignin catalyzed by nickel supported on zirconium phosphate. Green Chem 21:658–668

Mante OD, Thompson SJ, Soukri M, Dayton DC (2018) Isolation and purification of monofunctional methoxyphenols from loblolly pine biocrude. ACS Sustain Chem Eng 7:2262–2269

Mardiana S, Azhari NJ, Ilmi T, Kadja GTM (2022) Hierarchical zeolite for biomass conversion to biofuel: a review. Fuel 309:122119

McVeigh A, Bouxin FP, Jarvis MC, Jackson SD (2016) Catalytic depolymerisation of isolated lignin to fine chemicals: part 2 – process optimisation. Catal Sci Technol 6:4142–4150

Moxley G, Gaspar AR, Higgins D, Xu H (2012) Structural changes of corn stover lignin during acid pretreatment. J Ind Microbiol Biotechnol 39:1289–1299

Mukkamala S, Wheeler MC, Van Heiningen ARP, Desisto WJ (2012) Formate-assisted fast pyrolysis of lignin. Energy Fuels 26:1380–1384

Mullen CA, Boateng AA (2010) Catalytic pyrolysis-GC/MS of lignin from several sources. Fuel Process Technol 91:1446–1458

Mullen CA, Boateng AA (2013) Accumulation of inorganic impurities on HZSM-5 zeolites during catalytic fast pyrolysis of switchgrass. Ind Eng Chem Res 52:17156–17161

Mullen C.A. (2019) Upgrading of biomass via catalytic fast pyrolysis (CFP), In: Chemical Catalysts for Biomass Upgrading, Editor(s): Crocker M., Santillan-Jimenez E., Wiley.

Nolte MW, Saraeian A, Shanks BH (2017) Hydrodeoxygenation of cellulose pyrolysis model compounds using molybdenum oxide and low pressure hydrogen. Green Chem 19:3654–3664

Nowakowski DJ, Bridgwater AV, Elliott DC, Meier D, de Wild P (2010) Lignin fast pyrolysis: results from an international collaboration. J Anal Appl Pyrolysis 88:53–72

Ohta H, Feng B, Kobayashi H, Hara K, Fukuoka A (2014) Selective hydrodeoxygenation of lignin-related 4-propylphenol into n-propylbenzene in water by Pt-Re/ZrO2 catalysts. Catal Today 234:139–144

Onwudili JA, Williams PT (2014) Catalytic depolymerization of alkali lignin in subcritical water: influence of formic acid and Pd/C catalyst on the yields of liquid monomeric aromatic products. Green Chem 16:4740–4748

Partenheimer W (2009) The aerobic oxidative cleavage of lignin to produce hydroxyaromatic benzaldehydes and carboxylic acids via metal/bromide catalysts in acetic acid/water mixtures. Adv Synth Catal 351:456–466

Patel M, Hill N, Mullen CA, Gunukula S, DeSisto WJ (2020) Fast pyrolysis of lignin pretreated with magnesium formate and magnesium hydroxide. Energies 13:4995

Patil V, Adhikari S, Cross P, Jahromi H (2020) Progress in the solvent depolymerization of lignin. Renew Sustain Energy Rev, 133:110359

Paysepar H, Hu Y, Feng S, Yuan Z, Shui H, Xu C (2020a) Bio-phenol formaldehyde (BPF) resoles prepared using phenolic extracts from the biocrude oils derived from hydrothermal liquefaction of hydrolysis lignin. React Funct Polym 146:104442

Paysepar H, Venkateswara Rao KT, Yuan Z, Shui H, Xu C (2020b) Production of phenolic chemicals from hydrolysis lignin via catalytic fast pyrolysis. J Anal Appl Pyrolysis 149:104842

Pienihäkkinen E, Lindfors C, Ohra-aho T, Lehtonen J, Granström T, Yamamoto M, Oasmaa A (2021) Fast pyrolysis of hydrolysis lignin in fluidized bed reactors. Energy Fuels 35:14758–14769
Poveda-Giraldo JA, Solarte-Toro JC, Cardona Alzate CA (2021) The potential use of lignin as a platform product in biorefineries: a review. Renew Sustain Energy Rev 138:110688
Ragauskas AJ, Beckham GT, Biddy MJ, Chandra R, Chen F, Davis MF, Davison BH, Dixon RA, Gilna P, Keller M, Langan P, Naskar AK, Saddler JN, Tschaplinski TJ, Tuskan GA, Wyman CE (2014) Lignin valorization: improving lignin processing in the biorefinery. Science 344:1246843
Ralph J, Lapierre C, Boerjan W (2019) Lignin structure and its engineering. Curr Opin Biotechnol 56:240–249
Rawat S, Gupta P, Singh B, Bhaskar T, Natte K, Narani A (2020) Molybdenum-catalyzed oxidative depolymerization of alkali lignin: Selective production of Vanillin. Appl Catal A 598:117567
Renders T, Van den Bossche G, Vangeel T, Van Aelst K, Sels B (2019) Reductive catalytic fractionation: state of the art of the lignin-first biorefinery. Curr Opin Biotechnol 56:193–201
Riaz A, Verma D, Zeb H, Lee JH, Kim JC, Kwak SK, Kim J (2018) Solvothermal liquefaction of alkali lignin to obtain a high yield of aromatic monomers while suppressing solvent consumption. Green Chem 20:4957–4974
Rinaldi R, Jastrzebski R, Clough MT, Ralph J, Kennema M, Bruijnincx PC, Weckhuysen BM (2016) Paving the way for lignin valorisation: recent advances in bioengineering, biorefining and catalysis. Angew Chem Int Ed Engl 55:8164–8215
Roberts VM, Stein V, Reiner T, Lemonidou A, Li X, Lercher JA (2011) Towards quantitative catalytic lignin depolymerization. Chemistry 17:5939–5948
Rover MR, Hall PH, Johnston PA, Smith RG, Brown RC (2015) Stabilization of bio-oils using low temperature, low pressure hydrogenation. Fuel 153:224–230
San Miguel G, Domínguez MP, Hernández M, Sanz-Pérez F (2012) Characterization and potential applications of solid particles produced at a biomass gasification plant. Biomass Bioenergy 47:134–144
Sansaniwal SK, Pal K, Rosen MA, Tyagi SK (2017) Recent advances in the development of biomass gasification technology: a comprehensive review. Renew Sustain Energy Rev 72:363–384
Santos RB, Hart PW, Jameel H, Chang H (2013) Wood based lignin reactions important to the biorefinery and pulp and paper industries. BioResources 8:1456–1477
Schutyser W, Renders T, Van den Bosch S, Koelewijn SF, Beckham GT, Sels BF (2018) Chemicals from lignin: an interplay of lignocellulose fractionation, depolymerisation, and upgrading. Chem Soc Rev 47:852–908
Shafaghat H, Rezaei PS, Ro D, Jae J, Kim B-S, Jung S-C, Sung BH, Park Y-K (2017) In-situ catalytic pyrolysis of lignin in a bench-scale fixed bed pyrolyzer. J Ind Eng Chem 54:447–453
Shen R, Tao L, Yang B (2018) Techno-economic analysis of jet-fuel production from biorefinery waste lignin. Biofuels Bioprod Biorefin 13:486–501
Shu R, Long J, Yuan Z, Zhang Q, Wang T, Wang C, Ma L (2015) Efficient and product-controlled depolymerization of lignin oriented by metal chloride cooperated with Pd/C. Bioresour Technol 179:84–90
Shu R, Xu Y, Ma L, Zhang Q, Wang C, Chen Y (2018) Controllable production of guaiacols and phenols from lignin depolymerization using Pd/C catalyst cooperated with metal chloride. Chem Eng J 338:457–464
Shu R, Li R, Lin B, Wang C, Cheng Z, Chen Y (2020) A review on the catalytic hydrodeoxygenation of lignin-derived phenolic compounds and the conversion of raw lignin to hydrocarbon liquid fuels. Biomass Bioenergy 132:105432
Shuai L, Amiri MT, Questell-Santiago YM, Heroguel F, Li Y, Kim H, Meilan R, Chapple C, Ralph J, Luterbacher JS (2016) Formaldehyde stabilization facilitates lignin monomer production during biomass depolymerization. Science 354:329–333
Snowdon MR, Mohanty AK, Misra M (2014) A study of carbonized lignin as an alternative to carbon black. ACS Sustain Chem Eng 2:1257–1263

Song Y, Mobley JK, Motagamwala AH, Isaacs M, Dumesic JA, Ralph J, Lee AF, Wilson K, Crocker M (2018) Gold-catalyzed conversion of lignin to low molecular weight aromatics. Chem Sci 9:8127–8133

Stanzione J, La Scala J (2016) Sustainable polymers and polymer science: dedicated to the life and work of Richard P. Wool. J Appl Polym Sci 133:43794

Sun Z, Cheng J, Wang D, Yuan TQ, Song G, Barta K (2020) Downstream processing strategies for lignin-first biorefinery. Chemsuschem 13:5199–5212

Tarabanko VE, Petukhov DV, Selyutin GE (2004) New mechanism for the catalytic oxidation of lignin to vanillin. Kinet Catal 45:569–577

Teles CA, de Souza PM, Rabelo-Neto RC, Teran A, Jacobs G, Resasco DE, Noronha FB (2021) Hydrodeoxygenation of lignin-derived compound mixtures on Pd-supported on various oxides. ACS Sustain Chem Eng 9:12870–12884

Thring RW (1994) Alkaline degradation of Alcell@ Lignin. Biomass Bioenergy 7:125–130

Toledano A, Serrano L, Labidi J (2012) Organosolv lignin depolymerization with different base catalysts. J Chem Technol Biotechnol 87:1593–1599

Torr KM, van de Pas DJ, Cazeils E, Suckling ID (2011) Mild hydrogenolysis of in-situ and isolated Pinus radiata lignins. Bioresour Technol 102:7608–7611

Trinh TN, Jensen PA, Sárossy Z, Dam-Johansen K, Knudsen NO, Sørensen HR, Egsgaard H (2013) Fast pyrolysis of lignin using a pyrolysis centrifuge reactor. Energy Fuels 27:3802–3810

Tymchyshyn M, Rezayan A, Yuan Z, Zhang Y, Xu CC (2020) Reductive hydroprocessing of hydrolysis lignin over efficient bimetallic catalyst MoRu/AC. Ind Eng Chem Res 59:17239–17249

Van den Bosch S, Schutyser W, Koelewijn SF, Renders T, Courtin CM, Sels BF (2015a) Tuning the lignin oil OH-content with Ru and Pd catalysts during lignin hydrogenolysis on birch wood. Chem Commun (camb) 51:13158–13161

Van den Bosch S, Schutyser W, Vanholme R, Driessen T, Koelewijn SF, Renders T, De Meester B, Huijgen WJJ, Dehaen W, Courtin CM, Lagrain B, Boerjan W, Sels BF (2015b) Reductive lignocellulose fractionation into soluble lignin-derived phenolic monomers and dimers and processable carbohydrate pulps. Energy Environ Sci 8:1748–1763

Van den Bosch S, Renders T, Kennis S, Koelewijn SF, Van den Bossche G, Vangeel T, Deneyer A, Depuydt D, Courtin CM, Thevelein JM, Schutyser W, Sels BF (2017) Integrating lignin valorization and bio-ethanol production: on the role of Ni-Al2O3catalyst pellets during lignin-first fractionation. Green Chem 19:3313–3326

Vishtal A, Kraslawski A (2011) Challenges in industrial applications of technical lignins. BioResources 6:3547–3568

Voitl T, Rudolf von Rohr P (2008) Oxidation of lignin using aqueous polyoxometalates in the presence of alcohols. Chemsuschem 1:763–769

Wang H, Ruan H, Feng M, Qin Y, Job H, Luo L, Wang C, Engelhard MH, Kuhn E, Chen X, Tucker MP, Yang B (2017) One-Pot Process for Hydrodeoxygenation of Lignin to Alkanes Using Ru-Based Bimetallic and Bifunctional Catalysts Supported on Zeolite Y. Chemsuschem 10:1846–1856

Wang H, Wang H, Kuhn E, Tucker MP, Yang B (2018a) Production of jet fuel-range hydrocarbons from hydrodeoxygenation of lignin over super lewis acid combined with metal catalysts. Chemsuschem 11:285–291

Wang S, Shuai L, Saha B, Vlachos DG, Epps TH 3rd (2018b) From tree to tape: direct synthesis of pressure sensitive adhesives from depolymerized raw lignocellulosic biomass. ACS Cent Sci 4:701–708

Wang X, Du B, Pu L, Guo Y, Li H, Zhou J (2018c) Effect of particle size of HZSM-5 zeolite on the catalytic depolymerization of organosolv lignin to phenols. J Anal Appl Pyrolysis 129:13–20

Wang S, Li Z, Yi W, Fu P, Zhang A, Bai X (2021a) Renewable aromatic hydrocarbons production from catalytic pyrolysis of lignin with Al-SBA-15 and HZSM-5: synergistic effect and coke behaviour. Renew Energy 163:1673–1681

Wang T, Liu X, Liu H, He M (2021b) Synergistic effect of supercritical water and nano-catalyst on lignin gasification. Int J Hydrogen Energy 46:34626–34637

Wanmolee W, Laosiripojana N, Daorattanachai P, Moghaddam L, Rencoret J, del Río JC, Doherty WOS (2018) Catalytic conversion of organosolv lignins to phenolic monomers in different organic solvents and effect of operating conditions on yield with methyl isobutyl ketone. ACS Sustain Chem Eng 6:3010–3018

Xiang Q, Lee YY (2001) Production of oxychemicals from precipitated harwood lignin. Appl Biochem Biotechnol 91–93:71–80

Xu W, Miller SJ, Agrawal PK, Jones CW (2012) Depolymerization and hydrodeoxygenation of switchgrass lignin with formic acid. Chemsuschem 5:667–675

Xue Y, Sharma A, Huo J, Qu W, Bai X (2020) Low-pressure two-stage catalytic hydropyrolysis of lignin and lignin-derived phenolic monomers using zeolite-based bifunctional catalysts. J Anal Appl Pyrolysis 146:104779

Yan B, Lin X, Chen Z, Cai Q, Zhang S (2021) Selective production of phenolic monomers via high efficient lignin depolymerization with a carbon based nickel-iron-molybdenum carbide catalyst under mild conditions. Bioresour Technol 321:124503

Yohe SL, Choudhari HJ, Mehta DD, Dietrich PJ, Detwiler MD, Akatay CM, Stach EA, Miller JT, Delgass WN, Agrawal R, Ribeiro FH (2016) High-pressure vapor-phase hydrodeoxygenation of lignin-derived oxygenates to hydrocarbons by a PtMo bimetallic catalyst: product selectivity, reaction pathway, and structural characterization. J Catal 344:535–552

Zakzeski J, Jongerius AL, Bruijnincx PC, Weckhuysen BM (2012) Catalytic lignin valorization process for the production of aromatic chemicals and hydrogen. Chemsuschem 5:1602–1609

Zhang X, Zhang Q, Long J, Wang T, Ma L, Li Y (2014) Phenolics production through catalytic depolymerization of alkali lignin with metal chlorides. BioResources 9:3347–3360

Zhang B, Li W, Zhang T, Li X, Ogunbiyi AT, Chen K, Shen C (2021) Study on the removal and depolymerization of lignin from corn stover through the synergistic effect of Brønsted acid, Lewis acid and hydrogenation sites. Fuel 305:121509

Zhao Y, Xu Q, Pan T, Zuo Y, Fu Y, Guo Q-X (2013) Depolymerization of lignin by catalytic oxidation with aqueous polyoxometalates. Appl Catal A 467:504–508

Zhao L, Shi S, Liu M, Zhu G, Wang M, Du W, Gao J, Xu J (2018) Covalent triazine framework catalytic oxidative cleavage of lignin models and organosolv lignin. Green Chem 20:1270–1279

Zhao W, Li X, Li H, Zheng X, Ma H, Long J, Li X (2019) Selective hydrogenolysis of lignin catalyzed by the cost-effective Ni metal supported on alkaline MgO. ACS Sustain Chem Eng 7:19750–19760

Zhou XF (2014) Conversion of kraft lignin under hydrothermal conditions. Bioresour Technol 170:583–586

Zhou X-F, Lu X-J (2016) Co(salen) supported on graphene oxide for oxidation of lignin. J Appl Polym Sci 133:44133

Zhou S, Brown RC, Bai X (2015) The use of calcium hydroxide pretreatment to overcome agglomeration of technical lignin during fast pyrolysis. Green Chem 17:4748–4759

Zhou S, Xue Y, Sharma A, Bai X (2016a) 'Lignin valorization through thermochemical conversion: comparison of hardwood softwood and herbaceous lignin. ACS Sustain Chem Eng 4:6608–6617

Zhou M, Sharma BK, Liu P, Ye J, Xu J, Jiang J-C (2018) Catalytic in situ hydrogenolysis of lignin in supercritical ethanol: effect of phenol catalysts, and reaction temperature. ACS Sustain Chem Eng 6:6867–6875

Zhou M, Fakayode OA, Ahmed Yagoub AE, Ji Q, Zhou C (2022) Lignin fractionation from lignocellulosic biomass using deep eutectic solvents and its valorization. Renew Sustain Energy Rev 156:11986

Zhou X, Broadbelt LJ, Vinu R (2016b) Chapter Two - Mechanistic understanding of thermochemical conversion of polymers and lignocellulosic biomass. Adv Chem Eng 49:95–198

Zhu G, Qiu X, Zhao Y, Qian Y, Pang Y, Ouyang X (2016) Depolymerization of lignin by microwave-assisted methylation of benzylic alcohols. Bioresour Technol 218:718–722

Chapter 8
Material Applications of Lignin

Mandeep Poonia, Jeong Jae Wie, and Chang Geun Yoo

Abstract Recently, lignin has extensively been investigated for its material applications such as biocomposites, hydrogels, surfactants, and other materials. Owing to its aromatic nature, three-dimensional network structures, and abundance of functional groups, lignin is beneficial for replacing petroleum resources and is compatible with other commercial resins, expanding its applications. However, the applications of lignin-based materials are still at an early stage compared to cellulose-based materials due to the inherent molecular complexity and heterogeneity of lignin with technical challenges in its processing. This chapter introduces recent progress in lignin-based composites and hydrogels among various approaches. To better understand lignin-based materials, the impacts of lignin species, contents, synthesis/fabrication methods, and other co-components on the technical performance and characteristics of the lignin-based composites and hydrogels in recent studies are reviewed. In composite applications, lignin can enhance thermal, chemical, and dimensional stability, as well as biodegradability, UV blocking, antioxidant property, and antimicrobial activity. Similarly, lignin-based hydrogels demonstrate excellent biocompatibility with antibacterial and antioxidant properties. Current challenges and future perspectives of lignin-based materials are also discussed in this chapter.

M. Poonia · J. J. Wie · C. G. Yoo (✉)
Department of Chemical Engineering, State University of New York College of Environmental Science and Forestry, Syracuse, NY, USA
e-mail: cyoo05@esf.edu

J. J. Wie
Department of Polymer Science and Engineering, Inha University, Incheon, Republic of Korea

C. G. Yoo
The Michael M. Szwarc Polymer Research Institute, Syracuse, NY, USA

N. P. Nghiem et al. (eds.), *Biomass Utilization: Conversion Strategies*,
https://doi.org/10.1007/978-3-031-05835-6_8

8.1 Introduction

Recent resource scarcity and environmental concerns increased the demand for green technology to replace petroleum-based chemicals with renewable and sustainable resources. Biomass is one of the promising resources for diverse industrial applications from traditional construction, papermaking, and direct combustion energy source to biofuels, biochemicals, and renewable materials. Except for direct applications of whole biomass for construction and furniture, papermaking is the most typical biomass application. While cellulose has been highlighted as a promising renewable material for paper, nanocrystals, and nanofibers, lignin has been considered an undesired component in most biomass applications. Lignin is an aromatic macromolecule, mainly composed of syringyl, guaiacyl, and *p*-hydroxyphenyl units with different C–O and C–C bonds like β-O-4, β-β, β-5, α-O-4, and 5-5 linkages. Lignin provides structural strength to the plant fibers and protects polysaccharides in the plant cell walls from microbial attacks in nature. However, lignin is a major recalcitrance factor in biological conversion processes since it limits cellulose access to enzymes and deactivates enzymes (Yoo et al. 2020). Similarly, lignin has been removed from biomass during the pulping process to enhance the quality of paper. For these reasons, a certain amount of lignin is necessary to be removed from biomass to accomplish effective biomass utilization. Through mechanical, thermochemical, and biological treatments like pretreatment and pulping, unwanted fractions such as lignin and other extractives are separated from cellulose. The generated lignin has been considered a waste stream or a byproduct in many traditional biorefinery processes. However, lignin has great potential, due to its aromatic structure, relatively high energy content, and sustainability, for industrial applications. Unfortunately, only 2% or less of the generated lignin is used in limited applications such as wood adhesive, surfactant, and other chemical or material uses, while the rest is directly combusted as an energy source (Bajwa et al. 2019). Therefore, further investigation of lignin and its applications is necessary for effective lignin utilization. Valorization of lignin can contribute to the economic competitiveness as well as carbon neutrality of biomass-derived products including pulp products and cellulosic biofuels.

The importance of lignin has been highlighted in many recent publications. More than 64,000 lignin-related articles have been published during the recent four years (2018–2021) according to Web of Science.[1] These articles include fundamental lignin chemistry and biochemistry, lignin-related plant science, lignin utilization, and others. Among these lignin-related publications, articles on "lignin material" or "lignin-based material" were exported for the network map using the keyword co-occurrence analysis in 1006 scientific publications by the full counting method of VOSViewer (Perianes-Rodriguez et al. 2016). As Fig. 8.1 shows, lignin material applications in various fields such as polyurethane, activated carbon, electrodes, fibers, nanoparticles, nanofibers, and films, as well as composites and hydrogels are mainly covered in this chapter.

[1] Searching date: December 27, 2021.

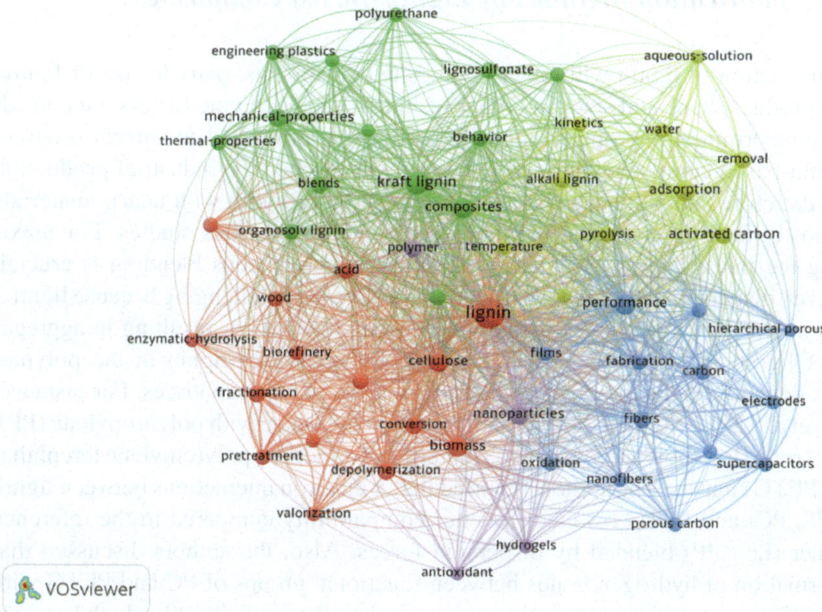

Fig. 8.1 The network map on "lignin material" or "lignin-based material" in the scientific publications from 2018 to 2021

8.2 Lignin-Based Composite Materials

For emerging technologies, the requirements for new materials are increasing with unique properties and multifaceted performance (e.g., lightweightness with high mechanical strength/toughness, flexibility, biodegradability, chemical/thermal resistance). Toward this end, composite materials have been investigated to achieve desirable performances not found in a single material. Typically, the discontinuous constituent is called filler (or reinforcement), and the continuous constituent surrounding the filler is called matrix. The tailored combination of constituent materials enhances their performances and/or imparts new properties with a facile tunability for a wide range of industrial applications. In particular, biocomposites composed of one or more phase(s) derived from a biological origin have been highlighted due to their environmental-friendly characteristics and sustainability.

8.2.1 Fabrication Methods of Lignin-Based Composites

The interaction of lignin with other composite constituents, particle size of lignin, target product shape and size, and other technical and economic factors are considered to determine the proper composite fabrication strategy. The current progress of lignin-based composite fabrication has not yet reached the industrial production level; therefore, among these factors, the interaction of lignin with matrix materials has mostly been discussed at the laboratory scale in previous studies. For maximizing the interfacial areas between materials, homogeneous blending is crucial. However, dispersion of lignin in matrix materials is often challenging because lignin–lignin interaction is stronger than polymer–lignin interaction, resulting in aggregation of lignin molecules. Hence, the dispersion quality of lignin in the polymer matrix depends on the interaction of lignin with the polymer matrices. For instance, Szabó et al. (2017) studied the interaction of lignosulfonate with polypropylene (PP), polystyrene (PS), polycarbonate (PC), and glycol-modified poly(ethylene terephthalate) (PETG) matrix. The authors reported that π electron interactions between lignin and PS, PC, and PETG resulted in better compatibility compared to the reference polymer (i.e., PP) blended by dispersion forces. Also, the authors discussed that the formation of hydrogen bonds between functional groups of PC and PETG with lignosulfonate makes the interactions stronger than the one with PS, which has only aromatic rings without functional groups (Szabó et al. 2017). In this chapter, the fabrication strategy for lignin-based composites is introduced by the dispersion methods: (1) solvent casting, (2) melt-mixing, and (3) extrusion methods (Table 8.1). The applications of lignin-based composites are also discussed with their advantages and disadvantages.

8.2.1.1 Solvent Casting

For the fabrication of lignin-based composite materials, lignin and matrix material(s) are dissolved in a solvent and transferred to substrates or a predefined three-dimensional (3D) mold (Ye et al. 2019). Filler is dispersed in a solvent along with polymer and has a large interparticle distance. The solvent is allowed to evaporate, which increases the solute concentration and thereby increases the viscosity of the composite solution. The increased viscosity reduces the diffusion of the fillers and eventually makes them immobilized in polymer matrices. For the solvent casting method, finding a common good solvent for both lignin and matrix materials is a prerequisite based on the solubilities of lignin and matrix materials. For instance, dimethylsulfoxide (DMSO) was used for the solvent casting of sulfonated poly(ether ketone) (SPEEK) and lignin (Ye et al. 2019), and dimethylformamide (DMF) was applied for that of Nafion and lignin (Ye et al. 2021). Alkali lignin was dissolved in water along with agar for the composite formation (Shankar et al. 2015). Some lignins were chemically modified prior to the dissolution in common solvents to increase

Table 8.1 Summary of fabrication methods and applications of lignin-based composites

	Lignin source	Lignin content	Matrix material	Method	Application	References
Solvent casting	Lignin (source unknown)	0–25 wt%	Sulfonated poly(ether ether ketone) SPEEK	Dissolve SPEEK and lignin in DMSO	Vanadium redox flow battery	Ye et al. (2019)
	Lignin (source unknown)	0–10 wt%	Nafion	Dissolve Nafion and lignin in DMF	Vanadium redox flow battery	Ye et al. (2021)
	Kraft lignin and acetylated kraft lignin	0–10 wt%	Poly(lactic acid)	Dissolve PLA and lignin in chloroform	Biodegradable composite materials	Kim et al. (2017)
	Alkali lignin	20 wt%	Chitosan	Mix chitosan-dispersed lactic acid solution and lignin-dispersed ethanol, and homogenize them for composite film-forming	Food packaging	Crouvisier-Urion et al. (2017)
	Alkali lignin (lignin copolymerized with β-butyrolactone and ε-caprolactone)	0 and 5 wt%	Poly(3-hydroxybutyrate)	Dissolve lignin and poly(3-hydroxybutyrate) in 1,1,1,3,3,3-hexafluoro-2-propanol and process by electrospinning	Biomedical applications	Kai et al. (2018)

(continued)

Table 8.1 (continued)

	Lignin source	Lignin content	Matrix material	Method	Application	References
	Steam pretreatment followed by DES, soda-anthraquinone, hydrotrope, organosolv lignin	0 and 5 wt%	Microcrystalline cellulose	Dissolve lignin and MCC in ionic liquid (DMAc/ LiCl solvent)	Advance packaging materials such as active drugs storage, long-time food storage, and biodegradable packaging	Guo et al. (2019)
	Organosolv lignin	0–50 wt%	Kraft birch pulp	Dissolve lignin and pulp in ionic liquid (1,5-diazabicyclo[4.3.0]non-5-enium acetate ([DBNH] [OAc]))	Low-cost carbon fibers	Le et al. (2021)
	Alkali lignin	0–10 wt%	Agar	Dissolve lignin, agar, and glycerol in distilled water	Food packaging	Shankar et al. (2015)
Melt-mixing	Organosolv hardwood lignin	0–40 wt%	Acrylonitrile-butadiene-rubber 41 & acrylonitrile–butadiene–styrene	Mix lignin and matrix material using Brabender Plasti-Corder Torque Rheometer	3D printing	Nguyen et al. (2018b)
	SW kraft and HW organosolv lignin	0–60 wt%	Nitrile-butadiene rubber, nylon 12, and carbon fibers	Mix lignin and matrix material at elevated temperature using Brabender Plasti-Corder Torque Rheometer	3D printing	Nguyen et al. (2018a)
	Hardwood kraft lignin	0–15 wt%	Poly(lactic acid)	Mix lignin and PLA in a high-speed mixer (MH50-H, DRAIS mixer)	Biodegradable food packaging	da Silva et al. (2019)
	Softwood kraft lignin	0–15 wt%	Poly(butylene succinate)	Mix lignin and PBS by Brabender plastograph internal mixing machine	Biomedical applications	Domínguez-Robles et al. (2020)

(continued)

Table 8.1 (continued)

	Lignin source	Lignin content	Matrix material	Method	Application	References
Extrusion	Softwood kraft lignin	0–3 wt%	Poly(lactic acid)	Extrude lignin and PLA by single screw extruder	Wound dressing	Domínguez-Robles et al. (2019)
	Hydrothermal lignin from eucalyptus	0–60 wt%	Poly(butylene adipate-co-terephthalate)	Extrude lignin and PBAT by twin screw extruder; maleic anhydride-graft-PBAT used as a compatibilizer	Packaging materials, and hygiene products	Xiong et al. (2020)
	Sodium lignosulfate	10 wt%	Polypropylene	Extrude lignin and PP by twin screw extruder; PP-grafted MAH used as a compatibilizer; Isophorone diisocyanate for covalent bonding	Biocomposites	Gil et al. (2019)
	High-temperature mechanical pretreatment followed by enzymatic hydrolysis of lignin	0–50 wt%	Polyhydroxybutyrate	Extrude lignin and PHB by twin screw extruder	3D printing	Vaidya et al. (2019)

their compatibility with the matrix materials. For instance, Kim et al. (2017) acetylated kraft lignin to improve its compatibility with poly(lactic acid) (PLA) and reduce the lignin aggregation by decreasing its hydrogen bond strength (Kim et al. 2017). In addition, Guo et al. (2019) used ionic liquids (i.e., N,N-dimethylacetamide/LiCl solvent) to dissolve both cellulose and lignin for the formation of the composite film (Guo et al. 2019). After the solvent casting, the resultant mixture was kept at ambient temperature to form a film, and then the film was washed with distilled water and dried.

Solvent casting method can also be applied with the mixture of two different solvents with homogenization treatment (Crouvisier-Urion et al. 2017). Crouvisier-Urion et al. (2017) individually dispersed chitosan in 1% (w/w) lactic acid aqueous solution and dispersed lignin in ethanol, and then formed the solution by mixing two solutions for solvent casting. For better dispersion of lignin, the blended solution was further homogenized. Light scattering and two-photon microscopy showed that the homogenization process reduced the particle size of lignin from 5 to 0.6 μm and enhanced the distribution of lignin in the chitosan matrix. The authors also discussed that the reduced lignin particle size via homogenization made the composite film more hydrophobic by exposing more hydrophobic groups like C–C groups of the subunits of lignin at the surface of the film.

The advantage of solvent casting is the superior mixing performance of lignin in the polymer matrix as the starting materials are solubilized at a molecular level prior to the fabrication of the composites. Relatively high lignin loading is applicable in the solvent casting method because of the better dispersion of lignin in the matrix materials. While solvents can enhance the blending of lignin and matrix materials, only limited solvents have good solubility for both lignin and matrix materials. Moreover, many of these solvents have a certain level of toxicity and require additional processing steps for solvent removal, which cause additional operation and equipment costs for the handling of the waste chemicals; therefore, minimal solvent use is desired.

8.2.1.2 Melt-Mixing

Lignin composite materials can also be fabricated by batch-type melt-mixing using an internal mixing machine (Jaafar et al. 2019). In this process, both the matrix material and the filler (i.e., lignin) are added to a mixing chamber of the internal mixing machine. Melting of the matrix material takes place due to the applied heat and friction between mixing components and the equipment rotor. A high-speed mixer (3000 rpm) was used for the fabrication of PLA-lignin composite with 0, 5, 10, and 15 wt% of hardwood kraft lignin loading (da Silva et al. 2019). Homogeneous composite was obtained after 1 min of mixing and molded into a film using a hot press. Nguyen et al. (2018a) fabricated nylon–lignin composite using a Brabender Plasti-Corder Torque Rheometer fitted with a mixing chamber and high-shear twin roller blades for melt-mixing of nylon and lignin (Nguyen et al. 2018a). Nylon 12 was pre-melted at 190 °C at 90 rpm for 5 min, then mixed with organosolv hardwood (HW) lignin (40–60 wt%) for 15 min to form the composites. Domínguez-Robles et al. (2020) also fabricated poly(butylene succinate)–lignin composite using a Brabender plastograph internal mixing machine. Softwood kraft lignin was melt-mixed with poly(butylene succinate) in the mixing chamber at 150 °C at 80 rpm for 10 min for the composite production (Domínguez-Robles et al. 2020).

Compared to the solvent casting method, melt-mixing does not require processing solvents for fabricating composite materials. However, the challenges of the melt-mixing method are its large energy consumption, high equipment cost, and potential thermal degradation of lignin and/or polymer matrix. Melt-mixing is performed in a batch process; hence, quality variation can occur with each batch, and the mass production of the composite materials is difficult.

8.2.1.3 Extrusion

The other lignin-based composite fabrication method is extrusion which utilizes elevated temperature and shear forces. As the temperature of the extruder increases, solid polymer pellets become polymer melt and form a continuous matrix, and lignin works as a reinforcement constituent. Polymer matrix coats the surface of lignin particles to form the composite. The temperatures of the extruder heating zones are adjusted according to the fluidity of the matrix material (Balasubramanian 2016). Domínguez-Robles et al. (2019) fabricated PLA–lignin composite using a single screw extruder (Domínguez-Robles et al. 2019). PLA pellets were mixed with softwood kraft lignin and castor oil and then extruded at 170–190 °C with an extruder speed of 5 rpm. Xiong et al. (2020) fabricated poly(butylene adipate-co-terephthalate) (PBAT)-lignin composite using a twin screw extruder (Xiong et al. 2020). Kraft lignin was obtained from eucalyptus and methylated with dimethyl carbonate to reduce the lignin aggregation. Maleic anhydride (MAH) grafted PBAT was used as a compatibilizer. The temperature profile from the feed port to the discharging hole of the extruder was: 130, 135, 142, 138, and 130 °C, with the feed port screw speed at 30 rpm and extruded twin screw at 50 rpm. The process of extrusion was repeated three times, and then the composite was dried at 60 °C for 12 h. Vaidya et al. (2019) fabricated polyhydroxybutyrate (PHB) composite filaments using biorefinery lignin, formed by mechanical pulping and enzymatic hydrolysis (Vaidya et al. 2019). A twin screw with a 40:1 length to diameter ratio (L/D) was used for extrusion at 300 rpm. Lignin was dry-blended with PHB powder at 10, 20, and 50 wt%. A reverse extruder barrel temperature profile, which has a hotter rear zone than the final downstream temperature zone, was used to avoid any potential thermal degradation of the polymer. The temperature profile of this study was: feeder to zone 3: 170 °C, zone 5 to zone 9: 160 °C, and zone 10 and die: 145 °C. The composite filament was cooled at room temperature and used for 3D printing applications with enhanced layer adhesion.

Extrusion has similar advantages (e.g., solvent-free processing) and disadvantages (e.g., high energy consumption, high equipment cost, potential thermal degradation of lignin or polymer matrix) as melt-mixing methods possess. However, extrusion is a continuous process that can fabricate composite materials with consistent quality. Also, 3D printer filaments and masterbatch pellets can be fabricated by the extrusion process.

8.2.2 *Impacts of Lignin on the Properties of Lignin-Based Composites and Their Applications*

The content and characteristics of lignin affect the composite properties. Ye et al. (2019, 2021) reported that the chemical stability of the composite increased with the lignin content. The authors fabricated the lignin-based composites by dissolving SPEEK and lignin in DMSO (Ye et al. 2019) and Nafion and lignin in DMF (Ye et al. 2021), and the resulted composites showed better chemical stabilities under acidic and oxidizing conditions. These composites were used for vanadium redox flow batteries as hydroxyl groups of lignin provided a transportation path for protons and suppressed vanadium ion permeability through the composite film (Ye et al. 2019, 2021). However, the impact of lignin on the hydrophobicity of the lignin-based composites varied. Vaidya et al. (2019) reported that the increase in biorefinery lignin content improved the hydrophobicity of the PHB-lignin composites (Vaidya et al. 2019). On the contrary, Shanker et al. (2015) reported a decrease in the hydrophobicity of the composite by blending alkaline lignin in the agar composite (Shankar et al. 2015). The authors discussed that the alkaline lignin was more hydrophilic than agar, resulting in the lower water contact angle (i.e., higher hydrophilicity) of the agar–lignin composite than that of the neat agar film. The effect of incorporation of lignin on the hydrophobicity of lignin-based composites depends upon several factors, including relative hydrophobicity of lignin compared to the polymer matrix and exposure of hydrophobic C–C groups on the composite film surface. Incorporating PLA matrix with kraft lignin increased the water vapor permeability (WVP) of the composite, while the incorporation with acetylated kraft lignin reduced the WVP because of the increased hydrophobicity of lignin by acetylation (Kim et al. 2017). The increase of lignin content in agar–lignin composite decreased WVP of the composite because the strong intermolecular interaction between agar and lignin led to good compatibility of these constituents and increased the tortuous path for water molecules (Shankar et al. 2015). The dimensional stability of the composite was also improved by increasing lignin content due to the intermolecular hydrogen bonding of lignin with polymer matrix, rigid benzyl structure of lignin (Ye et al. 2021), and less swelling nature of lignin (Shankar et al. 2015).

The impacts of lignin on the mechanical strength of the lignin-based composite materials were also investigated in previous studies. Ye et al. (2019) reported that the increase of lignin content (up to 15%) showed a positive impact on the mechanical strength of the SPEEK–lignin composite (Ye et al. 2019), while the addition of lignin reduced the tensile strength of the PLA–lignin composite (Kim et al. 2017).

Shanker et al. (2015) also reported that the tensile strength of the composite was improved by up to the inclusion of 3 wt% lignin in agar-lignin composite; however, the tensile strength decreased when the lignin content was increased further. The authors discussed that the lack of uniform dispersion and agglomeration of lignin beyond a certain lignin content reduced the tensile strength (Shankar et al. 2015). The mechanical strength of the composite material was also affected by the compatibility of lignin in the polymer matrix (Xiong et al. 2020). The compatibility can be increased by either chemical modification of lignin (e.g., methylation) or using a compatibilizer. For example, during the fabrication of PBAT–lignin composite, methylated lignin had less lignin aggregation due to the reduced intermolecular hydrogen bonding between hydroxyl groups of lignin molecules. This reduced lignin aggregation enhanced the compatibility of lignin in the composite. Compatibilizer in the polymer matrix also reduced lignin aggregation by increasing interfacial attraction between lignin and polymer matrix. Both methylation of lignin and the addition of a compatibilizer enhanced the tensile strength of the composite film compared to that of PBAT-unmodified lignin composite. The compatibility of lignin in the polymer composite can be further increased by employing primary chemical bonding between lignin and other constituent materials as compared to secondary intermolecular forces such as hydrogen bonding or π-π interactions. Gil et al. (2019) synthesized PP–lignin composite reinforced with glass fiber (GF) via reactive extrusion. Sodium lignosulfonate lignin is utilized with 10 wt% loading in the composites, and PP-grafted maleic anhydride (PP-g-MAH) is used as a compatibilizer. For primary bonding, isophorone diisocyanate (IPDI) was introduced in-situ during the extrusion process for the urethane reaction with hydroxyl groups of lignin and GF. An increase of the IPDI content to 7.5 wt% improved the tensile strength of PP–lignin composites from 28 MPa of 0 wt% IPDI counterpart to 35 MPa by the urethane formation. Above 10 wt% of IPDI content, the crosslinking reaction occurred between IPDI and urethane via allophanate reaction, which further increased the tensile strength to 46 MPa at 20 wt% IPDI content (Gil et al. 2019).

During FDM (fused deposition modeling) 3D printing, the high melt viscosity of polymers often causes clogging at the printing nozzle and reduces the deposition throughput. This problem can be solved by fabricating the composite with a shear-thinning profile. Nguyen et al. (2018a) fabricated the nylon–lignin composite, which showed a shear thinning behavior. At 230 °C and a shear rate of 100 rad/s, nylon 12 showed a high melt viscosity of 1050 Pa · s, while the composite containing 40, 50, and 60 wt% lignin showed a significantly lower melt viscosity of 150, 91, and 32 Pa · s, respectively (Nguyen et al. 2018a). The shear-thinning behavior of the composite was attributed to the formation of spherical lignin particles in the composite, which behaved like a lubricant phase and mobilized the nylon molecules. Vaidya et al. (2019) fabricated a lignin-based composite with biorefinery lignin in the PHB matrix (Vaidya et al. 2019). Melt rheology analysis showed that the melt viscosity of this composite was lower than that of pure PHB. The reduction in shear viscosity (shear thinning profile) enhanced the layer adhesion during 3D printing. However, due to the other impurities such as cellulose and hemicellulose, which have much lower thermal degradation temperatures than lignin and PHB, this biorefinery lignin-based composite material has limited application temperature.

The addition of lignin in the polymer matrix can improve the UV barrier and antioxidant properties of the composites, making them suitable for applications in food packaging. Domínguez-Robles et al. (2019) showed that increasing the lignin content in PLA–lignin composite enhanced the radical scavenging activity. The concentration of residual 2,2-diphenyl-1-picrylhydrazyl (DPPH) after 5 h reduced from 100% with PLA composite to 40, 35, 30, and 25% with 0.5, 1, 2, and 3 wt% lignin content in PLA–lignin composite, respectively (Domínguez-Robles et al. 2019). Biodegradability of the composite is also important in food packaging applications because the package is usually discarded in the waste. According to the biodegradability of PLA–lignin composite in garden soil over 180 days, neat PLA had a negligible reduction in mass, while PLA–lignin composite with 10 wt% lignin content showed a 5 wt% of mass reduction (da Silva et al. 2019).

As discussed earlier, lignin-based composites have various advantages in future material applications. However, there are still several technical challenges in lignin-based composite applications. The amount of lignin loading in the composite materials is currently limited because high lignin content composite showed a poor mechanical strength due to weak cohesive energy of phase separated lignin particles, operating as defect sites in the polymer composites. Also, the surface of composite materials became rough with an increase in lignin loading due to aggregation of lignin, which can cause abrasiveness in the composite. Lignin feedstock variability on the composite performance should be addressed with a deeper fundamental understanding. Further study is necessary to fully elucidate the unrevealed processing-structure–property-performance relationships in lignin-based composite toward practical use of lignin in future composite materials.

8.3 Lignin-Based Hydrogels

Hydrogels are 3D network structures having high water absorption and retention capacity. In the swollen state, hydrogels can retain water up to thousands of times their own dry weights (Meng et al. 2019b). Hence, hydrogels are useful in various applications such as water absorbents for soil water retention in arid and semi-arid regions, enzyme immobilization, drug delivery, wound dressing, tissue engineering, electrodes, biosensors, and food packaging. Petroleum-based monomers and polymers (e.g., acrylic acid, polyacrylamide) are widely used for hydrogel synthesis. However, these petroleum-based hydrogels are limited in biological degradation, and their decomposition compounds have biological toxicity (Song et al. 2020). Natural-based materials like cellulose and lignin have been explored for hydrogel synthesis due to their biodegradability, eco-friendliness, and sustainability (Teow et al. 2018; Song et al. 2020). For instance, cellulose-based hydrogel crosslinked with epichlorohydrin showed an excellent swelling capacity (4650%) and biodegradability (79.5% weight loss in 10 days) (Teow et al. 2018). Lignin also has great potential in this application due to its inherent characteristics such as antibacterial and radical scavenging activity (e.g., antioxidant) (Li et al. 2019; Zhang et al. 2019b). Lignin-based hydrogels showed antibacterial and antioxidant properties, whereas cellulose-based hydrogels need additional Ag nanoparticles or ZnO nanoparticles for these properties (Yadollahi et al. 2015). In this section, synthesis methods and applications of lignin-based hydrogels will be introduced with recent studies.

8.3.1 Synthesis Methods of Lignin-Based Hydrogels

Depending on the crosslinking nature, the synthesis of lignin-based hydrogels can be categorized into three approaches: (1) physical crosslinking, (2) chemical crosslinking, and (3) hybrid double crosslinking. Hydrogels prepared via physical crosslinking are also known as reversible hydrogels because some of them are thermally and mechanically reversible due to the relatively weak bond strength of the crosslinks (Morales et al. 2020). Chemically crosslinked hydrogels have an excellent mechanical strength; however, their synthesis often involves toxic chemicals, which can cause environmental issues and limit their applications in some fields like biomedical applications. Hybrid double-crosslinked hydrogels adopt the advantages of both chemical and physical crosslinking. Chemical crosslinks distribute the stress throughout the structure, and physical crosslinks dissociate to absorb more energy (Liu et al. 2020a). The hydrogel synthesis strategy can be determined according to the desired properties for the targeted applications. Table 8.2 summarizes lignin-based hydrogel synthesis methods with their potential applications.

8.3.1.1 Physical Crosslinking

Lignin-based hydrogels can be synthesized via several physical interactive forces such as hydrogen bonding, hydrophobic association, interpenetrating polymer networking, and electrostatic attraction (Thakur and Thakur 2015). Lignin can form hydrogel via intermolecular hydrogen bonding because of its abundant hydroxyl and carboxyl groups. Lignin-based hydrogel can also be formed by its π-π interaction with other hydrophobic materials. The hydrogel can be synthesized by partial interlacing of lignin with other polymers at the molecular scale in a polymer matrix through an interpenetrating polymer network (IPN) (Karak 2012). In addition, anionic groups of lignin like hydroxyl, carboxyl, and sulfonate groups can have an electrostatic attraction with other polymers having cationic groups like chitosan for the hydrogel synthesis (Ravishankar et al. 2019).

The mechanical strength of physically crosslinked hydrogels is relatively weak because of their low interactive forces. In general, the physical crosslinking approach for hydrogel formation has been applied through: (i) dissolution of lignin and other polymers such as cellulose, polyvinyl alcohol (PVA), chitosan, etc.; (ii) freeze–thaw cycles of the solution for physical crosslinking; and (iii) washing unreacted polymers

Table 8.2 Summary of lignin-based hydrogel synthesis methods and their applications

Crosslinking	Lignin source	Co-material	Crosslinker	Applications	References
Physical	Alkali lignin	Cellulose fibers	–	Hydrogel beads for immobilizing lipase	Park et al. (2015)
	Alkali lignin	Polyvinyl alcohol	–	Agriculture applications	Morales et al. (2020)
	Alkali lignin	Chitosan	–	Wound healing	Ravishankar et al. (2019)
	Lignocellulose from kraft pulping	–	–	Adsorption of lead and copper	Zhang et al. (2019a)
	Poly(ethylene glycol) methyl ether methacrylate grafted kraft lignin	–	–	Biomedical applications	Kai et al. (2015)
	Sodium lignosulfonate	Polyvinyl alcohol	–	Antibacterial application	Li et al. (2019)
	Sodium lignosulfonate	Sodium alginate, konjaku flour	–	Agriculture applications	Song et al. (2020)
	Sodium lignosulfonate	Chitosan, polyvinyl alcohol	–	Wound dressing	Zhang et al. (2019b)
	Cellulolytic enzyme lignin nanoparticles	Chitosan, polyvinyl alcohol	–	Biomedical applications	Yang et al. (2018)
	Aminated alkali lignin	Cellulose	–	Adsorption of dyes and heavy metal ions	Meng et al. (2019a)
Chemical	Alkali lignin	Microcrystalline cellulose	Epichlorohydrin	Alkali lignin fractionation from ethanol–water solution	Dai et al. (2019)
	Kraft lignin	Poly(ethylene glycol) diglycidyl ether	Poly(ethylene glycol) diglycidyl ether	pH-controlled actuator	Dai et al. (2020)

(continued)

Table 8.2 (continued)

Crosslinking	Lignin source	Co-material	Crosslinker	Applications	References
	Kraft lignin	N-isopropylacrylamide	N,N′-methylenebisacrylamide	Biomedical and environmental applications	Zerpa et al. (2018)
	Hydrothermally extracted lignin	Acrylamide	N,N′-methylenebisacrylamide	Catalyst backbone	Gao et al. (2021)
	Alkali lignin	Agarose	Epichlorohydrin	Personal hygiene products, underwater devices, water reservoirs for dry soils	Sathawong et al. (2018)
	Alkali lignin	Poly(ethylene glycol) diglycidyl ether	Poly(ethylene glycol) diglycidyl ether	Soil water retention	Mazloom et al. (2020)
	Sodium lignosulfonate	–	N,N′-methylenebisacrylamide	Removal of cadmium from the soil	Liu et al. (2020b)
	Alkali lignin	Polyacrylonitrile	Poly(ethylene glycol) diglycidyl ether	Flexible supercapacitor	Park et al. (2019)
Hybrid	Corncob lignin	Poly(ethylene glycol) diglycidyl ether	Poly(ethylene glycol) diglycidyl ether	Electrolyte in a flexible supercapacitor	Liu et al. (2020a)
	Alkali lignin nanoparticles	Polyacrylamide	N,N′-methylenebisacrylamide	Tissue engineering, artificial muscles, underwater antifouling materials	Chen et al. (2019)

to obtain the hydrogel. For example, Zhang et al. (2019b) individually dissolved the PVA in deionized water, chitosan in 2% acetic acid solution, and lignosulfonates in deionized water. These solutions were homogeneously mixed for casting (Zhang et al. 2019b). The casted solution in the mold was frozen at −18 °C for 8 h and thawed at room temperature. The freeze–thaw cycle was performed five times to obtain PVA–chitosan–lignin hydrogel, as described in Fig. 8.2. The hydrogel was formed via ionic bonding between amino groups of chitosan and sulfonic groups of lignin, hydrogen bonding between hydroxyl groups of PVA and chitosan, and hydrogen bonding between lignin and PVA as well as lignin and chitosan (not shown in Fig. 8.2). The increase of lignin content from 0 to 30 wt% improved the tensile strength of the hydrogel from 38.55 to 46.87 Mpa. Different freezing and thawing temperatures, time, and numbers of the cycle were investigated in previous studies (Yang et al. 2018; Morales et al. 2020). Yang et al. (2018) synthesized PVA–chitosan–lignin hydrogel with lignin nanoparticles (LNPs) formed by nanoprecipitation of lignin (Yang et al. 2018). The authors reported that the hydrogel formed with 1 wt% LNP had more uniform micropores and better swelling properties than the one with 0 and 3 wt% LNP. As compared to PVA-chitosan hydrogel, the swelling rate of hydrogel with 3 wt% LNP didn't improve despite more homogeneous pore structures.

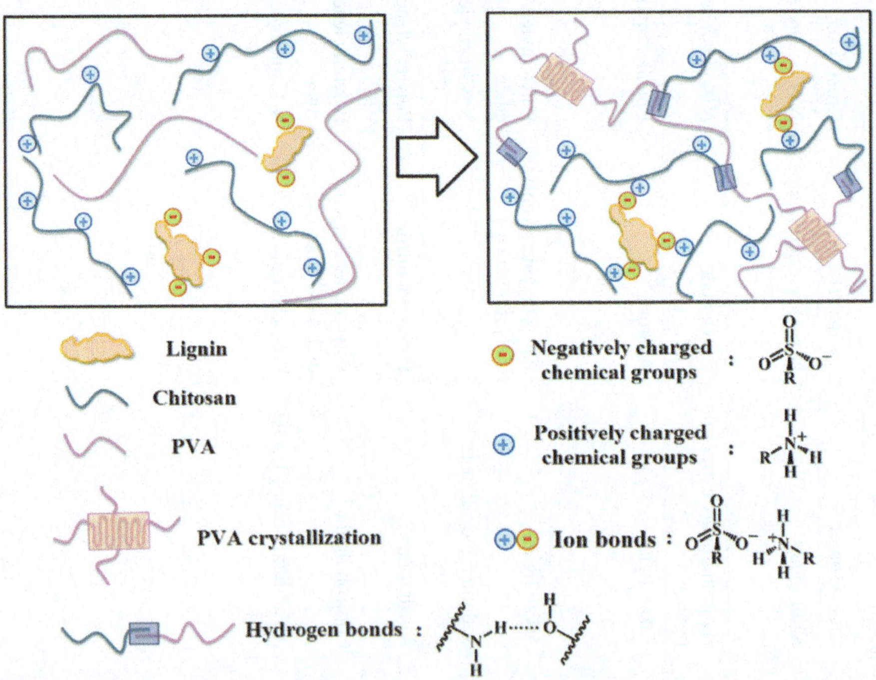

Fig. 8.2 Physical crosslinking mechanism of PVA-chitosan-lignin hydrogel (Reprinted from Mater. Sci. Eng., C 104, 110,002, Zhang, Y., Jiang, M., Zhang, Y., Cao, Q., Wang, X., Han, Y., Sun, G., Li, Y., Zhou, J., Novel lignin–chitosan–PVA composite hydrogel for wound dressing, Copyright (2019), with permission from Elsevier)

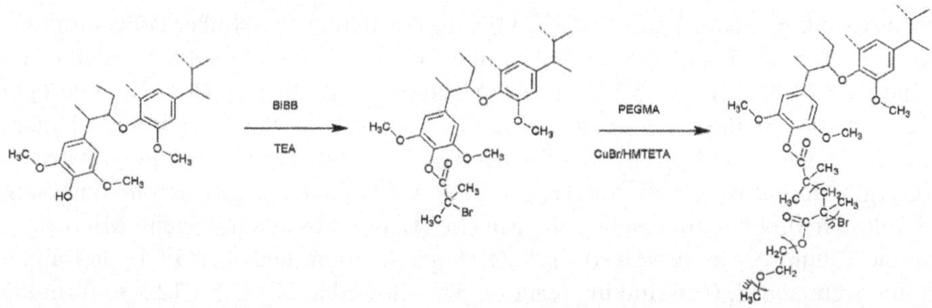

Fig. 8.3 Synthesis of polymer-grafted lignin via atom transfer radical polymerization (ATRP) (Reproduced with permission from Kai et al. (2015), Copyright 2015 American Chemical Society)

Lignin-based hydrogels were also synthesized via physical crosslinking by the inclusion complexation of lignin-based copolymers with α-cyclodextrin (α-CD) (Kai et al. 2015). As shown in Fig. 8.3, lignin-based copolymers were synthesized by atom transfer radical polymerization (ATRP) of kraft lignin with poly(ethylene glycol) methyl ether methacrylate (PEGMA). Firstly, lignin-based organic halide was synthesized, and it worked as a macroinitiator and a polymer backbone for ATRP reaction. In the second step, PEGMA-graft lignin copolymer was synthesized through the ATRP reaction. For the synthesis of hydrogel, PEGMA-lignin copolymer was dissolved in phosphate buffered saline (PBS) and mixed with the α-CD solution in PBS. The mixed solution was kept at room temperature for the formation of hydrogel via physical crosslinking by the inclusion complexation between PEGMA-grafted lignin and α-CD.

8.3.1.2 Chemical Crosslinking

Lignin has plenty of phenolic and aliphatic hydroxyl groups (Rico-García et al. 2020), which enable chemical reactions between lignin and crosslinking agents to synthesize lignin-based hydrogels. Strong covalent bonds in the polymeric networks make it difficult to change the shape of chemically crosslinked hydrogels; therefore, these are often referred to as permanent hydrogels (Thakur and Thakur 2015). Lignin-based hydrogels can be synthesized by chemical crosslinking of lignin with other polymers such as cellulose, PVA, chitosan, poly(ethylene glycol) diglycidyl ether (PEGDGE), polyacrylamide, and polyacrylonitrile with crosslinking agents such as epichlorohydrin, formaldehyde, PEGDGE. Also, unsaturated monomers such as acrylamide and N-isopropylacrylamide can be grafted on the lignin backbone and synthesize hydrogels by radical polymerization of monomers and crosslinkers. Dai et al. (2020) synthesized lignin-based hydrogel with hardwood kraft lignin and PEGDGE as a polymer as well as a crosslinker at 40 °C for 12 h (Dai et al. 2020). The phenolic hydroxyl groups of lignin reacted with epoxy rings of PEGDGE to form the hydrogel by chemical crosslinking. Mazloom et al. (2019, 2020) formed the hydrogel by crosslinking alkali lignin with PEGDGE at different crosslinker concentrations (0.3, 0.4, 0.5, 0.75, and 1.0 mmol/g of lignin) in three types of solvents (deionized water, 1.5 M NaOH, and 3 M NaOH) (Mazloom et al. 2019, 2020). The hydrogel was not formed in deionized water, while it formed in NaOH solutions. Higher swelling capacity (34 g/g-hydrogel) was observed with 1.5 M NaOH compared to 16 g/g-hydrogel with 3 M NaOH. Dai et al. (2019) synthesized lignin-containing cellulose hydrogel with epichlorohydrin (ECH) as a crosslinking agent. Microcrystalline cellulose was dissolved in NaOH/urea solution, and then ECH and alkali lignin were added. Crosslinking reaction was allowed at 60 °C for 12 h to form the hydrogel (Dai et al. 2019). Agarose–lignin hydrogel was also prepared with ECH as a crosslinker. Agarose was dissolved in water at 68 °C and crosslinked with kraft lignin for hydrogel formation (Sathawong et al. 2018). Free radical polymerization by grafting the unsaturated monomers on lignin was also applied for the lignin-based hydrogel synthesis (Meng et al. 2019b). In brief, free radicals can be generated by the thermal decomposition of azobisisobutyronitrile (AIBN) (Fig. 8.4a). Free radical is transferred to lignin to generate a phenoxy radical (Fig. 8.4b). The phenoxy radicals attack the carbon double bond of monomer N-isopropylacrylamide (NIPAAm) and crosslinker N,N′-methylenebisacrylamide (MBAAm) to form the initial propagating chain (Fig. 8.4c). The crosslinked structure of hydrogel is constructed by chain elongation (Fig. 8.4c) (Zerpa et al. 2018). The hydrogel is rinsed with acetone to remove the unreacted monomers. The lignin-based hydrogel had a more porous structure than synthetic hydrogel. Lignin was also modified with functional materials to tailor the hydrogel properties (Gao et al. 2021). Hydrothermally extracted lignin was modified with glycine, which contains amino and carboxyl groups. Glycine–lignin-based hydrogel was synthesized by free radical polymerization with acrylamide as a monomer, MBAAm as a crosslinker, and ammonium persulfate as an initiator.

8.3.1.3 Hybrid Double-Crosslinking

The performances of lignin-based hydrogels can be improved by synthesizing hybrid double-crosslinked hydrogels via a two-step network formation: (1) formation of covalent bonds through chemical crosslinking reaction; and (2) physical crosslinking formed by hydrophobic interactions, ionic interactions, and/or hydrogen bonding (Chen et al. 2019; Liu et al. 2020a). A schematic diagram of the hybrid double crosslinking is shown in Fig. 8.5 (Liu et al. 2020a). Chemical crosslinking reaction of ring-opening polymerization was carried out at 50 °C for 2 h with phenolic groups in lignin and epoxy rings of PEGDGE crosslinker for the formation of single crosslinked (SC) hydrogel. For physical crosslinking, the SC lignin hydrogel was immersed in 1 M H_2SO_4 for 12 h to obtain a double-crosslinked (DC) hydrogel. The protonation of unreacted phenol and carboxyl groups of lignin induced the hydrophobic interactions among lignin chains. The DC hydrogel resulted in 40-fold higher compressive mechanical strength (4.74 MPa) than that of SC hydrogel (0.12 MPa). The covalent crosslinks in the hydrogel efficiently distribute the stress throughout the network and

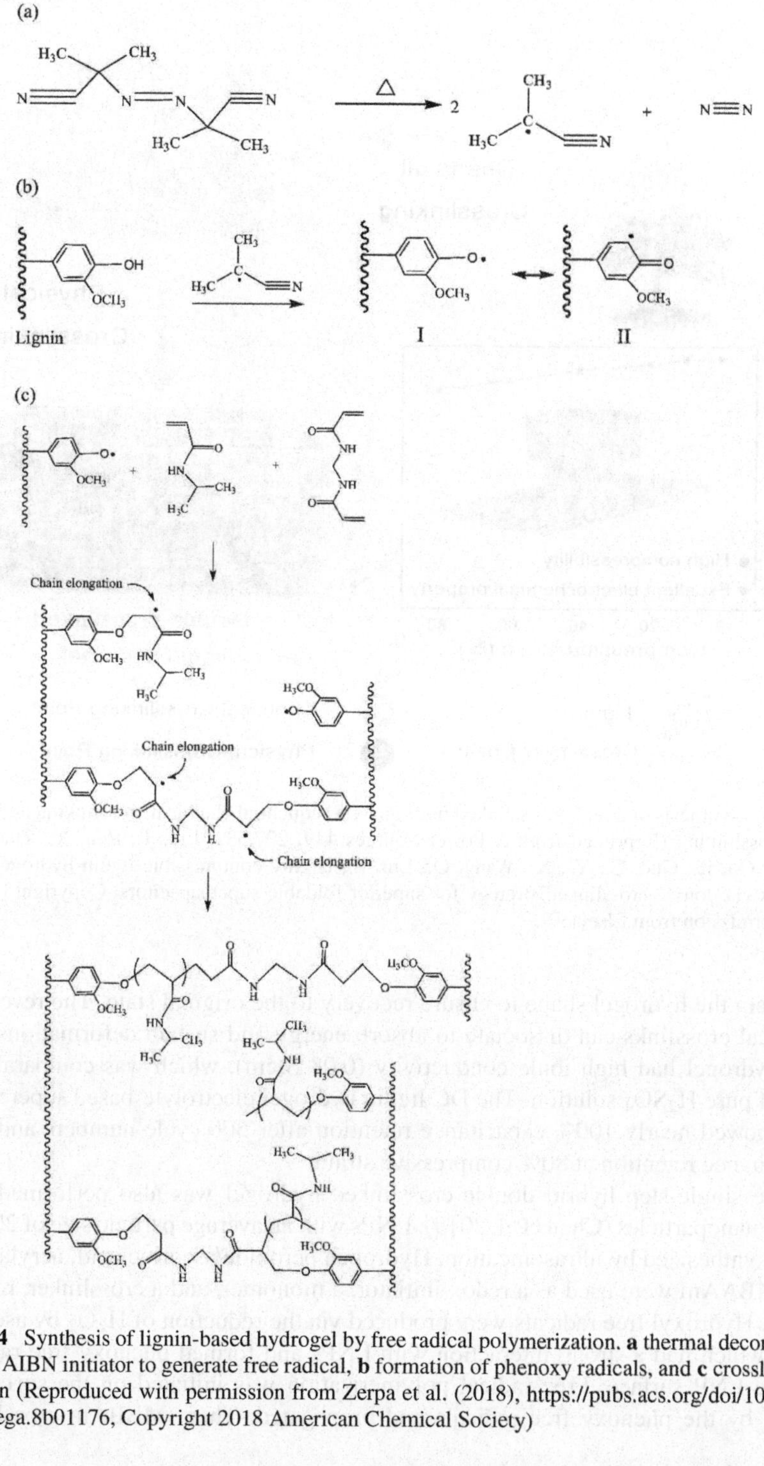

Fig. 8.4 Synthesis of lignin-based hydrogel by free radical polymerization. **a** thermal decomposition of AIBN initiator to generate free radical, **b** formation of phenoxy radicals, and **c** crosslinking reaction (Reproduced with permission from Zerpa et al. (2018), https://pubs.acs.org/doi/10.1021/acsomega.8b01176, Copyright 2018 American Chemical Society)

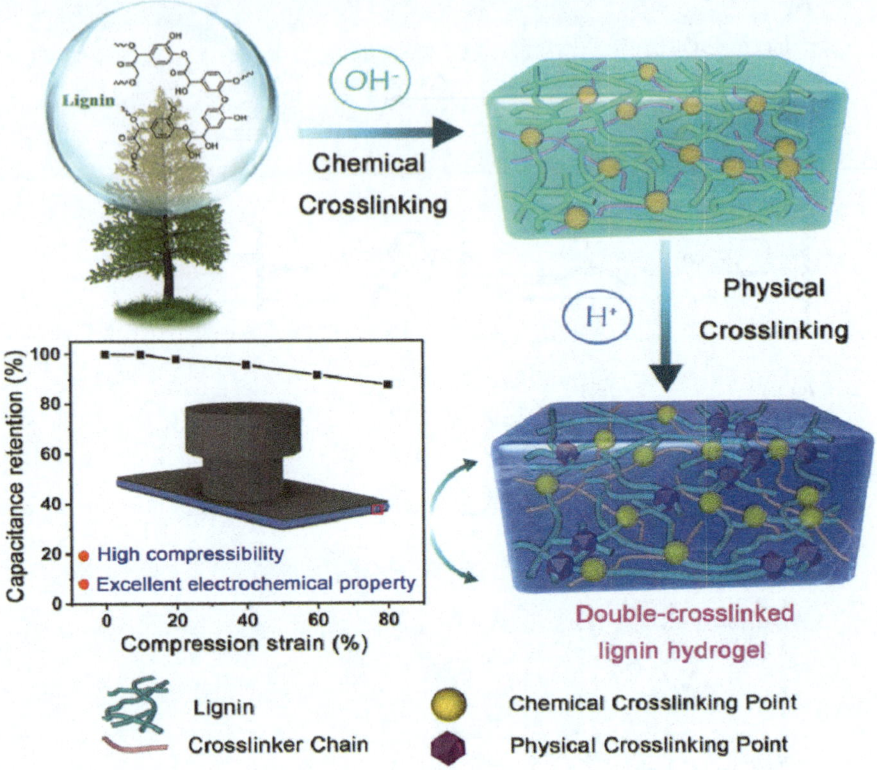

Fig. 8.5 Synthesis of double-crosslinked hydrogel via sequential chemical crosslinking and physical crosslinking (Reprinted from J. Power Sources 449, 227,532, Liu, T., Ren, X., Zhang, J., Liu, J., Ou, R., Guo, C., Yu, X., Wang, Q., Liu, Z., Highly compressible lignin hydrogel electrolytes via double-crosslinked strategy for superior foldable supercapacitors, Copyright (2020), with permission from Elsevier)

maintain the hydrogel shape to ensure recovery to the original state. The reversible physical crosslinks can dissociate to absorb energy and sustain deformations. The DC hydrogel had high ionic conductivity (0.08 S/cm), which was comparable to that of pure H_2SO_4 solution. The DC lignin hydrogel electrolyte-based supercapacitor showed nearly 100% capacitance retention after 500 cycle numbers and 85% capacitance retention at 80% compressive strain.

The single-step hybrid double-crosslinked hydrogel was also performed with lignin nanoparticles (Chen et al. 2019). LNPs with an average particle size of 200 nm were synthesized by ultrasonication. Hydrogen peroxide/ascorbic acid, acrylamide, and MBAAm were used as a redox initiator, a monomer, and a crosslinker, respectively. Hydroxyl free radicals were produced via the reduction of H_2O_2 by ascorbic acid, which had a strong interaction with LNPs and formed phenoxy free radicals on the LNP surface. Free radical polymerization was initiated on the surface of LNPs by the phenoxy free radicals and propagated. Then, MBBAm crosslinked the polyacrylamide (PAM) chains by connecting the neighboring LNPs. Some PAM chains were covalently bonded, and others were intertwined with LNPs via hydrogen bonding. PAM-LNP hydrogel showed higher toughness, elasticity, and stretchability as compared to PAM hydrogel. The inclusion of LNP into PAM hydrogel increased the fracture stress from 0.04 to 7.87 MPa, the toughness from 0.68 to 45.26 kJ/m^3, and the elongation at break from 190 to 750%. No serious deformation or strength degradation was observed after 100 loading–unloading cycles at 80% strain.

8.3.2 Applications of Lignin-Based Hydrogels

Lignin-based hydrogels have a wide range of applications such as soil water retention, adsorbent, flexible supercapacitor, food packaging, wound dressing, drug delivery, and tissue engineering because of their biodegradability, biocompatibility, antibacterial property, antioxidant activity, higher mechanical strength, and chemical and thermal stability.

8.3.2.1 Adsorption Applications

Lignin-based hydrogels showed a highly porous structure and negatively charged phenolic and aliphatic hydroxyl groups, making the lignin-based hydrogels ideal adsorbents. They were applied for the adsorption of both organic (Wang et al. 2017; Dai et al. 2019) and inorganic materials (Zhang et al. 2019a). Large varieties of dyes are used in textile dyeing, leather tanning, printing, and food packaging. These dyes are carcinogenic and mutagenic and need to be removed from wastewater before discharging into the environment. The lignin-based hydrogel was applied to adsorb methylene blue dye, a cationic dye, at pH 1–11 (Wang et al. 2017). The adsorption capacity of lignin-based hydrogel increased as the solution pH increased and leveled off when the pH reached 7 and higher. Protons at low pH competed with dye for adsorption sites, which resulted in poor dye removal. In contrast, high solution pH increased the negatively charged hydroxyl groups, which increased the dye adsorption by higher electrostatic attraction. The adsorption capacity of methylene blue dye on lignin-based hydrogel was up to 9.65 g/g of the hydrogel.

The chemically crosslinked cellulose–lignin hydrogel was also used for the fractionation of alkali lignin (Dai et al. 2019). The hydrogel was immersed in an alkali lignin solution, which had a weight average molecular weight (M_w) of 1054 g/mol with a polydispersity index (PDI) of 1.43. Lower molecular weight lignin (M_w of 499 g/mol) with PDI of 1.13 was adsorbed on the hydrogel, while the unadsorbed lignin showed M_w of 1220 g/mol with PDI of 1.21 in the filtered solution. This lignin-based hydrogel fractionated more uniform lignin fractions with narrow molecular weight distribution. The fractionation of lignin with the lignin-based hydrogel was attributed to smaller pore size and noncovalent π–π interactions between aromatic rings of alkali lignin and the lignin present in the hydrogel. Heteronuclear single quantum coherence (HSQC) nuclear magnetic resonance (NMR) and ^{31}P NMR results showed that the adsorbed and filtered lignin had very similar lignin composition and hydroxyl group contents.

Lignocellulosic (LC) hydrogels were used for the adsorption of copper and lead ions (Zhang et al. 2019a). Biomass with different lignin contents (6.5, 11.6, and 18.4 wt%) was obtained from poplar wood chips by the kraft pulping process using different alkali concentrations of cooking liquor. Lignin-free biomass was also obtained by bleaching the kraft pulp. These biomass solids were dissolved in N-methylmorpholine-N-oxide, and the hydrogels were obtained by solvent exchange with ethanol followed by water. As the lignin content increased up to 11.6 wt%, the adsorption of heavy metal ions (Cu^{2+} and Pb^{2+}) on lignin was improved, while the adsorption decreased when the lignin content was 18.4 wt%. The highest adsorption by the hydrogels with 11.6 wt% lignin content was attributed to the porous network of the hydrogel, which exposed more phenolic groups of lignin and hydroxyl groups of carbohydrates for the complexation with metal ions. However, further increase of lignin content in the hydrogel reduced the uniformity of pore structure due to the aggregation of lignin particles and led to the destruction of the network structure resulting in the decrease of active sites for heavy metals adsorption.

8.3.2.2 Biomedical Applications

Biocompatibility, antibacterial property, antioxidant activity, and low cytotoxicity of lignin (Kai et al. 2018; Domínguez-Robles et al. 2019; Zhang et al. 2019b; da Silva et al. 2019) make lignin-based hydrogels a potential material in biomedical applications such as wound dressing, drug delivery, and antimicrobial coatings. Physical hydrogels and hydrogels crosslinked by non-toxic crosslinkers have been investigated for biomedical applications (Larrañeta et al. 2018; Ravishankar et al. 2019; Zhang et al. 2019b). Hydrophobic drug loading was improved by the inclusion of hydrophobic materials in the hydrogel structure (Larrañeta et al. 2018). The lignin-based hydrogel was loaded with curcumin, a polyphenolic compound, for drug delivery applications. Higher loading of curcumin was available with higher lignin content in the hydrogel as both the lignin and curcumin contain aromatic rings.

Lignin-based hydrogels were also explored for wound dressing. The ideal requirements for a wound dressing material are a physical barrier with higher mechanical strength, antibacterial activity to avoid infections, high absorption capacity to clean up the metabolites, moisture-holding capacity, free radical scavenging, and antioxidant properties (Zhang et al. 2019b). Lignin-based hydrogels fulfill these requirements for wound dressing. For instance, PVA–chitosan–lignin hydrogel was prepared, which improved the mechanical strength from 38.55 to 46.87 MPa, protein adsorption capacity from 5 to 45 mg/g, and wound closure from 80% to nearly 100%, when lignin content increased from 0 to 30 wt% (Zhang et al. 2019b). Although the antibacterial activity of lignin has been demonstrated in previous studies (Larrañeta et al. 2018), Zhang et al. (2019b) reported that the antibacterial activity of the PVA–chitosan–lignin hydrogel was mainly due to cationic groups of chitosan. The authors reported that the increase of lignin content in the hydrogel reduced the antibacterial activity due to partial neutralization of cationic groups from chitosan with anionic groups of lignin. Chitosan–lignin hydrogel synthesized by physical crosslinking (electrostatic attraction) was also investigated for its wound dressing application (Ravishankar et al. 2019). Cytotoxicity of this hydrogel was evaluated in vitro and in vivo against mesenchymal stem cells and zebrafish, respectively. Hydrogels synthesized with alkali lignin were non-toxic, while those with sodium lignosulphonate were highly toxic. Cell viability was 114 ± 0.2% with alkali lignin, 99 ± 2% with chitosan, 99 ± 3% with chitosan–alkali lignin, while it was relatively low with sodium lignosulphonate (~60%).

8.3.2.3 Agricultural Applications

Hydrogels are used in arid and semi-arid regions for soil water retention and controlled release of fertilizers in the soil. However, synthetic materials-based hydrogels can cause soil pollution due to their limited biodegradability and/or biological toxicity of the degraded products. Lignin-based hydrogel, synthesized by crosslinking of lignosulfonate (L), konjaku flour (KJ), and sodium alginate (SA), was used for controlling the release of water and nutrients for tobacco plants (Song et al. 2020). The pore size of soil affects the saturated hydraulic conductivity of the soil. The addition of SA-KJ-L (sodium alginate-konjaku flour-lignosulfonate) hydrogel in the soil reduced the water loss during the infiltration process by blocking the soil pores and increasing friction between soil particles, hydrogel, and water. Saturated hydraulic conductivity of soil was reduced from 0.020 cm/min in no-hydrogel soil to 0.008 cm/min in 0.375% hydrogel soil, 0.003 cm/min in 0.650% hydrogel soil, and 0.002 cm/min in 0.975% hydrogel soil. The maximum water holding capacity (MWHC) of soil without hydrogel was 52.66% (i.e., 52.66 g of water in 100 g of dry soil). The MWHC increased to 61.63% with the addition of 0.975% lignin-based hydrogel. Nitrogen (N), phosphorus (P), and potassium (K) are the three main macronutrients in the fertilizers used for plant growth (Song et al. 2020). These nutrients entered the SA–KJ–L hydrogel molecular structure along with water which reduced the leaching and infiltration of nutrients in the soil. The addition of 0.975% hydrogel reduced the leaching of nitrate nitrogen, ammonium nitrogen, total phosphorus, and plant available potassium by 38.83%, 32.35%, 28.22%, and 26.89%, respectively, compared to the soil without hydrogel. Proline and reducing sugar are protective substances against tobacco stress, and their contents are used to identify the degree of drought in crops. The increase of proline content indicates the higher drought stress. Proline contents in plants treated with 0.375, 0.650, and 0.975% of SA–KJ–L hydrogel were lower (~0.40 μg/g, ~0.35 μg/g, and ~0.30 μg/g, respectively) than that of the untreated plants (0.88 μg/g).

Hydrogel crosslinked with alkali lignin and PEGDGE was applied to alleviate drought stress in maize and compared with the hydrogel synthesized with sodium polyacrylate (Mazloom et al. 2020). Maize plants were taller in soil amended with both hydrogels. In general, the P content in maize shoots declines with water shortage. In this study, the lignin-based hydrogel amended soil showed higher P contents (1.48–1.56%) in maize shoots than the content with no-hydrogel condition (0.44%). These were comparable to the P contents with the synthetic hydrogel amended soil (1.45–1.50%). Proline content in the maize grown in soil amended with lignin-based hydrogels was lower (0.6 and 1.4 μmol/g) than the contents with synthetic hydrogels (1.6 and 4.2 μmol/g) and no-hydrogel condition (4.8 μmol/g). The drought stress can also be identified by electrolyte leakage which occurs mainly due to cell membrane damage under drought-induced oxidative stress. The electrolyte leakage reduced from 71% in no-hydrogel condition to 62% with sodium polyacrylate hydrogel as well as lignin-based hydrogel treated soils due to alleviation in drought stress. Sodium polyacrylate hydrogel had a detrimental impact on soil chemistry due to the dissolution of sodium. The soil pH increased with sodium polyacrylate hydrogel, which interferes with the nutrient uptake by plants, while the pH was not increased with the lignin-based hydrogel because methoxyl, carbonyl, carboxyl, and hydroxyl functional groups of lignin contributed to soil buffering.

8.3.2.4 Other Applications

Lignin has a unique pH-stimuli-responding property currently used for the dissolution and precipitation of lignin at different pH (Dai et al. 2020). At low pH, hydroxyl and carboxyl groups of lignin remain protonated; hence reduction in repulsive forces decreases the swelling of lignin-based hydrogels. Conversely, the increase of pH deprotonates these groups and induces swelling due to electrostatic repulsion (Meng et al. 2019b). The hydrogel was synthesized by chemically crosslinking lignin with PEGDGE, and its pH-stimuli responsiveness was tested (Dai et al. 2020). The hydrogel filament was bent when immersed in an acidic solution, while the filament recovered its original shape in an alkaline solution. Mechanical switches and actuators can be synthesized by using the pH-responsive behavior of lignin-based hydrogels.

Although the above applications have been investigated, lignin-based hydrogels have not yet been successfully produced at industrial scales. Extraction of lignin at a certain molecular weight with narrow polydispersity is challenging. Most lignins are biocompatible and non-toxic, but some sulfur-containing technical lignins are not. For example, sodium lignosulfonate showed high cytotoxicity, which may not be proper for biomedical applications (Ravishankar et al. 2019). As mentioned earlier, physically crosslinked hydrogels have relatively low mechanical strength, while many chemically crosslinked hydrogels are still formed with toxic chemicals. Further study is required to form a lignin-based hydrogel with superior or at least comparable performances to synthetic hydrogels without toxic chemical uses.

8.4 Conclusion and Future Perspectives

Current lignin-based materials still have many technical limitations in their performances and processing methods. Lignin alone cannot meet the mechanical strength and/or flexibility; therefore, it is applied with other co-materials as a composite to overcome these challenges. Since two or more materials will be chemically and/or physically blended to form the materials, feedstock compatibility is a crucial factor. To improve the performances of lignin-based materials, refining or modification of lignin prior to the synthesis or processing of materials would be beneficial. Feedstock variability is another challenge in lignin utilization. Compared to cellulose, the properties and composition of lignin have relatively large variations depending on biomass species, growing environments, and processing methods (e.g., pulping and pretreatment) and conditions. Since the performance of lignin-based materials can be influenced by lignin, the understanding of lignin characteristics via proper analyses is essential. Machine learning can be applied with a sufficient experimental database to identify the best methodology based on the lignin source as well as the target performance of the products. Also, environmental aspects of lignin-based materials need to be considered further. One of the key features of lignin-based materials is carbon negativity; however, most studies simply discussed the alternation of petroleum resources with lignin. For developing true eco-friendly future materials, it is necessary to design carbon negative lignin-based materials on a cradle-to-grave life cycle basis.

Acknowledgements This work was supported by the National Science Foundation grant (CBET2027125) and the National Research Foundation of Korea (NRF) grant (NRF-2019R1A2C1004559).

References

Bajwa DS, Pourhashem G, Ullah AH, Bajwa SG (2019) A concise review of current lignin production, applications, products and their environmental impact. Ind Crops Prod 139:111526

Balasubramanian M (2016) Introduction to composite materials. In: Rana S, Fangueiro R (eds) Fibrous and textile materials for composite applications. Springer Singapore, Singapore, pp 1–38

Chen Y, Zheng K, Niu L, Zhang Y, Liu Y, Wang C, Chu F (2019) Highly mechanical properties nanocomposite hydrogels with biorenewable lignin nanoparticles. Int J Biol Macromol 128:414–420

Crouvisier-Urion K, Lagorce-Tachon A, Lauquin C, Winckler P, Tongdeesoontorn W, Domenek S, Debeaufort F, Karbowiak T (2017) Impact of the homogenization process on the structure and antioxidant properties of chitosan-lignin composite films. Food Chem 236:120–126

da Silva TF, Menezes F, Montagna LS, Lemes AP, Passador FR (2019) Effect of lignin as accelerator of the biodegradation process of poly(lactic acid)/lignin composites. Mater Sci Eng, B 251:114441

Dai L, Zhu W, Lu J, Kong F, Si C, Ni Y (2019) A lignin-containing cellulose hydrogel for lignin fractionation. Green Chem 21(19):5222–5230

Dai L, Ma M, Xu J, Si C, Wang X, Liu Z, Ni Y (2020) All-lignin-based hydrogel with fast pH-stimuli responsiveness for mechanical switching and actuation. Chem Mater 32(10):4324–4330

Domínguez-Robles J, Martin N, Fong M, Stewart S, Irwin N, Rial-Hermida M, Donnelly R, Larrañeta E (2019) Antioxidant PLA composites containing lignin for 3D printing applications: a potential material for healthcare applications. Pharmaceutics 11(4):165

Domínguez-Robles J, Larrañeta E, Fong ML, Martin NK, Irwin NJ, Mutjé P, Tarrés Q, Delgado-Aguilar M (2020) Lignin/poly(butylene succinate) composites with antioxidant and antibacterial properties for potential biomedical applications. Int J Biol Macromol 145:92–99

Gao C, Xiao L, Zhou J, Wang H, Zhai S, An Q (2021) Immobilization of nanosilver onto glycine modified lignin hydrogel composites for highly efficient p-nitrophenol hydrogenation. Chem Eng J 403:126370

Gil BM, Song SW, Lee JH, Jeon J, Lee KH, Wie JJ (2019) Introduction of primary chemical bonding in lignin-based PP composites for mechanical reinforcement via reactive extrusion. Compos B 165:510–515

Guo Y, Tian D, Shen F, Yang G, Long L, He J, Song C, Zhang J, Zhu Y, Huang C, Deng S (2019) Transparent cellulose/technical lignin composite films for advanced packaging. Polymers 11(9):1455

Jaafar J, Siregar JP, Tezara C, Hamdan MHM, Rihayat T (2019) A review of important considerations in the compression molding process of short natural fiber composites. Int J Adv Manuf Technol 105(7–8):3437–3450

Kai D, Low ZW, Liow SS, Abdul Karim A, Ye H, Jin G, Li K, Loh XJ (2015) Development of lignin supramolecular hydrogels with mechanically responsive and self-healing properties. ACS Sustain Chem Eng 3(9):2160–2169

Kai D, Chong HM, Chow LP, Jiang L, Lin Q, Zhang K, Zhang H, Zhang Z, Loh XJ (2018) Strong and biocompatible lignin /poly (3-hydroxybutyrate) composite nanofibers. Compos Sci Technol 158:26–33

Karak N (2012) Fundamentals of polymers. In: Vegetable oil-based polymers. Elsevier, pp 1–30

Kim Y, Suhr J, Seo H-W, Sun H, Kim S, Park I-K, Kim S-H, Lee Y, Kim K-J, Nam J-D (2017) All biomass and UV protective composite composed of compatibilized lignin and poly (lactic-acid). Sci Rep 7(1):43596

Larrañeta E, Imízcoz M, Toh JX, Irwin NJ, Ripolin A, Perminova A, Domínguez-Robles J, Rodríguez A, Donnelly RF (2018) Synthesis and characterization of lignin hydrogels for potential applications as drug eluting antimicrobial coatings for medical materials. ACS Sustain Chem Eng 6(7):9037–9046

Le N-D, Trogen M, Ma Y, Varley RJ, Hummel M, Byrne N (2021) Understanding the influence of key parameters on the stabilisation of cellulose-lignin composite fibres. Cellulose 28(2):911–919

Li M, Jiang X, Wang D, Xu Z, Yang M (2019) In situ reduction of silver nanoparticles in the lignin based hydrogel for enhanced antibacterial application. Colloids Surf B 177:370–376

Liu T, Ren X, Zhang J, Liu J, Ou R, Guo C, Yu X, Wang Q, Liu Z (2020a) Highly compressible lignin hydrogel electrolytes via double-crosslinked strategy for superior foldable supercapacitors. J Power Sources 449:227532

Liu Y, Huang Y, Zhang C, Li W, Chen C, Zhang Z, Chen H, Wang J, Li Y, Zhang Y (2020b) Nano-FeS incorporated into stable lignin hydrogel: A novel strategy for cadmium removal from soil. Environ Pollut 264:114739

Mazloom N, Khorassani R, Zohuri GH, Emami H, Whalen J (2019) Development and characterization of lignin-based hydrogel for use in agricultural soils: preliminary evidence. Clean (weinh) 47(11):1900101

Mazloom N, Khorassani R, Zohury GH, Emami H, Whalen J (2020) Lignin-based hydrogel alleviates drought stress in maize. Environ Exp Bot 175:104055

Meng Y, Li C, Liu X, Lu J, Cheng Y, Xiao L-P, Wang H (2019a) Preparation of magnetic hydrogel microspheres of lignin derivate for application in water. Sci Total Environ 685:847–855

Meng Y, Lu J, Cheng Y, Li Q, Wang H (2019b) Lignin-based hydrogels: A review of preparation, properties, and application. Int J Biol Macromol 135:1006–1019

Morales A, Labidi J, Gullón P (2020) Assessment of green approaches for the synthesis of physically crosslinked lignin hydrogels. J Ind Eng Chem 81:475–487

Nguyen NA, Barnes SH, Bowland CC, Meek KM, Littrell KC, Keum JK, Naskar AK (2018a) A path for lignin valorization via additive manufacturing of high-performance sustainable composites with enhanced 3D printability. Sci Adv 4(12):eaat4967

Nguyen NA, Bowland CC, Naskar AK (2018b) A general method to improve 3D-printability and inter-layer adhesion in lignin-based composites. Appl Mater Today 12:138–152

Park S, Kim SH, Kim JH, Yu H, Kim HJ, Yang Y-H, Kim H, Kim YH, Ha SH, Lee SH (2015) Application of cellulose/lignin hydrogel beads as novel supports for immobilizing lipase. J Mol Catal B: Enzym 119:33–39

Park JH, Rana HH, Lee JY, Park HS (2019) Renewable flexible supercapacitors based on all-lignin-based hydrogel electrolytes and nanofiber electrodes. J Mater Chem A 7(28):16962–16968

Perianes-Rodriguez A, Waltman L, van Eck NJ (2016) Constructing bibliometric networks: a comparison between full and fractional counting. J Informetr 10(4):1178–1195

Ravishankar K, Venkatesan M, Desingh RP, Mahalingam A, Sadhasivam B, Subramaniyam R, Dhamodharan R (2019) Biocompatible hydrogels of chitosan-alkali lignin for potential wound healing applications. Mater Sci Eng C 102:447–457

Rico-García D, Ruiz-Rubio L, Pérez-Alvarez L, Hernández-Olmos SL, Guerrero-Ramírez GL, Vilas-Vilela JL (2020) Lignin-based hydrogels: synthesis and applications. Polymers 12(1):81

Sathawong S, Sridach W, Techato K (2018) Lignin: isolation and preparing the lignin based hydrogel. J Environ Chem Eng 6(5):5879–5888

Shankar S, Reddy JP, Rhim J-W (2015) Effect of lignin on water vapor barrier, mechanical, and structural properties of agar/lignin composite films. Int J Biol Macromol 81:267–273

Song B, Liang H, Sun R, Peng P, Jiang Y, She D (2020) Hydrogel synthesis based on lignin/sodium alginate and application in agriculture. Int J Biol Macromol 144:219–230

Szabó G, Romhányi V, Kun D, Renner K, Pukánszky B (2017) Competitive interactions in aromatic polymer/lignosulfonate blends. ACS Sustain Chem Eng 5(1):410–419

Teow YH, Kam LM, Mohammad AW (2018) Synthesis of cellulose hydrogel for copper (II) ions adsorption. J Environ Chem Eng 6(4):4588–4597

Thakur VK, Thakur MK (2015) Recent advances in green hydrogels from lignin: a review. Int J Biol Macromol 72:834–847

Vaidya AA, Collet C, Gaugler M, Lloyd-Jones G (2019) Integrating softwood biorefinery lignin into polyhydroxybutyrate composites and application in 3D printing. Mater Today Commun 19:286–296

Wang Y, Xiong Y, Wang J, Zhang X (2017) Ultrasonic-assisted fabrication of montmorillonite-lignin hybrid hydrogel: highly efficient swelling behaviors and super-sorbent for dye removal from wastewater. Colloids Surf A 520:903–913

Xiong S-J, Pang B, Zhou S-J, Li M-K, Yang S, Wang Y-Y, Shi Q, Wang S-F, Yuan T-Q, Sun R-C (2020) Economically competitive biodegradable PBAT/lignin composites: effect of lignin methylation and compatibilizer. ACS Sustain Chem Eng 8(13):5338–5346

Yadollahi M, Gholamali I, Namazi H, Aghazadeh M (2015) Synthesis and characterization of antibacterial carboxymethyl cellulose/ZnO nanocomposite hydrogels. Int J Biol Macromol 74:136–141

Yang W, Fortunati E, Bertoglio F, Owczarek JS, Bruni G, Kozanecki M, Kenny JM, Torre L, Visai L, Puglia D (2018) Polyvinyl alcohol/chitosan hydrogels with enhanced antioxidant and antibacterial properties induced by lignin nanoparticles. Carbohydr Polym 181:275–284

Ye J, Cheng Y, Sun L, Ding M, Wu C, Yuan D, Zhao X, Xiang C, Jia C (2019) A green SPEEK/lignin composite membrane with high ion selectivity for vanadium redox flow battery. J Membr Sci 572:110–118

Ye J, Yuan D, Ding M, Long Y, Long T, Sun L, Jia C (2021) A cost-effective nafion/lignin composite membrane with low vanadium ion permeation for high performance vanadium redox flow battery. J Power Sources 482:229023

Yoo CG, Meng X, Pu Y, Ragauskas AJ (2020) The critical role of lignin in lignocellulosic biomass conversion and recent pretreatment strategies: a comprehensive review. Bioresour Technol 301:122784

Zerpa A, Pakzad L, Fatehi P (2018) Hardwood kraft lignin-based hydrogels: production and performance. ACS Omega 3(7):8233–8242
Zhang L, Lu H, Yu J, Fan Y, Ma J, Wang Z (2019a) Contribution of lignin to the microstructure and physical performance of three-dimensional lignocellulose hydrogels. Cellulose 26(4):2375–2388
Zhang Y, Jiang M, Zhang Y, Cao Q, Wang X, Han Y, Sun G, Li Y, Zhou J (2019b) Novel lignin–chitosan–PVA composite hydrogel for wound dressing. Mater Sci Eng C 104:110002

Chapter 9
Techno-Economic Analysis for Evaluating Biorefinery Strategies

Deepak Kumar, Tristan Brown, Shashi Kant Bhatia, and Vinod Kumar

Abstract Advanced biorefineries aiming to utilize all major components of biomass to produce a wide spectrum of bioproducts and fuel are seen to play a key role in implementing a biobased economy. However, due to the low technology readiness levels, it is critical to assess the commercial feasibility and economic competitiveness of these biorefinery approaches using techno-economic analysis (TEA) before making investment decisions and policy regulations. This chapter discusses the basic concepts of TEA and summarizes the studies on various advanced biorefinery concepts. Focus is given on three approaches: (i) biorefineries processing cellulosic feedstocks where sugars are converted to biofuel and part of the lignin is converted to fuel or high-value biochemical products, (ii) biorefineries converting hexose sugars to biofuel and pentose sugars to high-value chemicals, and (iii) biorefineries processing engineered lipid-producing energy crops.

9.1 Introduction

Increased energy demand due to rapidly growing population and industrialization, limited energy sources, and environmental concerns from fossil fuels necessitates the development of biobased energy technologies. The transportation sector accounts for

D. Kumar (✉)
Department of Chemical Engineering, State University of New York College of Environmental Science and Forestry, Syracuse, NY 13210, USA
e-mail: dkumar02@esf.edu

T. Brown
Department of Sustainable Resource Management, State University of New York College of Environmental Science and Forestry, Syracuse, NY 13210, USA

S. Kant Bhatia
Department of Biological Engineering, College of Engineering, Konkuk University, Seoul 05029, Republic of Korea

V. Kumar
School of Water, Energy and EnvironmentCentre for Climate and Environmental Protection, Cranfield University, Cranfield MK43 0AL, UK

N. P. Nghiem et al. (eds.), *Biomass Utilization: Conversion Strategies*,
https://doi.org/10.1007/978-3-031-05835-6_9

about one-third (29.84 trillion MJ) of the total energy used in the United States, and 95% of this is derived from petroleum (Kumar et al. 2021). Moreover, 28% (1854 Tg CO_2 equivalent) of anthropogenic greenhouse gas (GHG) emissions are generated from the transportation sector alone (Fasahati et al. 2019; Kumar et al. 2021). Bioethanol and biodiesel are promising alternatives to fossil liquid transportation fuels that are produced from renewable feedstocks, and they produce relatively low life cycle net GHG emissions. Bioethanol and biodiesel can be blended with gasoline and petroleum diesel, respectively, at various fractions (e.g., E10, E85, B5, and B20) and used in normal or flexible-fuel vehicles. Both biofuels are commercially produced in the United States, but mainly from food-based crops. The US produces more than 16 billion gallons (60 billion liters) of ethanol, but more than 90% of this is produced from corn starch. Similarly, out of 1.8 billion gallons of biodiesel produced in the US, more than 80% was produced by transesterification of vegetable oil obtained from soybean, canola, and corn.

Biofuel production from food crops is not a long-term sustainable solution due to several challenges, such as food versus fuel priority, capacity limitations, intensive agricultural inputs and cultivation costs, and land and freshwater use. Alternatively, the use of lignocellulosic feedstock, such as agricultural residues (e.g., corn stover, rice straw), agro-processing wastes (e.g., bagasse, shells, fruit pomace), forestry products and wastes, and dedicated energy crops (e.g., switchgrass, miscanthus, energycane) can address these challenges and provide a sustainable source of renewable biofuels and bioproducts. Considering the constraints of the limited availability of conventional oil crops (e.g., soybean, canola) and their high cultivation costs, significant efforts have been made in recent years to identify or develop alternative feedstocks that can provide oil/lipids for biodiesel production. One successful approach in that area is to use metabolic engineering and plant genetics to genetically modify the highly productive biomass crops like sugarcane, energycane, and sorghum to accumulate triacylglycerols (TAGs) in their vegetative tissues. It is a highly promising approach that could allow the production of far more industrial vegetable oil than previously possible. In addition to high productivities, most of these crops have relatively low cultivation costs and are water-use efficient and drought tolerant. For example, sorghum consumes about 33% less water compared to corn (Fasahati et al. 2019). Lignocellulosic biomass and engineered energy crops do not compete with food or feed, have a low economic value, and provide relatively superior environmental benefits, especially when used in a biorefinery approach. Similar to petrochemical refineries, biorefineries are based on the concept of producing multiple products from various constituents of the biomass using a series of advanced bioprocess and thermochemical technologies. The concept allows sustainable transformation of all fractions of biomass to biofuels and bioproducts, achieving maximum resource utilization and revenue. Lignocellulosic biomass is composed of three major constituents: cellulose (30–50%), hemicellulose (8–50%), and lignin (7–30%) (Baral and Shah 2017). Cellulose is a homopolymer of D-anhydro-glucopyranose units linked with β-1,4 glycosidic bonds. Hemicellulose is a heteropolymer of several sugars, including xylose, arabinose, glucose, mannose, and galactose. Lignin is a very complex 3-D heteropolymer of phenyl-propane units jointed by carbon–carbon and ether bonds. The cellulose and hemicellulose can be hydrolyzed to sugar monomers that can be subsequently fermented to biofuels and bioproducts. Lignin can be combusted for heat and power generation (conventional biorefinery) or could be recovered and used for the production of biofuels and high-value bioproducts (advanced biorefinery).

A wide array of biorefinery approaches have been developed and investigated on the lignocellulosic feedstocks and these engineered energy crops. Although most of these proof-of-concept biorefinery approaches look attractive based on the results from laboratory or pilot scale studies, the processing and manufacturing costs might be prohibitively high. It is critical to understand the commercial-scale viability and economic trade-offs between various options before their commercial deployment. In addition to the economic profitability, detailed material and energy balances of the whole system and individual unit processes included in the biorefinery are needed to determine product yields, scale of operation, raw materials requirement, and energy efficiency for successful implementation of these technologies. Techno-economic analysis (TEA) using process simulation models is used to conduct such investigations and determine the venture risk. The TEA can act as a connecting bridge between laboratory research and industrial application by allowing to foresee the process economics (Saini et al. 2020). This chapter discusses the TEA in the context of biorefinery systems. The main focus is for three types of biorefineries (i) biorefineries processing cellulosic feedstocks where both carbohydrate fractions are converted to ethanol (or other liquid fuel such as butanol) and part of the lignin is converted to fuel or high-value biochemical products, (ii) biorefineries similar to those in (i) except that the C5 sugars are converted to high-value chemicals instead of being used for ethanol production, (iii) biorefineries processing engineered lipid-producing energy crops.

9.2 Techno-Economic Analysis (TEA)

TEA is an effective approach to evaluate the commercial-scale feasibility (technical, energetic, and economics) of process technologies, especially the early-stage technologies. The analysis provides detailed process economics, such as equipment cost, capital investment, annual operational costs and their breakdown (e.g., labor, utilities, raw materials), rate of return, etc. Over the last two decades, the TEA has become an integral part of research both in industry and academia. In addition to the overall economic feasibility, TEA helps in identifying the bottlenecks/hot-spots and primary cost drivers in the process that can direct the research to improvise the process. For any biomass-based biorefinery, there are several possible technology choices. Even to produce the same main product, different process routes could be used. For example, during biomass to bioethanol production, there are numerous choices of pretreatment processes, hydrolysis (acid vs. enzymatic), lignin valorization (simple combustion vs. conversion to value-added products), and many other processes. Within any specific technology also, there is a choice of selecting the process parameters that affect the process yield. However, with that change in the product yield, there are always associated costs of energy and economics. Similarly, a trade-off between energy savings and capital investments is commonly encountered in biorefineries. TEA can help in understanding that trade-off and making the choice of technology and process parameters based on multiple constraints (product yield, economics, energy, etc.). Previous reports have demonstrated these trade-offs in the choice of technologies for various biorefineries (Kumar et al. 2021, 2018; Kumar and Murthy 2011; Kurambhatti et al. 2020; Somavat et al. 2018). A comprehensive TEA provides ample details of technical evaluation of biorefinery, such as equipment capacity requirement based on the scale of operation, heating and cooling energy requirement, power usage, skill requirement, etc. (Murthy 2022; Shah et al. 2016). In terms of economics, there are various accuracy levels of TEA: (i) Order of magnitude estimate (accuracy level: ±10–50%), (ii) sturdy estimate (accuracy level: ±30%), (iii) preliminary estimate (accuracy level: ±20%), (iv) definitive estimate (accuracy level: ±10%), and (v) detailed estimate (accuracy level: ±5%) (Murthy 2017). The choice is made based on the objective of the analysis, resources available, and the budget.

The analysis includes several steps, starting with the detailed process model design. The model is developed based on the actual biorefinery design (sequence of unit operations) and should take into account all technical and operational details of the essential unit operations and associated equipment and accessories. Figure 9.1 depicts the process model of a biorefinery co-producing ethanol and biodiesel

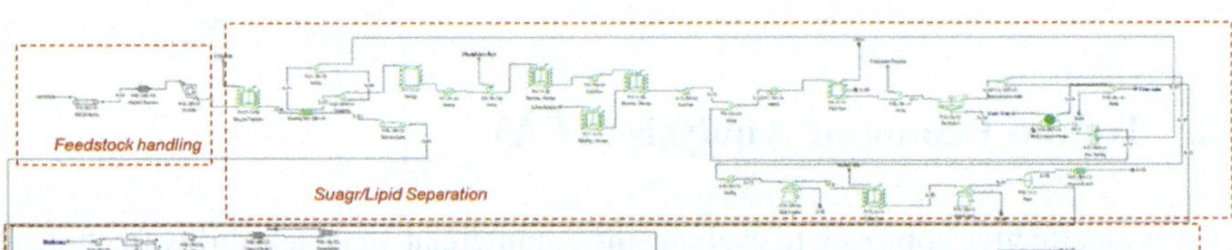

Fig. 9.1 Flowsheet of a process model of biorefinery co-producing ethanol and biodiesel from engineered energy cane (Adapted from Kumar et al. 2021)

from genetically engineered energycane. Various commercial (e.g., Aspen Plus, SuperPro Designer, CHEMCAD) and open-source platforms/tools (e.g., DWSIM, BioSTEAM, SIM42) can be used to develop these process models (Cortes-Peña et al. 2020). The data associated with the operating conditions and process efficiencies (e.g., enzyme dosage, solid loading, conversion efficiencies of cellulose and hemicellulose during enzymatic hydrolysis) of all unit processes can be obtained from lab and pilot scale studies, mathematical models, and commercial facilities, which is input into the process model. The simulation models (both commercial and open-source) allow conducting a mass and thermodynamically rigorous energy balance to determine the equipment design specifications and their capacities, product yields (from an individual unit operation and overall process), utilities (amounts of heating and cooling agents, electricity), and waste streams. Comprehensive reports are generated from the model simulations, and the data is further analyzed to obtain more meaningful results.

The monetary value estimations of the process include capital cost, operational costs, and revenue, which are further used to calculate the overall profitability in terms of return on investment (Shah et al. 2016; Towler and Sinnott 2012). The total capital investment consists of direct fixed capital (DFC), startup cost, and working capital. Out of these three costs, DFC is the main cost that accounts for designing and installing a plant. DFC includes equipment purchase costs and several associated direct (e.g., installation, piping, insulation, electrical facilities) and indirect costs (e.g., engineering and construction). The costs of equipment for the respective capacities (estimated from the process model simulations) can be obtained from various sources, including vendor quotations, trade journals, literature, and built-in cost models in the simulation software (Murthy 2017). Since the equipment capacities vary among various options, the cost of the equipment for a specific capacity can be calculated from the extrapolation of the exponential scaling equation (Eq. 9.1). The value of "exp" in the equation varies between 0.4 and 0.9 depending on the type of equipment and industry, but an average value of 0.6 is commonly used in the biorefinery TEA studies (Shah et al. 2016). The associated direct and indirect costs are dependent on the type of industry and can be determined using multiplication factors obtained from the previous reports or process modeling textbooks in Chemical Engineering. These values can be estimated on an individual context or could be lumped into a single factor, known as the Lang factor. Both approaches have been used in the published biorefinery TEA studies, and DFC value is typically estimated 3 to 4 times that of the equipment purchase cost (Huang et al. 2016b; Kumar et al. 2021, 2018; Kumar and Murthy 2011; Somavat et al. 2018). The startup cost is a one-time investment to prepare a new refinery for operation. The working capital corresponds to funds required to operate the biorefinery for initial some time till revenue is generated. Most of the TEA studies estimate working capital as 5% of DFC (Huang et al. 2016a; Khounani et al. 2019; Kurambhatti et al. 2020; Ntimbani et al. 2021; Shen et al. 2019; Somavat et al. 2018).

$$\text{New Cost} = \text{Base Cost} * \left(\frac{\text{new size}}{\text{base size}}\right)^{\text{exp}} \tag{9.1}$$

Annual operating cost is calculated by accounting for several expenditures associated with raw materials, utilities, labor, quality analysis, plant maintenance, and waste disposal. The quantities of raw materials and demand of utilities and labor are determined from process simulations and are multiplied by the respective purchase prices to calculate the total cost. The unit cost of the product (e.g., $/L biofuel) is calculated by dividing operational costs by the product yield. Revenue is calculated from the sale of the main product and co-products. The quantities (annual production) of products are calculated from the process models and are multiplied by the selling price to estimate total revenues. As biorefineries produce multiple products, there are various approaches to calculate the cost of one product. Most of the studies identify one output as the main product and calculate its cost based on the net operational costs (total operational cost—co-product revenue) (Kumar et al. 2018; Somavat et al. 2018). In another approach, the production cost of all products produced from biorefinery is calculated based on their amounts and market values (Huang et al. 2016a). After calculating capital costs, annual operational costs, and revenues, a detailed cash flow analysis is conducted using these cost numbers to determine the overall profitability of the process. Considering the scope of this chapter, only a brief discussion on the TEA methodology is provided here, and the readers can refer to other book chapters that provide basics and details of the TEA methodology (Murthy 2022, 2017; Shah et al. 2016).

9.3 TEA of Lignocellulosic Biomass-Based Biorefineries

Considering the need for analysis, various studies have investigated the TEA of biomass-based biorefineries. The conventional biorefineries TEA considered the conversion of structural carbohydrates (cellulose and hemicellulose) to bioethanol and the burning of the lignin-rich stream for in-house steam and electricity productions. This is the common practice used to utilize lignin in the pulp and paper industries with the burning of more than 95% of total extracted lignin (50 million tons annually) (Baral and Shah 2017; Shen et al. 2019). Most of the TEA studies of conventional biorefineries studies concluded that the energy produced from the burning of lignin was higher than that of biorefinery requirement and the excess steam was converted to electricity and sold to the grid (Hasanly et al. 2018; Humbird et al. 2011; Kazi et al. 2010; Kumar and Murthy 2011; Rajendran and Murthy 2017; Sanchez et al. 2013). Kumar and Murthy (2011) conducted a TEA of cellulosic refineries for four pretreatment technologies using grass straw as feedstock and found that lignin energy was sufficient to meet the energy requirement of biorefinery in all the cases. The excess electricity ranged from 0.77 to 1.78 kWh per L of ethanol production (Kumar and Murthy 2011). Similar observations have been reported for the biorefineries producing butanol (Baral and Shah 2017). Although the direct combustion of lignin stream is the current practice, several recent studies have emphasized the importance of converting lignin to high-value chemicals and bioproducts instead of burning it. The low oxygen-to-carbon ratio (O/C ratio) and ring structure of lignin provide an opportunity to produce various high-value aromatic compounds, biochemical, and biomaterials with a wide application to replace petroleum-derived counterparts (Shen et al. 2019). Some examples of these products include epoxies, phenolic resins, adhesives, biopolymers (PHAs; polyhydroxyalkanoates), vanillin, jet fuel, etc. Moreover, there is a large capital investment associated with steam and electricity production system. According to a comprehensive TEA study conducted by Humbird et al. (2011), the steam and electricity generation system (boiler/turbogenerator) accounted for more than 25% of the total installed equipment cost. In this chapter, the studies where not all of the lignin was burnt for thermal energy needs but part of it was used for biofuel and bioproduct production are discussed. The details of these studies are summarized in Table 9.1.

To keep the biorefineries self-sustainable, a part of lignin can be burnt to produce steam and electricity to meet the energy demand of the biorefinery and the remaining lignin can be valorized to produce high-value chemicals and bioproducts. Bbosa et al. (2018) evaluated the feasibility of biorefinery valorizing lignin fraction (80% of lignin from the ethanol process) through hydrothermal liquefaction (HTL) process and found the process highly profitable, with the minimum ethanol selling price (MESP) of \$1.03/gal (\$0.27/L), significantly lower than the reference ethanol price (selling price in Iowa of \$1.94/gal (\$0.51/L)). Although the production cost was low, the capital cost was relatively higher due to the high equipment cost associated with the lignin valorization section (~40% of the total installed equipment cost). They also conducted a comprehensive uncertainty analysis and concluded that the uncertainty level is high due to the low maturity level of lignin valorization technology. Shen et al. (2019) conducted a TEA of biorefinery co-producing bioethanol (from carbohydrates) and jet-fuel (from lignin) using corn stover as a feedstock. The lignin fraction from the distillation column of the bioethanol process was converted to jet fuel using the hydrodeoxygenation (HDO) route. Three scenarios for various ratios of lignin stream distribution between power generation (100, 76.1, and 23.6%) and jet-fuel were evaluated. The capital investment for the scenarios producing jet fuel from lignin was up to 8% higher compared to the base case (lignin burning). Correspondingly, the minimum ethanol selling price (after accounting for co-product credit from electricity and/or jet-fuel) was also slightly higher (\$2.88/gal (\$0.76/L) vs. \$2.85/gal (\$0.75/L) ethanol). It is important to note that although there was not much benefit in terms of process economics, it is critical to establish jet-fuel production technologies using feedstock other than conventional oil crops. It was also reported that the ethanol and jet-fuel co-producing biorefinery could be more profitable with the improvement in the jet-fuel production yields and increase in jet-fuel market price (Shen et al. 2019). Baral and Shah (2017) compared the techno-economic feasibility utilization of stillage (lignin stream from distillation) from cellulosic butanol refinery in two pathways: direct combustion and fast pyrolysis to produce bio-oil. They concluded that although significant quantities of bio-oil and biochar can be produced through the fast pyrolysis route, the direct combustion process is more economically feasible. The stillage processing cost in fast pyrolysis was found to be 12% lower compared to the direct combustion route. By conducting the sensitivity analysis, they found that the cost can be reduced, and the pyrolysis process can be

Table 9.1 Techno-economic analysis of various biorefineries using lignocellulosic biomass

Biomass ($ year)	Plant capacity	Process	Products	Capital investment ($ million)	Minimum selling price	Remarks	References
Corn stover (2017 $)	2000 dry MT/day biomass	Integrated Cellulosic ethanol and HTL (valorizing lignin)	Ethanol, catechol, cresols, acetic acid, furfural, formic acid, phenol, and acetaldehyde	624.5	$0.272/L ethanol	80% lignin for HTL and 20% for steam and power	Bbosa et al. (2018)
Corn stover (2014 $)	2000 dry MT/day biomass	Integrated Cellulosic ethanol and jet-fuel (valorizing lignin)	Ethanol, jet-fuel, electricity	463.6–501.5	$0.75 –0.76/L ethanol	Various scenarios of lignin fraction utilization between jet-fuel and power production	Shen et al. (2019)
Corn stover (2017 $)	792 MT/day biomass (657.4 dry MT/day)	Integrated ethanol and xylitol/furfural process	Ethanol, xylitol/furfural, electricity	148–173	$0.7–1.65/L ethanol	Steam explosion pretreatment, hemicellulose sugars converted to xylitol/furfural	Giuliano et al. (2018)

(continued)

Table 9.1 (continued)

Biomass ($ year)	Plant capacity	Process	Products	Capital investment ($ million)	Minimum selling price	Remarks	References
Sugarcane bagasse and harvest residues (2018 $)	1560 dry MT/day biomass	One-stage furfural and cellulosic ethanol co-production	Ethanol, furfural, electricity	272–322	−$0.02 to $0.42/L ethanol (for co-production) $0.595/L (solo-ethanol)	Mixture of bagasse (70%) and harvesting residues (30%); Steam explosion pretreatment, hemicellulose sugars converted to furfural; fraction of biomass burnt to supply heat and electricity	Ntimbani et al. (2021)
Switchgrass (2017 $) $68/ton	2000 dry MT/day biomass	one-pot biomass fractionation and furfural production	Ethanol, furfural, lignin, electricity	445.4	$625/ton Furfural	Use of a biphasic solvent comprising aqueous choline chloride (ChCl) and methyl isobutyl ketone (MIBK)	Zang et al. (2020)

profitable as the technology is matured and the process conditions of all operations are optimized considering the process intensification (Baral and Shah 2017).

Considering the challenges of fermenting C5 sugars to ethanol using current industrial yeast strains, some TEA studies have investigated the feasibility of valorizing C5 sugars to chemicals instead of fermenting them to ethanol. Giuliano et al. (2018) conducted one such analysis in which they compared the process economics of three scenarios: (i) ethanol production from both glucose and xylose, (ii) ethanol from glucose and xylitol from xylose, (iii) ethanol from glucose, and furfural from xylose. Due to the high value of xylitol (three times the selling price compared to ethanol), the payback selling price of ethanol for the third case was 51% lower compared to the base case (only ethanol) (Giuliano et al. 2018). Similar results were reported by Ntimbani et al. (2021) with the TEA of one-stage furfural and cellulosic ethanol from the mixture of sugarcane bagasse (70%) and harvesting residues (30%). The minimum ethanol selling price for integrated biorefinery scenarios (−$0.02 and $0.42/L) was significantly lower than the sole-ethanol scenario ($0.595/L) at the internal rate of return (IRR) used in the study. Due to relatively large capital investments ($305-$327 million) compared to the sole-ethanol process ($294 million), the MSEP was highly dependent on the IRR value. The integrated biorefinery was more profitable at an IRR of 10%. However, at an IRR of 15% or higher, the MSEP from integrated biorefinery was at least 37% higher ($1.14–$2.23/L) compared to the sole-ethanol process ($0.83/L) (Ntimbani et al. 2021). Rajendran and Murthy (2017) performed a TEA of bana grass and energy cane-based biorefineries where C5 sugars were converted to ethyl acetate through a process of producing acetic acid from the C5 sugars followed by reactive distillation with ethanol. Although the ethanol production was lower in these scenarios (due to C5 diversion and using some ethanol for reactive distillation), the return on investment was found to be significantly higher (~5% vs. −9.0% for bana grass and ~5% vs. −9.3% for energy cane) compared to the base case (only ethanol production). Zang et al. (2020) conducted a TEA of an integrated biorefinery co-producing ethanol, furfural, and lignin from switchgrass through a one-pot biomass fractionation process using a biphasic solvent (aqueous choline chloride (ChCl) and methyl isobutyl ketone (MIBK)). Processing of 1 MT of feedstock (20% moisture) using this integrated system resulted in the production of 124 kg ethanol, 109 kg furfural, and 97 kg lignin. Unfermented carbohydrates and unrecovered lignin were burnt to produce steam and electricity to partially meet the energy demands of the biorefinery. The process was found to be economically feasible, with the minimum selling price of furfural at $625/ton, which was about 35% lower compared to the market price (Zang et al. 2020).

Although several of the studies discussed above highlighted the importance and benefits of converting lignin and C5 sugar fractions into high-value compounds, further research is needed to improve these technologies. The decision on the utilization of lignin and C5 sugars for the production of high-value products will depend on the type of feedstock and the technology used because in some cases, the quantities of lignin or C5 sugars might not be sufficient to justify the capital investment on the technologies required for their valorization. Also, the market price of potential products from lignin valorization would play a role in making these decisions. It would be important to point out that the market price and selling price assumptions can vary among studies and that variation causes inconsistency. The purchase or selling prices of raw materials keep fluctuating, and the fluctuation could be very high for biofuels and bioproducts. For example, the wholesale price of ethanol ranged between $1.09 and $3.51/gal ($0.29–$0.93/L) during 2010–2020 (Kumar et al. 2021). The assumptions of the selling price of products and co-products would directly affect the TEA results. Although several studies use the current year price for the analysis, using a historical average (e.g., 10-year average) could make the analysis more rigorous. Moreover, sensitivity analysis or uncertainty analysis could be conducted to understand the feasibility of biorefineries under various potential scenarios (e.g., considering extreme low and high values of selling prices). More detailed discussion can be found in a published paper (Brown 2018), which focused on the effect of commodity price uncertainty on the TEA feasibility study of cellulosic biorefineries.

9.4 TEA of Engineered Lipid-Producing Energy Crops-Based Biorefineries

Considering this new technology, only a handful of studies have been conducted to evaluate the commercial-scale feasibility of engineered lipid-producing energy crops-based biorefineries (Table 9.2). As these crops are under development stage and there is uncertainty about the actual composition of these feedstocks, most of these studies evaluated various scenarios by varying the TAG levels in the biomass. The chemical composition of feedstocks under these scenarios was estimated based on energy balance (Fasahati et al. 2019; Huang et al. 2016a, b; Kumar et al. 2018). This approach assumes that sucrose and lipids are two major energy reserves, and after substituting sucrose with lipids, the remaining mass ends up as fiber in the biomass (Huang et al. 2016a; Kumar et al. 2018). Based on this assumption, Huang et al. (2016a) considered that the fiber content in the lipid-producing sugarcane (with 20% lipids) was 73% (dry basis) compared to 43.3% for normal sugarcane. Using a similar assumption, Fasahati et al. (2019) assumed about an 8% increase in cellulose (16.21 vs. 15.02%) in engineered sorghum (5% lipids) compared to non-engineered sorghum.

The first study on the TEA of engineered lipid-producing energy crops-based biorefinery was conducted by Huang et al. (2016a) to evaluate the feasibility of co-producing ethanol and biodiesel from lipid-producing sugarcane, also known as lipid-cane. The biorefinery was found economically feasible, with an IRR of 13.7–24.0% compared to 15% for the soybean biodiesel process. An inverse relationship was observed between the process economics and lipid content in the biomass. With an increase in lipid content from 2 to 20% (dry basis), the biodiesel production cost decreased by more than 30% ($0.59/L for 20% lipids vs. $0.89/L for 2% lipids) (Huang et al. 2016a). Similar observations were made by Kumar et al. (2018) from

Table 9.2 Techno-economic analysis of various biorefineries using engineered energy crops

Biomass ($ year; biomass cost)	Plant capacity	Process	Products	Capital investment ($ million)	Production cost	Minimum selling Price	Remarks	References
Lipid-producing sugarcane (2013 $) – $35/ton (wet basis)	8000 MT/day (200 operational days/year)	Integrated ethanol and biodiesel production	Ethanol, biodiesel, electricity	158.5–199	$0.59–$0.89/L biodiesel; $0.40–$0.46/L ethanol	–	Ethanol from sucrose, biodiesel from lipids; bagasse conversion to power generation; 2–20% lipid in the feedstock	Hung et al. (2016a)
Lipid-producing sugarcane (2016 $) – $35/ton (wet basis)	8000 MT/day (200 operational days/year)	Integrated ethanol and jet fuel production	Ethanol, jet-fuel, electricity	238.1–351.2	$0.73–$1.79/L jet fuel	$1.4–$3.63/L jet-fuel	Ethanol from sucrose, jet fuel from lipids using HEFA process; bagasse conversion to power generation; 5–20% lipid in the feedstock	Kumar et al. (2018)

(continued)

Table 9.2 (continued)

Biomass ($ year; biomass cost)	Plant capacity	Process	Products	Capital investment ($ million)	Production cost	Minimum selling Price	Remarks	References
Lipid-producing sugarcane and sorghum (2013 $) – $35/ton lipid-cane (wet basis), $32/ton sorghum (wet basis)	8000 MT lipid-cane /day (200 operational days/year), 7200 MT sorghum/day (60 additional days)	Integrated ethanol and biodiesel production	Ethanol, biodiesel, electricity	–	$0.56–$0.84/L biodiesel; $0.38–$0.48/L ethanol	–	Ethanol from sucrose, biodiesel from lipids; bagasse conversion to power generation Lipid-cane processing for 200 days, sorghum processing in the same plant for 60 days in the lipid-cane off-season	Huang et al. (2016b)
Lipid-producing sorghum (2014$) – $48/ton (wet basis)	2000 MT/day dry sorghum	Integrated ethanol and biodiesel production	Ethanol, biodiesel, electricity	499.7	–	$0.65/L ethanol	Ethanol from structural carbohydrates, biodiesel from lipids	Fasahati et al. (2019)
Lipid-producing sugarcane Energy cane (2019 $) – $35/ton (wet basis)	8000 MT/day (200 operational days/year)		Ethanol, biodiesel, electricity	238.7–406.6	$0.69–$0.81/L biodiesel; $0.35–$0.45/L ethanol	–	Ethanol from sucrose and biodiesel from lipids; bagasse conversion to power generation or cellulosic ethanol (2G)	Kumar et al. (2020)

the TEA of biorefinery co-producing ethanol and jet fuel from this engineered sugarcane. The jet fuel production cost decreased from \$1.79/L to \$0.73/L with an increase in the lipid content of sugarcane from 5 to 20%. However, at the same time, the capital investments were estimated to be relatively higher to process biomass with high lipid levels. Kumar et al. (2018) reported that 48% higher capital investment (\$351.2 vs. \$238.1 million) is needed to process 20% lipid-containing sugarcane compared to 5% lipid-containing sugarcane. This increase was due to capital-intensive jet fuel production technology. As the biodiesel production process is not that capital-intensive, Huang et al. (2016a) reported only an 11.6% increase in capital investment to process 20% lipid-containing sugarcane compared to 5% lipid-containing sugarcane. Both studies considered the utilization of bagasse to produce steam and electricity. In both cases, the energy from bagasse burning was sufficient to meet the biorefinery demand and the excess electricity was sold to the grid. The surplus electricity for 20% lipid-containing sugarcane was estimated at 217.2 and 156.9 kWh/MT biomass processed for ethanol-biodiesel biorefinery and ethanol-jet fuel biorefinery, respectively (Huang et al. 2016b; Kumar et al. 2018). Moreover, considering the high productivity of sugarcane, the biodiesel yield per unit of land was estimated up to 13 times higher than that of soybean (Huang et al. 2016a). Considering the harvesting time (6–7 months/year) of sugarcane, both of these studies assumed a 200-day annual operational time in the process model development. In a follow-up study, Huang et al. (2016b) investigated the TEA of a biorefinery processing lipid-producing sugarcane and lipid-producing sorghum in the same plant. Since the harvesting time of sugarcane and sorghum is different, with integration of sorghum processing into the lipid-cane biorefinery, the plant could be operated for additional 60 days without adding more capital costs. This integration resulted in a decrease in the production cost of ethanol and biodiesel by \$0.02/L and \$0.03/L, respectively. The IRR was calculated to be 29.2% compared to 24% for the lipid-cane-only biorefinery (Huang et al. 2016b). In another TEA study on lipid-producing sorghum-based biorefinery, Fasahati et al. 2019 reported that for the base case (10% lipids in sorghum), the minimum selling price of ethanol (MSPE) from this advanced biorefinery was about 20% lower (\$0.65/L vs. \$0.82/L ethanol) compared to the ethanol-only process (conventional process). From the sensitivity analysis, they concluded that the MSPE could meet the market price if engineered sorghum contains at least 13% oil and costs less than \$65/ton (dry basis).

In the most recent study, Kumar et al. (2021) investigated the TEA of biorefinery processing lipid-producing energy cane for co-production of ethanol and biodiesel. However, unlike previous TEA studies assuming ethanol production only from the sucrose in the juice, this study evaluated the feasibility of biorefinery under two scenarios: (i) biodiesel from TAGs and ethanol production only from the sucrose in the juice and bagasse burning for power generation, (ii) biodiesel from TAGs and ethanol production from both sucrose and structural carbohydrates in bagasse (Kumar et al. 2021). The cellulosic ethanol production analysis was important in this case because energy cane contains almost twice the amount of fiber compared to sugarcane. For the same reasons, the maximum oil content in energy cane was estimated at 7.7% compared to 20% in the case of sugarcane. The biorefinery was found economically viable and more profitable compared to the soybean biodiesel process, with biodiesel production costs in all scenarios (at various levels of TAGs in biomass) lower ($0.66/L–$0.9/L vs. $0.91/L) compared to soybean diesel cost. Due to the lower maximum possible oil content (7.7 vs. 20%) compared to lipid-producing sugarcane, the biodiesel production per unit of land was relatively lower but was still five times that of soybean. Based on the Monte Carlo simulations, it was concluded that the biorefinery was profitable with a 29–65% probability (Net present value; NPV > 0), mainly controlled by the biodiesel market price and feedstock composition. This study also highlights the importance of TEA in understanding the trade-off of energy, economics, and productivity among various processing options. For example, the total biofuel produced in scenario-2 was about five times higher compared to scenario-1. But at the same time, capital investments were also up to 60% higher than in scenario-1 ($394–$406 million vs. $238–$244 million). Moreover, the scenario-1 biorefinery was self-sustainable in energy and produced surplus electricity that could avoid the use of fossil-based energy and power. Whereas the in-house steam and electricity productions are not sufficient to meet the biorefinery demands in scenario-2. Using these TEA results, a detailed life cycle assessment could further help in understanding the environmental benefits trade-off among these technology options.

Overall, all of these studies demonstrated that the engineered lipid-producing biomass-based biorefineries could be cost-competitive and provide significant advantages in terms of fuel production per unit land compared to conventional oil crops. Moreover, considering the potential of growing these crops in marginal lands with low agricultural outputs, the development and processing of these crops could play a critical role in solving the energy issues in the United States and possibly other places, and meeting the renewable fuel mandates. At the same time, these TEA studies concluded that high oil levels need to be achieved in these crops before large investments can be made.

9.5 Summary

Through the integration of thermochemical and biological processes, the biorefinery approach aims to maximize resource utilization by converting all fractions of biomass to fuel, power, and bio-products. Considering the low technology readiness levels of these concepts, techno-economic analysis using process simulations could play a critical role in determining the economic viability and associated risks at the commercial-scale application. The analysis becomes even more crucial for the biorefineries based on engineered-energy crops that are currently in the conceptual phase only and lack the actual experimental data. The TEA can guide the research and development by identifying the technical bottlenecks and target levels (in terms of biomass composition or product yields) required to achieve economic viability and shorten the time between lab to commercialization.

References

Baral NR, Shah A (2017) Techno-economic analysis of utilization of stillage from a cellulosic biorefinery. Fuel Process Technol 166:59–68

Bbosa D, Mba-Wright M and Brown RC (2018) More than ethanol: A techno-economic analysis of a corn stover-ethanol biorefinery integrated with a hydrothermal liquefaction process to convert lignin into biochemicals. Biofuels Bioprod Biorefin 12(3):497–509

Brown TR (2018) Price uncertainty, policy, and the economic feasibility of cellulosic biorefineries. Biofuels Bioprod Biorefin 12(3):485–496

Cortes-Peña Y, Kumar D, Singh V, Guest JS (2020) BioSTEAM: a fast and flexible platform for the design, simulation, and techno-economic analysis of biorefineries under uncertainty. ACS Sustain Chem Eng 8(8):3302–3310

Fasahati P, Liu JJ, Ohlrogge JB, Saffron CM (2019) Process design and economics for production of advanced biofuels from genetically modified lipid-producing sorghum. Appl Energy 239:1459–1470

Giuliano A, Barletta D, De Bari I, Poletto M (2018) Techno-economic assessment of a lignocellulosic biorefinery co-producing ethanol and xylitol or furfural. In: Computer aided chemical engineering, vol 43. Elsevier, pp 585–590

Hasanly A, Talkhoncheh MK, Alavijeh MK (2018) Techno-economic assessment of bioethanol production from wheat straw: a case study of Iran. Clean Technol Environ Policy 20(2):357–377

Huang H, Long S, Singh V (2016a) Techno-economic analysis of biodiesel and ethanol co-production from lipid-producing sugarcane. Biofuels Bioprod Biorefin 10(3):299–315

Huang H, Long SP, Clemente TE, Singh V (2016b) Technoeconomic analysis of biodiesel and ethanol production from lipid-producing sugarcane and sweet sorghum. Ind Biotechnol 12(6):357–365

Humbird D, Davis R, Tao L, Kinchin C, Hsu D, Aden A, Schoen P, Lukas J, Olthof B, Worley M (2011) Process design and economics for biochemical conversion of lignocellulosic biomass to ethanol: dilute-acid pretreatment and enzymatic hydrolysis of corn stover. National Renewable Energy Lab. (NREL), Golden, CO (United States)

Kazi FK, Fortman JA, Anex RP, Hsu DD, Aden A, Dutta A, Kothandaraman G (2010) Techno-economic comparison of process technologies for biochemical ethanol production from corn stover. Fuel 89:S20–S28

Khounani Z, Nazemi F, Shafiei M, Aghbashlo M, Tabatabaei M (2019) Techno-economic aspects of a safflower-based biorefinery plant co-producing bioethanol and biodiesel. Energy Convers Manage 201:112184

Kumar D, Murthy GS (2011) Impact of pretreatment and downstream processing technologies on economics and energy in cellulosic ethanol production. Biotechnol Biofuels 4(1):1–19

Kumar D, Long SP, Singh V (2018) Biorefinery for combined production of jet fuel and ethanol from lipid-producing sugarcane: a techno-economic evaluation. Gcb Bioenergy 10(2):92–107

Kumar D, Long SP, Arora A, Singh V (2021) Techno-economic feasibility analysis of engineered energycane-based biorefinery co-producing biodiesel and ethanol. GCB Bioenergy 13(9):1498–1514

Kurambhatti C, Kumar D, Rausch KD, Tumbleson ME, Singh V (2020) Improving technical and economic feasibility of water based anthocyanin recovery from purple corn using staged extraction approach. Ind Crops Prod 158:112976

Murthy GS (2017) Techno-economic assessment. In: Li Y, Khanal SK (eds) Bioenergy: principles and applications, vol 1. Wiley, pp 507–520

Murthy GS (2022) Techno-economic assessment. In: Biomass, biofuels, biochemicals. Elsevier, pp 17–32

Ntimbani RN, Farzad S, Görgens JF (2021) Techno-economic assessment of one-stage furfural and cellulosic ethanol co-production from sugarcane bagasse and harvest residues feedstock mixture. Ind Crops Prod 162:113272

Rajendran K, Murthy GS (2017) How does technology pathway choice influence economic viability and environmental impacts of lignocellulosic biorefineries? Biotechnol Biofuels 10(1):1–19

Saini R, Osorio-Gonzalez CS, Hegde K, Brar SK, Magdouli S, Vezina P, Avalos-Ramirez A (2020) Lignocellulosic biomass-based biorefinery: an insight into commercialization and economic standout. Curr Sustain/Renew Energy Rep 1–15

Sanchez A, Sevilla-Güitrón V, Magaña G, Gutierrez L (2013) Parametric analysis of total costs and energy efficiency of 2G enzymatic ethanol production. Fuel 113:165–179

Shah A, Baral N, Manandhar A (2016) Technoeconomic analysis and life cycle assessment of bioenergy systems. In: Advances in bioenergy, vol 1. Elsevier, pp 189–247

Shen R, Tao L, Yang B (2019) Techno-economic analysis of jet-fuel production from biorefinery waste lignin. Biofuels Bioprod Biorefin 13(3):486–501

Somavat P, Kumar D, Singh V (2018) Techno-economic feasibility analysis of blue and purple corn processing for anthocyanin extraction and ethanol production using modified dry grind process. Ind Crops Prod 115:78–87

Towler G, Sinnott R (2012) Chemical engineering design: principles, practice and economics of plant and process design. Elsevier

Zang G, Shah A, Wan C (2020) Techno-economic analysis of an integrated biorefinery strategy based on one-pot biomass fractionation and furfural production. J Clean Prod 260:120837

The manufacturer's authorised representative in the EU is Springer
Nature Customer Service Centre GmbH, Europaplatz 3, 69115 Heidelberg,
Germany. If you have any concerns regarding our products, please
contact ProductSafety@springernature.com

Printed and bound by CPI Group (UK) Ltd, Croydon, CR0 4YY
28/09/2026
02224073-0006